Alien Ocean

Alien Ocean

Anthropological Voyages in Microbial Seas

STEFAN HELMREICH

University of California Press

BERKELEY   LOS ANGELES   LONDON

University of California Press, one of the most distinguished university presses in the United States, enriches lives around the world by
advancing scholarship in the humanities, social sciences, and natural
sciences. Its activities are supported by the UC Press Foundation and
by philanthropic contributions from individuals and institutions. For
more information, visit www.ucpress.edu.

University of California Press
Berkeley and Los Angeles, California

University of California Press, Ltd.
London, England

Library of Congress Cataloging-in-Publication Data

Helmreich, Stefan
    Alien ocean : anthropological voyages in microbial seas / Stefan
    Helmreich.
        p.   cm.
    Includes bibliographical references and index.
    ISBN 978–0-520-25061-1 (cloth : alk. paper)
    ISBN 978–0-520-25062-8 (pbk. : alk. paper)
    1. Marine microbiology—Research.   2. Marine biologists.   3.
Human ecology.   I. Title.

QR106. H45 2008
578. 77—dc22   2008020955

18   17   16   15   14   13   12   11
10   9   8   7   6   5   4   3   2

Contents

Illustrations

Moorings

The ocean is strange. For those of us settled in down-to-earth common sense and facts-on-the-ground science, the ocean symbolizes the wildest kind of nature there is. It represents a contrast to the cultivated land and even, sometimes, to the solid order of culture itself. Although many people have tried to capture this sea—whalers, painters, poets, politicians—marine biologists have offered some of the most authoritative accounts of the ocean and the life it sustains, particularly for publics compelled and captivated by the explanatory stories of science.

Marine biologists' visions of the ocean are today in transformation. These scientists are learning to see the sea not only as the home medium for marine mammals, fishes, and seaweeds but also as a realm inhabited, maintained, and modulated by an extraordinary mix of microbes, many of which live at astonishing extremes of light, temperature, pressure, and chemistry. Using molecular biological techniques, gene sequencing, bioinformatics, and remote sensing, marine biologists are coming to view the ocean as a web of microbial life joining the sunniest surface waters to the dimmest depths of the sea floor. Novel configurations of technology and theory are leading them to conceptualize the ocean as a site in which the object of biology—life—materializes as a networked phenomenon linking the microscopic to the macrocosmic, bacteria to the biosphere, genes to globe. Microbes are key figures in this new scientific ocean, pointers to the origin of life, climate change, and promising biotechnologies.

This book offers an anthropological account of how one cluster of marine biologists, marine microbiologists, are making such microbes meaningful—to themselves, to other scientists, and to broader publics. It examines how marine microbes are becoming items of interest and contest among such varied players as environmentalists, biotech entrepreneurs,

Acknowledgments

My debts are deep.

I embarked on my ethnographic work at the Monterey Bay Aquarium Research Institute (MBARI), where Judith Connor, director of information and technology dissemination, made my presence possible. I am grateful to Ed DeLong for inviting me to participate in the work of his MBARI lab, to Steven Hallam for leading me through the technical and ethical seascapes of today's marine biology, and to Pete Girguis for directing me aboard the research vessel *Point Lobos*. I also thank Aaron Cozen, José de la Torre, Shana Goffredi, John Graybeal, George Matsumoto, Chris Preston, Chris Scholin, and Bob Vrijenhoek. Kim Solano, Nathan Sawyer, and J.P. made me feel at home in the hamlet of Moss Landing, and Matt McCarthy provided a retreat in Santa Cruz.

During my time on Oʻahu, Michael Cooney was an incisive guide into the world of marine biotechnology. Bob Bidigare, Mark Goldman, Jo-Ann Leong, Pat Takahashi, Georgia Tien, and Jian Yu helped me understand cyanobacteria past, present, future. Dave Karl provided the best explanation of microbial oceanography I could hope to hear, clarifying my course in this book immensely. For conversations in Hawaiʻi about microbes, algae, molluscs, and politics, I thank Isabella Abbott, Pauline Chinn, Kaipo Faris, Ben Finney, Margaret McManus, Rebecca Most, Grieg Steward, and Mililani Trask.

When I joined the faculty of the Massachusetts Institute of Technology, Penny Chisholm became an invaluable interlocutor about all aspects of ocean and earth system science. Tracy Mincer and Martin Polz gave me evolutionary insights into marine microbial genetics. Matt Sullivan and Virginia Rich were always ready to comment on chapter drafts and game to view a good (or bad) ocean documentary. At Harvard, Colleen Cavanaugh answered questions about vent science with wisdom and wit.

Penny Chisholm enabled me to sign on to an expedition into the Sargasso Sea, on which voyage I did my best to reciprocate by collecting marine viruses. Brian Binder and Liz Mann were a pleasure to travel with and learn from, and the rest of the science party and the crew of R/V *Endeavor* made the journey—which included a zag into the Bermuda Triangle—a remarkable experience. I thank Cornelia Bailey for her hospitality and narrative insight on Sapelo Island, Georgia.

My trip to the Juan de Fuca Ridge on R/V *Atlantis* and in the deep submergence vehicle *Alvin* was made possible by the University of Washington's Deborah Kelley, who welcomed me onto her expedition with wry humor and catching enthusiasm. Among the science party, I particularly thank Deb Glickson, Jim Holden, and Kris Ludwig. I dived in *Alvin* with geologist John Delaney and pilot Bruce Strickrott, extraordinary guides into an extraordinary realm.

Back on land, I thank Lynn Margulis for an eye-opening fieldtrip to the Sippewissett marsh on Cape Cod and also for her microscopic reading of my writing on astrobiology. From Woods Hole, I thank Mitch Sogin, Andreas Teske, and Carl Wirsen. I thank Ford Doolittle for multiple readings of my arguments about the tree of life. I thank marine biotechnologist Dominick Mendola of Encinitas, California, for sharing his life story. From farther flung worlds I thank Loka Bharathi, Chuck Fisher, Stephen Hourdez, Karen Nelson, Lata Raghukumar, Frank Robb, and Art Yayanos. Deborah Day, archivist at the Scripps Institution of Oceanography, helped me scout out odd documents in marine biological history.

Friends, colleagues, and students from the recombinant worlds of science studies and anthropology—as well as from media studies, literary criticism, and history—helped me get my bearings amid a wash of topics that threatened to swamp me entirely. Among those whose soundings proved essential for outlining the shape of the book were Bill Maurer, whose anthropologies of fishes—cryptogenic, star-shaped, and lateral-lined—kept my social theory suitably at sea; Sarah Franklin, who first shepherded my ruminations on sea science into print; Mike Fortun, who read the beta version of this manuscript and made salutary suggestions for resequencing my data; Rich Doyle, whose transmissions reassured me about the aliens; Cari Costanzo Kapur, who helped me navigate the choppy waters of Hawaiian politics; Dmitry Portnoy, who read across lines of anthropological and genetic text for surface tensions and submarine meanings; Cris Moore, whose grappling with time, space, and scaling in theoretical computer science helped me unknot loops in my underwater arguments; Hannah Landecker, who asked clarifying questions when I got muddled up; and

Chris Kelty, who taught me to think "against networks." Hillel Schwartz provided a steady signal of citations and observations. Jane and George Collier intervened at an early stage to make sure I knew where I was heading. I also thank Nadia Abu El-Haj, Samer Alatout, Pamela Ballinger, Debbora Battaglia, David Bjarnason, Tom Boellstorff, Geoff Bowker, Laurel Braitman, Tony Crook, Marianne de Laet, Carol Delaney, David Derrick, Virginia Dominguez, Gary Downey, Joe Dumit, Troy Duster, Ron Eglash, Erika Flesher, Rayvon Fouché, Peter Galison, Ilana Gershon, Sharon Ghamari-Tabrizi, Sherine Hamdy, Evelynn Hammonds, Donna Haraway, Cori Hayden, Eva Hayward, Deborah Heath, Aída Hernández, David Hess, Linda Hogle, Mimi Ito, Sarah Jain, Sarah Jansen, Sheila Jasanoff, Henry Jenkins, Natalie Jeremijenko, Carina Johnson, Douglas Kahn, Wyn Kelley, Eduardo Kohn, Wen-Hua Kuo, Anthony Lioi, Margaret Lock, Theresa MacPhail, Robert Markley, Emily Martin, Sally Engle Merry, Lisa Messeri, Zara Mirmalek, Hiro Miyazake, Michael Montoya, Lynn Morgan, Chandra Mukerji, Robin Nagle, Alondra Nelson, Diane Nelson, Damien Neva, Julie Olson, Susan Oyama, Gísli Pálsson, Trevor Pinch, Marcelle Poulos, Paul Rabinow, Hugh Raffles, Ronald Rainger, Rayna Rapp, Jenny Reardon, Peter Redfield, Annelise Riles, Lars Risan, Michael Rossi, Dan Segal, Bill Shackford, Elta Smith, Levent Soysal, Stefan Sperling, Jared Stark, Hallam Stevens, Ajantha Subramanian, Karen-Sue Taussig, Rebecca Thomas, Charis Thompson, Miranda von Dornum, Cathy Waldby, Wendy Walker, Charles Watkinson, Kath Weston, and Lambert Williams.

The anthropology program at MIT is a lively, congenial place to think and work. For their close readings of chapters, I thank Jim Howe, Jean Jackson, Erica James, Susan Silbey, and Chris Walley. Hugh Gusterson and Susan Slymovics, though now posted elsewhere, remain the best mentors one could hope to have. Rosie Hegg and Amberly Steward, program administrators, provided essential aid in getting this manuscript out the door. Colleagues in MIT's broader program in History, Anthropology, and Science, Technology, and Society also engaged with *Alien Ocean* early on, in part or entirety. I thank Michael Fischer, Deborah Fitzgerald, David Jones, Evelyn Fox Keller, Vincent Lépinay, Ken Manning, David Mindell, Harriet Ritvo, and Sherry Turkle. I thank graduate students in Biogroop, a reading collective with whom I had many vital conversations. Etienne Benson, Natasha Myers, Sophia Roosth, and Sara Wylie will find in this book many transductions of our discussions.

Portions of *Alien Ocean* have been presented to a variety of academic audiences. I thank listeners at Amherst, Bowdoin, Cornell, Harvard, MIT, the Monterey Bay Aquarium Research Institute, Mount Holyoke, the New

School for Social Research, Rensselaer Polytechnic Institute, Rice, the School of American Research, UC Berkeley, UC Irvine, UC Santa Cruz, the University of Hawai'i at Mānoa, the University of Iceland, the University of Wisconsin at Madison, and Wesleyan.

Financial support for this project was provided by Grant no. 6993 from the Wenner-Gren Foundation for Anthropological Research and by awards from the office of MIT's Dean of Humanities, Arts, and Social Sciences. The 2006 James A. and Ruth Levitan Prize in the Humanities, administered through MIT, helped me finish the manuscript. My editors at University of California Press, Stan Holwitz and Elizabeth Berg, made production proceed swimmingly. Copyeditor John Thomas tamed all manner of grammatical and typographical monsters.

To gather up, finally, the most traditional knots of kinship, I thank my parents, Mary and Gisbert Helmreich, who remind me that the ocean waits for me in Southern California. My in-laws, Tom and Judi Paxson, continue to amaze me with their selfless dedication to kith, kin, and social justice. My wife and partner in all things anthropological, Heather Paxson, made the research and writing of this book possible, traveling with me to obscure ports and places, joining her own thinking on human and microbial cheese cultures with mine on seas, and reading tirelessly through endless rafts of chapter drafts. Heather made this book thinkable, for it was she who imagined the title, *Alien Ocean*. Her love is a constant current that keeps me going. Our toddler son, Rufus Paxson Helmreich, soaking up multiple viewings of *Yellow Submarine* as I finished this book, filled my eleventh-hour sallies through the sea of science with happily alien sing-alongs.

Before embarking on *Moby-Dick*, Hermann Melville wrote to his friend and father-in-law, Judge Lemuel Shaw, "It is my earnest desire to write those sort of books which are said to 'fail.'" Although I cannot hope to fail as spectacularly as Melville, this book does press against limits in my abilities to comprehend and articulate all that crossed my anthropological path. I only hope such failures as eddy through these pages are, like aliens, usefully diagnostic of today's difficulties in apprehending the laws of nature at sea.

Cambridge, Massachusetts,
and Encinitas, California, 2008

Arguments I originally developed in the following articles are reprinted with permission:

Trees and Seas of Information: Alien Kinship and the Biopolitics of Gene Transfer in Marine Biology and Biotechnology. *American Ethnologist* 30(3): 340–58. Published University of California Press, ©2003 American Anthropological Association.

How Scientists Think; About 'Natives,' for Example: A Problem of Taxonomy among Biologists of Alien Species in Hawaii. *The Journal of the Royal Anthropological Institute, Incorporating MAN* 11(1): 107–28. ©2005 The Royal Anthropological Institute.

The Signature of Life: Designing the Astrobiological Imagination. *Grey Room* 23(4): 66–95. ©2006 The MIT Press.

Blue-Green Capital, Biotechnological Circulation, and an Oceanic Imaginary: A Critique of Biopolitical Economy. *BioSocieties* 2(3): 287–302. ©2007 Cambridge University Press and the London School of Economics and Political Science.

An Anthropologist Underwater: Immersive Soundscapes, Submarine Cyborgs, and Transductive Ethnography. *American Ethnologist* 34(4): 621–41. Published University of California Press, ©2007 American Anthropological Association.

Introduction: Life at Sea

Dropping off the edge of California, the undersea canyon of Monterey Bay falls 4,000 meters into an ink-dark world. It is a realm of tubeworms, clams, and the odd whale skeleton, invisibly scrimshawed now by microbes burrowing for minerals in bones broken up and breaking down. Above this abyssal district fly fleets of tuna, salmon, and sharks as well as sea lions, harbor seals, and porpoises. Closer to shore, Silicon Valley scuba divers practice their submarine yoga amid forests of kelp and smacks of jellyfish. Oceanographers, suspended on the surface, haul jugs of seawater onto the decks of research vessels. In the nearby coastal town of Moss Landing, DNA sequencing machines are set to scan for signs of life, primed to read the genes of marine microorganisms floating within these microcosmic seas.

Sitting in a lecture hall overlooking the bay on a June day in 2000, I am listening to microbiologist Ed DeLong as he addresses patrons of the Monterey Bay Aquarium. He informs his audience that life on Earth likely originated in swarming seawater, descending perhaps from a crew of microbes named the *Archaea*, or "ancient ones," the most famous of which reside at high-pressure, high-temperature, sulfur-spitting volcanic vents on the seafloor. These extraordinary microorganisms, DeLong says, might reveal the upper temperature limits of life and even suggest the outlines of life forms on lightless alien worlds, like Jupiter's satellite Europa, which may host hydrothermal activity. Microbial extremophiles—lovers of extremes—are ubiquitous on Earth, integral to the maintenance of this ocean planet.

Though they are vanishingly small, marine microbes range across myriad ecological contexts and operate at global scales. DeLong tells us: "Microbes are responsible for the health of the oceans. They shape the chemistry of the sea and the atmosphere. These organisms that we can't even *see* are extremely important. These little guys *control* the biogeochemistry of our world. They are the stewards of our planet." Microbes are pivotal players in those processes through which atmospheric carbon, nitrogen, phosphorous, and sulfur are converted into the bodies of earthly organisms and back into elemental substance. Detailing the dizzying multiplicity of microbial life—from heat- to cold- to acid-loving archaea, to bioluminescent bacteria, to everyday phytoplankton floating on the sea surface—DeLong guides us on a slide show through what he calls "microscopic forests of the sea," a phrase that conjures images of rain forests, potent terrestrial symbols of ecological diversity. In DeLong's rendering, marine microbes provide vital bonds to biotic forces past and present.

And future. DeLong catalogs the potential of extremophilic marine microbes to supply materials to biotechnology. Heat-loving microbes in deep-sea vents offer enzymes that enable the copying of DNA in the lab, a high-temperature process crucial to gene sequencing. Cold-loving bacteria are promising sources for pharmaceuticals. Microbes thriving on petroleum hydrocarbons, toxic to most creatures, might digest oil spills. In DeLong's depiction, the microbial sea emerges as a storehouse of curative powers that humans might harness, transforming the most alien lifestyles into allied forces to heal people and planet.

Marine microbiology—long the province of seasick scientists struggling to isolate single cells in unstable culture media on unstable boats—is undergoing something of a renaissance. Led in the United States by DeLong and colleagues in Monterey Bay, at MIT, and the University of Hawai'i, the field is increasingly described as *microbial oceanography*, a phrasing that makes mapping microbial life coincident with mapping the sea itself, that suggests that microbes are not just *in* the sea but, in an important sense, *are* the sea.[1] DeLong offers an emblem for the revolution in a striking PowerPoint slide: an image of the whole Earth, seen from space, upon which sits superimposed Leonardo da Vinci's *Vitruvian Man* (ca. 1485–90), the perfectly proportioned figure proposed by the ancient Roman architect Vitruvius as a metric for the construction of temples and resurrected in the 1990s as a symbol of the Human Genome Project (figure 1). DeLong displays this icon to argue that the genetic techniques aimed at decoding human biology might be extended to what he calls "this other beast, our living planet." Gene sequencing and DNA databasing—genomics and bioinformatics—afford microbiologists fresh ways of scrutinizing the planet's membrane of marine microbes, linking genomes to biomes.[2]

The power of DeLong's image derives from pasting the iconography of early modern science into the frame of satellite imaging, from reaching

FIGURE 1. DeLong's PowerPointing Vitruvian Man. "Earth: The Blue Marble," image courtesy of NASA, created by Reto Stockli with Alan Nelson, under Fritz Hasler, for NASA's Visible Earth Project. Vitruvian Man photo by Luc Viatour, reproduced under GNU Free Document-ation Licence. Composite image, after DeLong, by Michael Rossi.

across scales of human and planetary embodiment, from placing the bio-diversity-rich Amazon to Vitruvian Man's left and a hurricane—harbinger of global warming—to his right, from juxtaposing Species Man with Mother Earth, and from presenting a distant but dazzling view of an Earth at once extraterrestrial and familiar. Reflecting on the slide, DeLong offers, "Earth is a misnomer. The planet should be called Ocean—or maybe it should be called Life or even Ocean Life." Invoking the Greek goddess whose name has become synonymous with the hypothesis that our planet is a self-regulating system, he adds, "Gaia fits."

I am transfixed by this image, not least because my own discipline, anthropology, also places humanity at its center, but more, because the man in the planetary petri dish points me to questions about the cultural coordi-nates of DeLong's vision of the microbial ocean. How does the microbial sea restage or reconfigure older views of the ocean as a primordial life-giving liquid, as a sublime wilderness, or as a social, economic, and scientific frontier? How does this new portrait provide fresh language to projects of environmental stewardship or reengineering? How might marine microbes be imported into secular, civic, or even spiritual apprehensions of the ocean world? How does the microorganismic, microcosmic sea reinforce or upend interpretations of the ocean as an extraterritorial space across which public, private, or national interests might be projected? How do framings of the "global" and "local" transform in its light? What cast do these visions take when diffracted through the techniques of genomic science? This book tracks how marine microbiologists seek to make new scientific knowledge about biological *life forms* meaningful to social and cultural *forms of life*. It follows how they scale from microscopic to macroscopic, how they define parameters of social and scientific relevance.

I approach my questions through an account of anthropological research I conducted from 2000 through 2005 among scientists in the United States, where much of the new marine microbiology is emerging. Microbial oceanography is a growing, still amorphously bounded field, made up of people who identify as microbiologists (concerning themselves with creatures too small to see, including bacteria, single-celled eukaryotes like yeast, and viruses), marine biologists (examining how organisms make their living in the sea), and biological oceanographers (studying plankton, from seaweed on down to diatoms and smaller). Strong traditions in microbiology and oceanography exist in France, the United Kingdom, Germany, Russia, and Japan, but I choose the United States because it is the country in which these two fields are most explicitly converging, and also because setting out from this national address permits me to inspect old and new tensions between the national and international, public and private, aspirations of ocean science. Microbial oceanography comes of age in the wake of post–cold war shifts in funding for marine science from physical oceanography to biological oceanography; as military monies for sounding the sea have declined, there has been a rise in state funding for high-tech, genomic bioscience and in venture capital for biotechnology.[3] Launching my investigation from the United States also allows me to look at the relation of a particular nation to the ocean, a task that in this book presses marine microbiology up against histories of the Pacific frontier and colonial Atlantic. Because the claims of U.S. science are often global, even universal, in ambition, they also cry out for placement in a more parochial frame. At the same time, because of their power, such claims demand to be taken seriously as preparing new futures and cosmopolitics. *Alien Ocean* examines how scientists working in the United States craft portraits of an oceanic nature they believe both enables and constrains human biological and social life.

In his 1948 foreword to marine biologist Ed Ricketts's *Between Pacific Tides*, novelist John Steinbeck—who took Ricketts as the model for "Doc" in *Cannery Row*—wrote incisively about marine biological investigations of the ocean realm: "The world is being broken down to be built up again, and eventually the sense of the new worlds will come out of the laboratory and penetrate into the smallest living techniques and habits of the whole people."[4] I am interested in those processes through which marine micro-biologists are breaking down the ocean world to build it up again today. Because of the sea's symbolic association with the origin and perpetuation of vitality, and because of its fluid capacity to link the smallest microorganism to the largest ecosystem, I am interested, too, in the ocean as a medium through which to explore shifting limits of the category of *life* in biological sciences. If the way scientists view the ocean is transform-ing, the way they understand life is mutating, too. This book explores what the concept of life looks like once it has been broken down into net-works of marine microbial genes to be built up again into new forms like biotechnologies—or into novel webs of ecological relation linked to Gaia, the frame James Lovelock in the 1960s proposed for Earth seen as a self-regulating, dynamically equilibrated system continually sustaining an assemblage of geological, chemical, and biological feedback systems that allow life to survive.

Bruno Latour, in *The Pasteurization of France*, documents how bacteri-ologists pushed microbes to the center of nineteenth-century hygiene pro-grams. "In redefining the social link as being made up everywhere of microbes," writes Latour, "Pasteurians and hygienists regained the power to be present everywhere."[5] Their project was so successful that one com-mentator in 1896 pronounced, "Society can exist, live, and survive only thanks to the constant intervention of microbes, the great deliverers of death, but also dispensers of matter."[6] Marine microbiologists such as DeLong similarly seek to persuade their publics that the entwined orders of nature and society cannot exist without microbes, though they empha-size that microbes are mostly allies to be understood rather than enemies to be defeated. Microbial oceanographers argue that marine microbes are central to life on Earth, that the lowly microbe constitutes a force of leviathan significance. If the whale was a key figure for the nineteenth-century sea—as a symbol of work, trade, and natural history—and the dolphin one emblem of the twentieth-century sea—as a mascot for environ-mental science and inspiration—the marine microbe is being readied for its twenty-first-century turn on the stage of technoscientifically fueled imaginings of the sea.

LIFE FORMS AND FORMS OF LIFE

For marine microbiologists, the ocean teems with newly discovered life forms. By "life forms," I mean those embodied bits of vitality called *organisms*, variously apprehended as ranged into species (durable, but changeable genealogical kinds) or as sorted into types occupying spaces of physical, metabolic, or ecological possibility (e.g., photosynthesizers, deep-sea dwellers).[7] In a more expansive sense, however, I also mean the relations of creatures with one another, following here those biologists who regard organisms as inextricably situated in ecologies. When I write of "forms of life," I mean those cultural, social, symbolic, and pragmatic ways of thinking and acting that organize human communities. I adapt this phrase from Ludwig Wittgenstein, who defined it in his *Philosophical Investigations* as a frame of reference within which linguistic action becomes meaningful.[8] Whereas historians and sociologists of science have employed the term to speak to the coherence of bounded worldviews, my usage emphasizes a plurality of forms of life—scientific, religious, economic, ethical—as well as their uneven, contested, and overlapping character.[9] The relation between life forms and forms of life today entails questions of how best to construe and construct the link between the biological and social, a problem that can be discerned, for example, in debates about whether human embryonic stem cells constitute proper materials for medical research.[10] Marine microbiologists' meditation on the meanings of microbes as long-lost relatives, environmental stewards, and biotech workers is animated by a contemporary preoccupation with how to connect life forms to forms of life.

The question of how to relate life forms to forms of life is of recent vintage. The very idea that life is a property that manifests in forms is modern, coming into being in coordination with the rise of biology as a discipline, solidifying around 1800. Michel Foucault, the omnivorous archeologist of ideas, argues that prior to the nineteenth century, "if biology was unknown, there was a very simple reason for it: that life itself did not exist. All that existed was living beings, which were viewed through a grid of knowledge constituted by *natural history*."[11] In other words, life had not yet been cordoned off into the domain known as biology. Such would come with texts like the German naturalist Gottfried Treviranus's programmatic

1802 *Biologie*, which offered that the "objects of our research will be the different forms and phenomena of life."[12]

Life itself in the nineteenth century became an at once metaphysical and scientifically underwritten property uniting humans, animals, plants, and microorganisms, a view that crystallized with Darwin's model of evolution, which named natural selection as the invisible hand that gave form to life. Once life had been rendered abstract, theoretical, the management of life forms within a specifically biological logic could become a question for social thought and practice. Foucault argued that the governance of biological processes emerged with such activities as public health and population control, which made it possible for nation-states to bring "life and its mechanisms into the realm of explicit calculations,"[13] to organize human life forms according to the social imperatives of forms of life—of national belonging and colonial expansion, for example. Eugenics is only the most extreme example of such endeavors. Foucault named these practices *biopolitics*.

To translate this history into the classic idiom of anthropology, I write here of the relation of nature to culture. Nature, importantly, has a genealogy distinct from biology, conveying into apprehensions of life forms additional meanings. Nature is not just that province scientists seek to describe but also a topology that retains a remnant of the mythic. In Western epistemology, nature has been imagined as a force to be dominated, tamed, struggled against. At the same time, nature has carried a strong (Judeo-Christian) religious charge, as that which is moral, inevitable, God-given, perhaps even rationally or harmoniously designed. Placed in this second tradition, biology, understood as a genre of nature that grounds culture, has often been a reference point for legitimating social relations—for naturalizing power, authorizing forms of life with reference to the moral sturdiness of life forms.[14] Think only of appeals to biology—in social Darwinism, sociobiology, evolutionary psychology—to rationalize gender and racial hierarchy.

In the age of biotechnology, genomics, cloning, genetically modified food, and reproductive technology, however, when human enterprise rescripts and reengineers biotic material, a founding function for nature is not so easily discernable. Culture and nature no longer stand in relation as figure to ground. Life forms cannot unproblematically anchor forms of life. "Life," suggests anthropologist Michael Fischer, "is outrunning the pedagogies in which we have been trained."[15] Biopolitics no longer only refers to the management of biotic things but now includes their reengineering as well, now encompasses the ways culture eddies into nature to transform it—through transferring deep-sea microbial genes into corn to make

INSIDE

■ Gore announces
efforts to protect
oceans/Page 3A

■ Panel addresses
education's essential role
for ocean science/Page 3A

■ Clinton's visit brings
together small army
of law enforcement/Page 3A

Delegates agree: Sea is life

**Conference
crystallizes
the need
for change**

By Laura Helmuth

Vice President Al Gore guards his notes from gusty winds at the U.S. Coast Guard Pier Thursday

**A source
of climate,
travel, food**

By Leslie Harris

FIGURE 3.   Front page, *Salinas Californian*, June 12, 1998.

state of the oceans are one."[19] In a conference pamphlet, first lady Hillary Clinton put the bond between humans and the sea in a personal, sentimental register, pronouncing that "seventy-one percent of our planet is ocean, and seventy-one percent of our body is salt water. . . . There is this extraordinary connection between who we are as human beings and what happens in this magnificent body of water."[20]

Clinton's analogy—evocative though asymmetrical (it compares surface area to volume)—revives a premodern theory of correspondences between the world and our bodies, giving a numinous humanist warrant to interest in the sea.[21] Earle anchors such analogical reasoning in evolutionary time: "Our origins are there, reflected in the briny solution coursing through our veins."[22] Seawater, likened to blood, a powerful symbol of relatedness, becomes a substance securing human kinship with aqueous Earth. What historian of biology Donna Haraway has diagnosed as a contemporary ecological "yearning for the physical sensuousness of a wet and blue-green Earth" is rendered experientially accessible in our own flesh.[23] One year-of-the-ocean book gives such castings an amniotic spin, enfolding the biotic world in a maternal embrace: "The sea is mother to all life on the planet."[24]

Such declarations connect individual humans to the planet in ways that both call upon and bypass evolutionary history, offering a one-step program of communion with the sea. These formulations construe the human body as a reflection of an ordered, harmonious whole—a cosmos. We are not presented here with a biopolitical call for human populations to reorganize their relations with the ocean. Neither are we called upon to inhabit a molecular biopolitics in which we take responsibility for, say, projects of genetic engineering that transform our life chances—an activity that anthropologist Paul Rabinow, discussing the rise of genetically articulated identity and community politics (around, e.g., hereditary conditions such as dwarfism or Huntington's disease), has named *biosociality*.[25] We are exhorted to think of our individual connection not to a population, not to our genes, but to the planet's ocean and to Gaia. This is not rhetoric of biosociality, but of *gaiasociality*.

While National Ocean Conference participants revered the sea as the matrix from which life emerged, like Ed DeLong they also tied it to the future. Earle: "Our survival is utterly dependent on the existence of life on Earth—of biodiversity."[26] If biodiversity exists at scales ranging from genetic to ecosystemic to global, then the life for which it serves as a proxy becomes a sliding signifier speaking at once of DNA, of life forms, of Gaia. With biodiversity a resource for resilience, a fountain of rejuvenation, it serves as a placeholder for the future.

Conferees saw this future as one that might be safeguarded by enlightened national stewardship. A report on the conference published by President Bill Clinton's cabinet in 1999 was titled "Turning to the Sea: America's Ocean Future." This future was futuristic: national efforts to protect marine ecosystems and genetic diversity would be hitched to projects of high technology. The report elaborated on the promises of bioengineering: "The tools of marine biotechnology have been applied to solve problems in the areas of public health and human disease, seafood safety and supply, new materials and processes, and marine ecosystem restoration and remediation. Many classes of marine organisms demonstrate a wide variety of compounds with unique structural features that suggest medicinal, agricultural, and industrial applications."[27]

The quest for chemical compounds and genes in the bodies of such critters as deep-sea vent microbes rewrote the ancient life-giving ocean as a technoscientific frontier to be explored with a can-do commitment to comprehending, taming, and commercializing a vast wilderness. This future, unlike earlier projects of resource extraction, was envisioned as environmentally sensitive, sustainable. A historically aggressive and quasi-religious American pioneer narrative might be rewritten in the service of a scientifically oriented project dedicated to preserving life on Earth—a sentiment implicit in the biblical resonance of the acronym for the National Oceanographic and Atmospheric Administration—NOAA (pronounced "Noah")—which suggests that Spaceship Earth is an ark and the United States its steward.

The Center of Marine Biotechnology at the University of Maryland—an institution examining marine microbes as materials for pharmaceutical and industrial products—posed the project of bioengineering this way: "Biotechnology allows us to tap the potential of the oceans without depleting them as a resource. . . . [T]he tools of biotechnology allow researchers to clone . . . genes, reproduce them, and produce desired substances in the laboratory, leaving the organisms where they belong—in the environment."[28] In America's ocean future, the biological resources of the planet can be called upon to help Earth heal itself, to turn to a natural, as yet unfathomed reserve of life. NOAA and its academic and industrial partners seek to save Gaia according to her own logic. Haraway diagnoses the vacillation between apocalypse and promise animating such narratives: "Belief in advancing disaster is actually part of a trust in salvation, whether deliverance is expected by sacred or profane revelations, through revolution, dramatic scientific breakthroughs, or religious rapture."[29] "Turning to the sea," then, suggests a return to origins that is also a journey into the future—and also a voyage into what Ocean Conference discussion papers, drawing on a timeworn American trope, called an "unexplored frontier."[30]

All this talk of life demands a comment about death. The celebration of oceanic diversity at the Ocean Conference unfolded alongside a keen sense that it was under threat. Beneath the "sea is life" rhetoric lurked the shadow fear that humans were visiting death upon the oceans. Global warming, coral bleaching, and contamination refer us to older images of the sea as a space of drowning, death, and shipwreck. This is only the surface of the submerged history. American visions of the ocean as a space of healing, therapy, and recreation wash over the twin history of the sea as a space of imperialism, the Middle Passage, submarine warfare, and radioactive waste.

A vision of the ocean as not only endangered but dangerous dominated a second workshop I attended, four years after the National Ocean Conference, a gathering devoted to discussing "Marine Bioinvasions," the human-mediated transport of creatures from one marine ecosystem to another, an activity that frequently has deleterious effects on destination waters. Organisms transported beyond regular ranges—through passage in discharged ballast water from ships, or through the travel of farmed and aquarium fish—are often called "alien species." The conference, held at the Scripps Institution of Oceanography in La Jolla, California, in March 2003, on the eve of the U.S. invasion of Iraq, was a political lifetime after the sunny 1998 Ocean Conference and could not escape uneasy resonance with post–September 11 fears of invisible alien others set to use biology as a weapon. One speaker worried that microbial communities in ballast water tanks might, through exchanging genes with one another while ships were under way, develop resistance to antibiotics. A new category of invasives emerged: "In the case of bacteria, the presence of numerous genetic variants and their associated phenotypic expressions may change our frame of reference from 'invasive species' to 'invasive genotypes.'"[31] Paul de Kruif's 1926 *Microbe Hunters*, a popular (and still in print) account of microbiology, pitched microbes as "silent assassins that murdered babes in warm cradles."[32] Nowadays, the association of microbes with disease and death— their most prominent public face starting with Pasteur's nineteenth-century war on anthrax—is framed at an ecological scale. Conferees discussed the possibility of bioterrorist attacks arriving in disease-dosed container ships, a view ratcheted to sensationalist pique in William Langewiesche's 2004 *The Outlaw Sea*, which depicts the ocean as "an anarchic expanse" and "a harbinger of a larger chaos to come" at the hands, perhaps, of an elusive "al Qaeda Navy" sailing under ever-changing flags of convenience.[33]

I learned at the Scripps conference that supernutrification of ocean waters from fertilizer runoff and corporate animal feed lot waste may lead to massive microbial blooms smothering larger sea creatures. In the Gulf of Mexico, the flushing of the American toilet down the Mississippi River has created swaths of sea called "dead zones" where algal growth blocks light from reaching levels below and oxygen is sucked out of the seawater, suffocating fish.

Scripps paleontologist Jeremy Jackson cautions that blooms of microbes might lead to a "loss of structure" in ocean food chains, a change he glosses as "the rise of slime." Here, the ocean becomes abject, a substance akin to pus—at once of and not of us. Time turns upside down, with oceanic others tagged as throwbacks, primitives; Jackson suggests that "dead zones reverse the achievements of more than half a billion years of evolution to take us back to the Precambrian Era before the rise of animals."[34] Human biocultural practices flow into the putatively natural zone of the ocean, scrambling nature and culture, life forms and forms of life.

Mentions of biodiversity at this conference were delivered with a reminder that it was under threat from bioinvasions with ultimately anthropogenic triggers, which could be traced to patterns of global commerce and politics. We witness a resurgence of an apocalyptic notion that the oceans will not wash away our sins but rather drown us in them.

The National Ocean Conference and the Marine Bioinvasions conference provide a double vision that persists in this book, for marine microbes mimic the double life of the ocean. On one side, webs of microbial genes tie the globe into a ball of interdependent relationship that finds a key symbol in the homeostatic and briny blue Earth. On the other, microbes lurk as a source of danger.[35] The title of a PBS television special on microbes, *Intimate Strangers*, captures the paired discourses.[36] Marine microbes morph from stewards of the planet to threatening pathogens and, often, through the magic of biotechnology, back again. Microbiologist Howard Shapiro summarizes:

> Different people see microorganisms from different perspectives. To evolutionary and molecular biologists, microbes are relatives, with whom we set up correspondence. To biotechnologists, they are workers, to be employed and, perhaps, exploited. To environmental microbiologists, they may be merely scenery, or analogous to canaries in coal mines, but they are generally viewed as good neighbors if we have good fences. To clinical, food, and sanitary microbiologists, and to the defense establishment, microorganisms are enemies to be tracked, contained and killed, and to leaders of rogue states and terrorist organizations, they are useful tools which are much easier to get through airports than are firearms or explosives.[37]

With such pluripotentiality, marine microbes might be installed, in the language of Latour, as "obligatory points of passage"[38]—necessary nodes—in nascent networks of research about the ocean and its relation to human welfare.[39] We enter the realm of what anthropologist Heather Paxson calls *microbiopolitics*, "the creation of categories of microscopic biological agents; the anthropocentric evaluation of such agents; and the elaboration of appropriate human behaviors vis-à-vis microorganisms."[40] Such evaluations and elaborations of categories and behaviors are compound, scalar—biological embodiment is always multiple, manifold—as the capacity of oceans to conjoin micro and macro might remind us (consider, e.g., the title of the National Research Council's *From Monsoons to Microbes: Understanding the Ocean's Role in Human Health*).[41] As evolutionist Lynn Margulis has shown, microbes are implicated through symbiosis and cell fusion in the emergence of biotic variety at all scales, in a process named *symbiogenesis* (the incorporation of once free-living mitochondria and chloroplasts into animal and plant cells, respectively, is exemplary).[42]

Microbiopolitics, then, are joined by *symbiopolitics,* the governance of relations among entangled living things.

The alternation between promise and apocalypse marked out by visions of the sea as life or as life-threatening has long featured in Euro-American conceptions of the nature of the ocean. For Christian Europeans, the sea was a feared vestige of primordial chaos, host to demonic monsters. For medieval and early modern seafarers, it became a wilderness against which sailors might test their faith and mettle, and with the Romantics it was configured as an element with which lone individuals might seek to merge, reconnecting with an ocean now imagined as the nourishing matrix of life itself. The life-taking and life-giving ocean also embodied a dualistic femininity, alternately maternal and witchlike.[43] By the middle of the nineteenth century, the ocean had become a master symbol of the sublime, of the awesomely beautiful and terrifying, of the natural that exists on such an overwhelming scale as to suggest that it ultimately partakes of something supernatural. Freud's *oceanic feeling*—his name for a human nostalgia for a lost communion with the universe—draws on this sentiment, though it adds a fear of slipping into the undertow of the id, into the slime that hides within the sublime.[44] The fusion of the strange and the sublime sea in the age of genomics, biotechnology, and microbial assemblages is the newest and most vivid form of what I call the alien ocean.

ALIEN OCEAN

If the wild and wondrous sea belongs to a zone of being beyond a steady and grounded self—if it belongs in part to what anthropologists call the order of the Other—today's microbial marine world can profitably be seen from the science fiction–spangled angle of the alien. If marine mascots have scaled down from the nineteenth-century whale to the twentieth-century dolphin to, now, the emerging microbe, it may not be surprising that the imagery aimed at apprehending this creature—neither charismatically mammoth nor wet and cuddly—reaches toward the sensationally odd and not-quite other.

I employ the figure of the alien because marine biologists so often invoke it as they describe the unfamiliar universe of marine microbes. Education-minded researchers, for example, appeal to the alien to invite kids to consider microbiology as a career: "Being a microbiologist is like being an explorer in a vast, unseen world full of weird, alien-like creatures."[45]

Alien associations abound when scientists speak of the strangeness of marine biota like hydrothermal vent microbes, the foreignness of invasive species, and the possibility that marine microbes might serve as models for extraterrestrial life forms (a connection brought into pop culture by James Cameron in a 2005 documentary about extremophiles called *Aliens of the Deep*). The word *alien* also flags the space-age imagery that informs scientific invocations of the term; not only are microbial realms often compared to outer space, but microbes inhabit contexts and scales—worlds, microcosms—inaccessible to prosthetic-free human experience, zones hard to apprehend as connected to our own forms of life. Microbes are life forms that exist at scales "unperceived by ordinary human experience," in contexts that push to their limits (genetic, metabolic, ecological) what biologists have imagined living systems capable of enduring or enacting.[46] Microbiologists often describe these creatures in a biotechnological idiom, as devices solving engineering problems, devices deploying the logic of information processing to propagate their kind in extended, subaqueous webs. Microbes' status as novel scientific objects, and as objects of wonder and anxiety, makes science fiction an inviting vernacular in which to describe them.[47]

The figure of the alien materializes, I contend, when uncertainty overtakes scientific confidence about how to fit newly described life forms into existing classifications or taxonomies, when the significance of these life forms for forms of life—and particularly, for secular, civic modes of governance—becomes difficult to determine or predict. In *Alien Ocean*, we encounter a few kinds of aliens. Most straightforward, we find the alien to be a revelatory, funhouse reflection of particular selves. James Cameron's 1988 science fiction film *The Abyss* offers an ideal-typic illustration. In this movie, the crew of an undersea oil rig encounters a submarine settlement of creatures resembling extraterrestrials; one of the protagonists, voicing a sense of an alien presence at the bottom of the sea, insists to her compatriots that "there's something down there—something not us." This "not us" has lessons for "us"; the ovoid-faced jellylike submarine creatures of this cold war allegory find themselves, like us terrestrials, wavering between war and peace. The oceanic other's uncanny ability in *The Abyss* to sculpt water into technologies of useful energy as well as of careless destruction mirrors humanity's own fateful facility with atomic power. The alien, here as more generally, is that which, as Jodi Dean has it, "reminds us that nothing is completely other (and everything is somewhat other), that the very border between 'like' and 'unlike' is illusory."[48] Aliens are figures through which, as Kathleen Stewart and Susan Harding phrase it, "the imaginary 'them' [becomes] the surreal 'us.'"[49] In other words, aliens are strangers, entities not yet—or not fully ever—friend or enemy, self or other.[50] Stealing a phrase from Stuart Hall, the alien could be glossed as constituted by "difference which is not pure 'otherness.'"[51] Aliens are life forms whose place in our forms of life is yet to be determined.

We meet other sorts of aliens in this book, too: oblivious unfamiliars, for example, like the methane-eating microbes of chapter 1, survivors from an earlier, oxygen-thin Earth whose out-of-the-ordinary metabolisms keep greenhouse gases out of the atmosphere, protecting aerophiles like us from being poisoned. Then there are the alien kin of chapters 2 and 6, hyper-thermophiles thriving in the extreme heat and pressure of deep-sea volca-noes. These creatures may be revealed through gene sequencing to be our aboriginal relations—or may turn out to be difficult to place because of their habit of exchanging genes laterally, confounding linear genealogies. They remind us that aliens often mess up lineages and confuse the unfold-ing of chronology itself.

Microbes traveling across ports in ballast water, like those I treat in chapter 4, confound space and place. Invasive species are emissaries of what Jean and John Comaroff call "alien-nature," a biotic world of illegitimate, inundating flows called forth by the shifting and contradictory dynamics of globalizing social forces.[52] Alien species are often dressed up in imageries of the primitive and rootless, tokens for anxieties about those ethnically or racially other to the people describing them. These are aliens defined as the opposite of natives and, often, as those outside nations as such. *Alien Ocean* may recall for readers the phrase "alien nation," beloved by science fiction screenwriters as well as paranoid antiimmigration editorialists.[53] An alien nation might be a nation that comprises aliens. Or, more radically, it might undo the idea of nation as such, suggesting an entity constituted by strangers to the very thing to which they belong. Alien oceans are oceans formed of already mixed-up lineages or of those outside, beyond, or within known oceans.

Some readers may object that I have not written *Our Oceans, Ourselves*, a book that would highlight human intimacy with the sea, that would emphasize a sense of oceanic communion. That other book is indeed here, though submerged within *Alien Ocean*, which, as I was writing, absorbed that kindly text. *Alien Ocean* is skeptical of any simple identification with the sea, pessimistic about whether scientific knowledge alone about the ocean is enough for making sense of it (let alone protecting it), and insis-tent that all accounts of the sea are partial and that therefore there can be no such self-evident category as "our oceans." I do not mean that people do not or should not cultivate affinities with oceans, that we should not be concerned about worrisome ecological changes in Earth's waters. I ask that readers pause to consider the ocean's difference from humanity as well as critically examine how, by whom, and with what effects that difference is narrated.

ANTHROPOLOGY AT SEA

An oft-neglected arm of anthropology, maritime anthropology, has described how people imagine and encounter the sea, frequently examining the lives of fisher people—how they view the ocean, navigate sea surfaces, acquire fish for food, adjudicate property relations in fluid domains.[54] The ocean of nineteenth-century whalers is not that of twentieth-century surfers, the ocean of Icelandic fishers not that of Hawaiian canoe clubs, the ocean of Darwin on the *Beagle* not that sea explored by biologists looking for hyperthermophiles in *Alvin,* the noted oceanographic research submarine.

Gislí Pálsson offers that anthropology itself emerges out of colonial engagement with oceans: "As a result of voyages by sea, different and isolated worlds were connected into a global but polarised network of power-relations. Prior to these voyages, the idea of anthropology did not exist. In a very real sense, then, anthropology, the study of humanity, is as much the child of seafaring as of colonialism."[55]

Many inaugural anthropological journeys began with sea voyages. Bronislaw Malinowski, a founding figure in British anthropology, who in the 1920s described fishing magic in *Argonauts of the Western Pacific,* offers a fieldwork arrival scene in which the history of colonialism and anthropology melt into one another: "Imagine yourself suddenly set down surrounded by all your gear, alone on a tropical beach close to a native village, while the launch or dinghy which has brought you sails away out of sight."[56] Malinowski's sojourn in the Trobriand Islands, off New Guinea, during World War I has an important watery condition of possibility. Malinowski was an Austrian subject working in Britain when the war broke out and was permitted to substitute a stay in the western Pacific for internment as an enemy alien in the United Kingdom.[57] His ocean was a space between places.

Oceanographers have taken this space as their central subject, seeking to comprehend the ocean within the strictures of natural science. Anthropologists want to know whether and how scientific accounts of the ocean might also be crosscut by currents of belief and practice that link laboratories and research vessels to wider cultural sentiments about the ocean—as, for example, a strange and sublime wilderness, a zone of unclaimed or common-use resources, a site of coastal leisure, or a stage revealing the carelessness of humanity. Certainly, some of these seas are familiar to me—having grown up first in New England and then, after my parents' midlife crises brought them west, on the beaches of Southern California (where my father and I promptly almost drowned). My recent visits to Encinitas, where I graduated from a high school that offered class credit for surfing, have been marked not only by a wistful sense of a beach-going adolescence passed but by a growing awareness that the sea may never have been as pristine as I had imagined.[58]

The question of how sentiment and science about the sea inflect one another is the question this book tackles for contemporary marine biology, a task which—aside from examinations of applied fisheries biology—has not been approached by maritime anthropologists. Social scientists have conducted studies of shipboard interaction, relations between scientists and funding agencies, and the complexities of mapping the ocean.[59] The earliest approximation of an ethnographic account of marine biology comes to us from the assistant ship's steward on Britain's HMS *Challenger* during the first oceanographic voyage (1872–76).[60] Joseph Matkin wrote in letters home about living "at sea with the scientifics."[61] Today, as marine science sees fewer boat-borne naturalists handling submarine oozes and more lab-lubbing geneticists sequencing DNA from the sea, tracking the making of the twenty-first-century ocean requires the ethnographer to follow people to shore, to labs, and onto the Web to see how the marine world is being envisioned anew.

Tacking between Web, lab, and sea well describes the fieldwork I undertook researching this book. I began by surfing the Web, reading scientific papers, attending lectures, going to conferences, speaking with marine biologists, visiting labs. In 2000, I got in touch with DeLong and other scientists at the Monterey Bay Aquarium Research Institute (MBARI) in Moss Landing, a private research center founded and funded by computer magnate David Packard. I returned to MBARI for four months in 2003, participating in everyday labor in DeLong's lab and on institute ships. This work initiated contacts with marine biologists around the United States. I spent three months in Hawai'i in summer 2003, speaking with marine biotechnology researchers, evolutionary biologists, and native activists on O'ahu. Starting in fall 2003, when I joined the MIT faculty, I attended the biweekly lab meetings of Penny Chisholm on a regular basis for a year and a half, and

I made frequent trips to the Woods Hole Oceanographic Institution on Cape Cod. Contacts at MIT allowed me to sign on for ten days to an oceanographic research trip to the Sargasso Sea on the research vessel *Endeavor* as well as to join an eighteen-day expedition on the research vessel *Atlantis*, with the deep submergence vehicle *Alvin*, to the hydrothermal vent fields of the Juan de Fuca Ridge, 2,200 meters below the surface of the eastern Pacific. I also attended a variety of research conferences—on deep-sea science, marine biotechnology, taxonomy, Earth system science.[62] I conducted formal and open-ended interviews (some fifty-six in all, primarily with biologists, though also with physical oceanographers, geologists, chemists, computer scientists, marine technicians, ship crew), asking people about their work, how they arrived at their interests, whether new technologies had transformed their methods or questions, what they considered open debates, how they saw their work contributing to a wider sense of the ocean world. I also participated in the everyday life of institutions I visited, not only attending talks and workshops but also joining scientists for lunch, dinner, drinks, recreation. To get to know aspects of marine microbiological practice well, I assisted with lab work, on land and at sea.[63] For historical background I could not get from interviews, the library, or the Web, I conducted archival research at the Scripps Institution of Oceanography, at Woods Hole, and the Hawai'i State Archives. Because the community working in marine microbiology is small, I have decided not to use pseudonyms for established researchers who would be easily identifiable from any description I could give. People named have given approval for my use of quotations. I have given pseudonyms to graduate and undergraduate students.

Each site allowed me to encounter a different group of scientists, from people working at private research foundations, to those in the biotech industry, to academics in universities, and from disciplinary elders to college students. A distinct array of microbial life came attached to each site as well—from greenhouse gas–eating methanotrophs in Monterey Bay, to biotechnologically intriguing cyanobacteria and invasive species off the coast of Hawai'i, to environmentally important phytoplankton in the Sargasso Sea, to saltmarsh microbes in Georgia, to hyperthermophiles off Washington State, and to extraterrestrial analogs in marshes near Woods Hole. Each episode allowed a weave of encounters with contemporary technical and political enunciations of life, signaled by the ubiquitous and proliferating prefix *bio-:* biogeochemistry, bioinformatics, biodiversity, bioprospecting, biotechnology, bioinvasion, biohazards, and, finally, breaking the morphological rule, astrobiology. Although my research was centered in the United States, reports on international conferences make this study more cosmopolitan than its geographic parameters initially suggest. More, fieldwork I undertook in the Sea Islands of Georgia and in the Hawaiian Islands—locations where the American seascape runs directly into the nation's slave-trading legacy in the Atlantic and its frontiering history in the Pacific—pushed this research up against the margins of the American marine dream.

Oceanographers are accustomed to talking across academic disciplines. When I approached marine biologists as a curious colleague from anthropology, they accommodated my questions with relative ease.[64] Much of our interaction took the character of a conversation. To be sure, there were moments when exchanges stalled, or someone would instruct me about which side of a debate was correct—and whenever someone told me to discount another position on biological theory, ecology, or politics, my anthropological attitude sent me to speak directly with an exponent of the other position. I gave several talks outlining my anthropological questions to audiences of marine biologists—in Moss Landing, Honolulu, Cambridge, and even on board a research ship. My interest in transformations in marine biology became itself an object of notice among marine biologists. In one presentation I attended, MIT's Penny Chisholm gave as evidence of a sea change in her field the very fact that an anthropologist was interested in what microbiologists were up to.

Oceanography and ethnography have much in common. The convergences came to me most vividly when I was at sea one day in Monterey Bay. I was on a trip with scientists using a remotely controlled robot to gather marine microbial samples. Our presence at sea, it occurred to me, was fieldwork for both marine scientists and myself. If marine microbiologists are engaged in a kind of oceanography—a word that speaks to the immensity, if not impossibility, of writing down the ocean—ethnography shares the similarly difficult task of writing up an account of a people's practices (and certainly not always those of a unified ethnos) that can speak beyond an instance of fieldwork. Equally striking to me during this oceano-ethno-graphic encounter were our shared activities and metaphors of investigative, one-step-removed presence: marine biologists' immersion of devices, like their robot, in the deep sea, my immersion for a time in their social practice and language; their remote readouts of deep dynamics, my semidetached participant-observation. The more I thought about it, though, the stranger *fieldwork* seemed as a word for what we were doing: marine biologists were assaying an underwater environment in motion; I was following activities that would take me to labs and classrooms not always so clearly fenced off from my everyday academic life.[65] And just as marine microbiologists are edging away from the need to place their organisms "in culture"—seeing them as networks of genes, as "assemblages," instead—so anthropologists now resist seeing cultures as self-contained, coherent objects, centering our attention rather on crosscutting social, political, and economic difference, contradiction, and displacement, even as we do not abandon our attempts to write it all down. A better description of the common practice joining marine biology and anthropology might be that we each engage in representing entities—marine worlds, cultural beliefs and practices—subject to transformations of boundaries and substance, boundaries and substances called forth in part by our own representations and interventions.[66]

Lest this way of putting things suggest a flat aesthetic in which the anthropologist now represents marine biologists representing the sea, let me say that I am in search of what Latour calls more "indirect, crosswise, and crablike"[67] conversation between myself and the people of whom I write—what anthropologist Bill Maurer has named a *lateral* mode of engagement, where threads of connection between the lives of scientists and of anthropologists get tied in knots, do not exist solely in parallel or linear reflection but take on "other spatiotemporal or faunal formations, like the radiality of, say, starfish or sand dollars."[68] My anthropological work and that of my marine biologist interlocutors is entangled, like kelp in Monterey Bay—or better, like consortia of distantly related marine microbes pressed into association by common circumstance. We share puzzlements about how to conceptualize the links between the biological and the social, between life forms and forms of life, links not preformed but also performed and deformed by such puzzling.

To take just one example, both marine microbiologists and anthropologists have been concerned with the relation of natural classification to social practice; the genetic materials that hold the attention of marine biologists are not simply bits of newly visible information but also symbols of unexpected evolutionary relation, even kinship. Delivering explanations of microbial communities involves conceptual conundrums similar to those anthropologists face in drawing up social accounts; microbiologists Farooq Azam and Alexandra Worden write that assessing "functional diversity [i.e., what genes *do*] depends on the environmental context of microbial expression. The task concerns an age-old theme in ecology: scale, both spatial and temporal."[69] Anthropologists are also concerned with context and scale, examining the relevant relationships and social frames—households, neighborhoods, nation-states, or, in the present study, laboratories, disciplines, funding structures, epistemological communities—within which their subjects act.

ATHWART THEORY

In *Alien Ocean,* I read anthropological and oceanographic materials and theories through one another. In this process, I treat theories—whether in anthropology or marine biology—both as tools for explaining worlds and as phenomena in the world to be examined. I think of such tacking back and forth as working *athwart theory.* This is not the anti–philosophy of science offered by Paul Feyerabend in *Against Method* in which "anything goes"; it is precisely a method, one that does not take for granted the difference between things and forms of explanation or abstraction, tracing instead how these items exist in alignment and tension.[70] Neither is this working *against theory,* a practice proposed by Steven Knapp and Walter Benn Michaels to argue that all interpretations—whether they offer schemes for understanding everything or deny the idea of correct interpretation at all—founder on a "single mistake": the assumption that problems set by theoretical frames are themselves real.[71] Working athwart theory, I not only cross-wire ethnographic and marine microbiological theories (of, e.g., space, place, and community) but also claim that such transverse, oblique, operations can produce compelling renderings of a real world.

The aesthetic of representation that has animated the social sciences has often asked for a map, a scale model, of social reality, founded on representative sample sizes mobilized in the name of generalization. Many anthropological accounts, meanwhile, follow an aesthetic of perspective—seeking to capture various vantage points on a social situation—a pluralized collection of what Malinowski called "the native's point of view." Implicit in this vision has been the promise that different lines of sight might, through juxtaposition, triangulate upon a social terrain, producing a reliable map without necessarily covering all the spots in the territory.

Working athwart theory asks not for the isomorphism of direct representation, nor for the second-order objectivity of triangulation, but rather for an empirical itinerary of associations and relations, a travelogue which, to draw on the nautical meaning of *athwart,* moves sidewise, tracing the contingent, drifting and bobbing, real-time, and often unexpected connections of which social action is constituted, which mixes up things and their descriptions. Such an approach operates through not taking for granted a context within which a text or event will sit but rather creating and inhab-iting contexts along the way. It is not dissimilar to what Clifford Geertz called "thick description," though in this instance the anthropologist's multilayered account is interleaved with—even, sometimes, scribbled over and rewritten by—the interpretations of her or his interlocutors. As Marilyn Strathern observes, "The anthropologist's contexts and levels of analysis are themselves often at once both part and yet not part of the phe-nomena s/he hopes to organize with them. Because of the cross-cutting nature of the perspectives they set, one can always be swallowed by another."[72] "An anthropological voyage," affirms Hortense Powdermaker, "may tack and turn in several directions."[73] The itinerary offered in this book traces the convergences, complicities, and crosscutting incorporations that connect oceanographic and ethnographic concerns.

In crafting my crabwise narrative, I take a cue from the manifold prac-tices of microbes, making *associations*. Uncultured microbes of the sort studied by DeLong, after all, are often known only through association—and more particularly through *associates,* like clams at the bottom of the Monterey Bay that live in symbiosis with bacteria that process hydrogen sulfide. Throughout this book, microbes appear not only in connection with those scientists who would represent them but also alongside whales, jellyfish, tubeworms, coral, algae, sand dollars—associates I think of, fol-lowing Haraway, as mess mates, companion species.[74] These organisms are significant others both to microbes and to those of us who, when we think of the sea, have only peripheral associations with microbes and think more about big fish eating little fish, ice caps melting, or beached jellyfish sting-ing our feet. I mean association to retain a resonance with the social, even as I take it not to be simply reducible to human relations. As Latour argues in his account of the microbiology of Pasteur, association—a word used in biology to speak of symbioses—recognizes novel kinds of networked agents, human and nonhuman, in the drama of the sciences.[75] Importantly, not all are companions; some are *stranger species;* both are players in sym-biopolitics.

Assaying the alien ocean also requires the traditional sociological labor of following associations of scientists with one another. Such associations in this book feature the storyteller, me, the anthropologist, as one agent of connection. But my association with marine microbiologists is not simply that of microbiologist impersonator/alien auditor—what we in the anthro biz call a participant-observer. If "anthropology as social science is the study of alien encounters,"[76] I have not only met marine biologists in what UFO enthusiasts call a "close encounter of the fourth kind"—being taken on board their ships as a visitor—but consider that the work we both do is a practice of encountering that which we do not yet know.

I do not claim that the concept of the alien ocean captures all there is to say about how the ocean looks to contemporary marine biologists, in the United States or elsewhere. This ethnography makes not a general claim but an interpretative, heuristic one, offering ethnographically inspired associations and the frame of the alien ocean as a means for asking questions and discovering methods for arriving at answers.

HEADINGS

The chapters of *Alien Ocean* may be thought of as headings, not only in the bookish sense but also in the maritime sense: pointers in particular directions. Each chapter is also a channel, a passage into an aspect of the alien ocean. This book can be read linearly or channel-surfed.

Chapter 1, "The Message from the Mud," details the basics of marine microbiological seagoing and laboratory work and opens with a journey into Monterey Bay with members of Ed DeLong's Moss Landing lab. I follow MBARI microbiologists as they gather methane-eating microbes from the seafloor using a remotely operated robot, sequence these organisms' genes back on shore, and interpret microbial DNA for messages about climate, a process that has them work across a mix of media: seawater, televisions, gene amplifiers, petri dish cultures, computers. If fisheries management stood for the relevance of ocean science in the early twentieth century, climate monitoring plays that role now (it is fitting that Al Gore led the Ocean Conference; as I wrote this book, I came to think of marine biologists and myself as occasionally phasing into the lost universe in which Gore became president). Deep-sea microbes materialize as unfamiliar life forms whose activities have implications, even moral meanings, for forms of social life organized around fossil fuel use.

Chapter 2, "Dissolving the Tree of Life," investigates whether marine microbes might point to the genesis of life on Earth. The heat-loving archaea of deep-sea hydrothermal vents star in this chapter. I scrutinize debates about whether the genes of hyperthermophiles can draw lines back to the root of what biologists call the "tree of life," and I pay attention to the possibility that widespread lateral gene transfer—often glossed as the travel of "alien genes"—may undo this branching structure. A new form of thinking about biogenetic kinship may be in the making, a form that recognizes horizontal relation. At issue for biologists are narratives of origins and futures—as well as how lay publics, including aquarium goers and creationists, engage such stories.

Chapter 3, "Blue-Green Capitalism," tracks marine biologists in Hawai'i who have staked high hopes on employing microorganisms from Hawaiian waters as resources for industrial and pharmaceutical biotechnology. I analyze the lab and legal instruments through which scientists seek to turn microbes such as cyanobacteria—blue-green algae—into money, charting the rise and fall of the University of Hawai'i's Marine Bioproducts Research Engineering Center, a hybrid organization funded by the National Science Foundation and biotech startups. Drawing on ethnography on O'ahu in 2003, I trace the transfer of biological and intellectual property between academia and industry. I argue that comprehending how marine organisms are conjured into capital requires analysis of scientists' sentiments about the sea. In Hawai'i, Euro-American associations of the ocean with vacation, rejuvenation, and frontier space underwrite what I term "blue-green capitalism," where blue stands for the freedom of the open ocean and sky-high speculation, and green for biological productivity and ecological sustainability.[77] Such visions have run into resistance from Native Hawaiians concerned that "bioprospecting" the Islands' waters reinvigorates colonial practices through which they were alienated from their territory.

Chapter 4, "Alien Species, Native Politics," examines scientific accounts of introduced organisms in Hawaiian waters, juxtaposing these with Native Hawaiian parsings of ecological boundaries and integrity. Leaping up the planktonic size scale, from cyanobacteria to seaweed, "alien algae" are our guides into these worlds. Unlike microbes, visible algae densely entangle scientific and everyday interactions with sea creatures (and thwart, for a chapter, this book's fixation on microbes). Drawing on interviews with evolutionary biologists and Hawaiian activists, I show that the taxonomy and politics of alien and native species are slippery, especially in the sea. Examining worries about planktonic globalization, I show how contexts of locality and globality are themselves created.

The making of a global microbial ocean is the theme of chapter 5, "Abducting the Atlantic," which stays in sunny surface waters, though joins visions of the ocean as life-giving with conceptions of the sea as death-dealing. I analyze the entry into marine microbiology of J. Craig Venter, the man credited in 2000 with completing the human genome. In 2004, Venter set out in his private yacht to produce a genetic profile of microbial assemblages around the world, suggesting that he would sequence "the ocean's genome."[78] I compare Venter's work with the efforts of MIT-trained marine biologists who seek to place genetic "text" in ecological context and also place marine viruses in the mix with light-loving cyanobacteria. The chapter documents a trip I joined in May 2004 to the Sargasso Sea. Contrasting Venter's plan to snapshot the world's genome to the efforts of scientists trained by Penny Chisholm to chart a baroque network linking genes to organisms to ecologies, I examine how each encounters the ocean as a space to be broken down and built up again—on boats, in freezers, in databases. I describe this as the abduction of the ocean, where *abduction* refers to a logical operation in which information is understood in relation to contexts yet to come—such as a *global genome* or a microbially inflected model of Gaia. Uploading marine gene sequences to databases deterritorializes ocean ecologies, renders them virtual, a cyberspatialized domain that does not always map back onto ocean space. The chapter concludes with a visit to Georgia's Sea Islands, home to Geechee descendents of enslaved Africans and now also the site of the Sapelo Island Microbial Observatory. The Sea Islands reverberate with another history, the Atlantic slave trade, which still echoes through the racial economy of science, where African Americans remain underrepresented. Arrangements of place, space, and race reconfigure as informatically enabled marine biology remaps seascapes.

Chapter 6, "Submarine Cyborgs," returns to hydrothermal vents. I document my June 2004 dive in *Alvin*, the three-person submersible deployed by R/V *Atlantis*, to visit the Juan de Fuca Ridge, a system of undersea volcanoes 200 miles off the coast of Washington State. I employ the figure of the cyborg to describe *Alvin* and also to make sense of how biologists depict microbes as "little living machines." In my *Alvin* narrative, I veer away from a tale about what I saw, attending rather to what I heard. Examining the sound of my dive, I offer the metaphor of *transduction*—the transfer of signals across media—to think through how immersive cyborg presence—and, indeed, ethnographic presence—is produced. At chapter's end, I return to themes of bioprospecting, though now in waters outside national jurisdiction, beyond those 200-mile Exclusive Economic Zones that have fringed coastal nations since 1982's UN Convention on the Law of the Sea. I report on a meeting at the United Nations about governance of deep-seabed bioprospecting and argue that a cybernetic, self-correcting model of science infuses regulatory thinking on this matter. In the stateless sea, where the relation of life forms to territorial forms of life is in flux, I detect a political form of the alien ocean.

Chapter 7, "Extraterrestrial Seas," explores how marine microbes are treated as analogs for possible life on other worlds—aliens in a more cosmic sense—drawing on a 2005 Woods Hole workshop on astrobiology I attended at which marine microbiological findings fed into questions about life on Mars and Europa. I revisit themes of this book and push on the limits of life as a category of analysis. If the chapter you now read opens with images of microcosmic seas in drops of water, the final chapter does not so much scale discussion up to the macrocosmic seas of space as offer a vision of alien oceans as heterotopias, baroque topologies that mix the utopian and dystopian, the hospitable and the hostile.

Alien Ocean asks after realms of the sea usually out of sight and reach—the subvisible world, the deep sea, and areas outside national sovereignties. The microbe is our guide into these zones. In asking how to represent these realms, *Alien Ocean* also asks how to imagine human presence—ethnographic, oceanographic—in or in relation to these realms. Indeed, this book wonders whether "life itself" is any longer based on the presence of organisms to themselves. It also looks over the shoulder of marine biologists who wonder whether humans will continue to be present on Earth, whether the absent fishes of the world's seas are premonitions of our own vanishing.

The idea of an alien ocean filled with surprising life forms offers a disturbed mirror for terrestrial concerns, for uncertainties about the viability of contemporary forms of life, particularly those forms, like science, founded on the promise of the ordering power of representation. In this sense, the aliens sighted in this book are symbols—of insecurity, though also wonder and amazement—and this book constitutes a contribution to symbolic anthropology, the study of what Geertz called "webs of significance," "the layered multiple networks of meaning carried by words, acts, conceptions, and other symbolic forms."[79] However, in investigating the symbolism of the alien ocean, I am careful not to get tugged into the undertow of limitless, off-in-all-directions interpretation. Considering the ocean—so often described as the most protean of places—to be endlessly symbolic consigns it to the empire of the elusive other. More, the symbolic itself, associated at least since Hegel with the ambiguous—one symbol can have many meanings, many meanings more than one symbol—has often signified the very essence of otherness and unfathomable alterity.[80] Symbols must be interpreted with respect to the forms of life within which they become meaningful. Let me offer just two historical, admittedly provocative, examples of how one might read ocean organisms as symbols. Toni Morrison argues that Melville's *Moby-Dick* is about America's vexed pursuit of whiteness on the eve of the Civil War, with the white whale standing as the unearthly symbol of that obsession.[81] More recently, cold-war, space-age interests in communicating with dolphins, advocated by scientists involved in the Search for Extraterrestrial Intelligence, can be read as standing for an optimism that other intelligences might be wiser, and less warlike, than our own.[82]

What, then, might contemporary microbial oceanography and the figure of the marine microbe—alien, fascinating, strange, dangerous—tell us about American preoccupations today? In 2001, when I began planning research for this book from a postdoctoral post at New York University, I charted possible directions for my anthropological investigations. My imagination was abruptly arrested after witnessing the collapse of the Twin Towers, and for a long while, living and teaching in the disquietingly named "frozen zone" below 14th Street in Manhattan, I was unable to rescue any narrative line, let alone significance, for my project. Months later, I opened the pages of *Moby-Dick* and came upon a passage in which Melville's Ishmael reflects on the inconsequence of his heading to sea. The eerie, oracular character of this passage jolted me back into a salutary sense of history. Ishmael imagines his journey "a sort of brief interlude and solo between more extensive performances"—thrown into relief by the events of the day, in the 1840s, which Melville conjured as

Grand Contested Election for the Presidency of the United States.
WHALING VOYAGE BY ONE ISHMAEL.
BLOODY BATTLE IN AFGHANISTAN.[83]

These headings, bizarrely prefiguring headlines in 2000 and 2001, speak to the history of governance and colonial adventure, and Ishmael's small story is, of course, part of the drama of his day: the imperial depredation of the seas, the passing of the great age of cetaceans, pushed to their limits. Melville pushed me to ask: In what fresh histories are today's ocean organisms and the people who study them enmeshed? Which histories might be repeating themselves? How might it matter that the scientists of whom I write struggle to make sense of their research in institutions sited in the United States—and in a cultural and political milieu that radically transformed in the wake of the contested election of George W. Bush, the historical fracture known as 9/11, ongoing battles in Afghanistan and Iraq, and White House–led challenges to scientific expertise in arenas reaching from evolution, to climate change, to biotechnology?

To anticipate and hypothesize: the United States is a nation in a tug-of-war about the relation between its sociopolitical forms of life and what will count as life forms—as a glance at debates about creationism, evolution, abortion, racialized medicine, global warming, and stem cell therapy might suggest. If the bony and blubbery body of the whale was a surface upon which were written nineteenth-century stories of nature and race, the genes of marine microbes are texts upon which marine scientists seek now to inscribe commitments to reasoned thinking and teaching, responsible ecological stewardship, and promising biotechnology. The apparition of the alien signals uncertainty about the classification of life forms, about the stability of any national (or, indeed, international) framing of marine science, and therefore introduces and bespeaks an indeterminacy about how to make such unsettled life forms meaningful for forms of life.

1 The Message from the Mud

Making Meaning Out of Microbes
in Monterey Bay

Call me distracted, divided, but as I head to sea this Friday in March 2003, I already wonder how I will render my oceangoing experience into text. I am not alone. The marine biologists swaying back and forth around me are also preoccupied with representations to come, with setting sampling and recording devices in place to aid in piecing together a portrait of the waters we will visit today. We are all a bit groggy in the morning fog, just finding our feet on the drizzly deck of the *Point Lobos*, a small oceanographic vessel operated by the Monterey Bay Aquarium Research Institute (MBARI). The ship is gliding out of Moss Landing, California, a tiny coastal hamlet between Santa Cruz and Monterey. The town is harbor to fishing boats with names like *Desperado* and *Baits Motel*, home to antique stores cluttered with compasses and harpoons, and hub to marine research devoted to mapping both the undersea Monterey Canyon and the genes of its resident marine life. I have joined the *Lobos* to conduct anthropological fieldwork about and alongside marine biological fieldwork.

This will be a one-day expedition to scout for microscopic ocean life crucial to the balance of the biosphere. The chief scientist on board, marine biologist Pete Girguis, briefed me a few days ago over a preparatory calamari lunch. He told me we would be looking for deep-sea microbes that eat methane, a potent greenhouse gas. Residing in muddy methane-rich zones of the seafloor known as "cold seeps," these microbes are often found in sediment beneath vesicomyid clams and vestimentiferan tubeworms, themselves thriving in intimate relation with bacteria that live off compounds poisonous to most creatures. The goal of our trip today is to dredge up sediment from the bottom of Monterey Canyon, muck heavy with methane-metabolizing organisms, so that, as Pete puts it, he and his colleagues can decode "the message from the mud." This message, stowed in the cells and genes of methanotrophic (methane-eating) microbes, will eventually offer these scientists insights into how such teensy creatures may be linked to global biogeochemical processes, those interconnected biological, geological, and chemical cycles that sustain the Earth system. It will tell a story about how the smallest scales of marine biology affect the largest lattices of life.

In assembling the materials and equipment that will make the message from the mud intelligible, these scientists work with a variety of media: seawater, cameras, computers, deep-diving robots, petri dishes, DNA libraries. By *media,* I mean substances, channels, or instruments through which forms of action are propagated. These marine biologists engage with *mediation*—watery, televisual, digital, biotechnological—at every step in their journey, from data collection to analysis. As I follow Pete and company from sampling sludge to sequencing genes, from their Monterey Bay boat to their Moss Landing lab—the trajectory I trace in this chapter—I will see the scientific sea manifest as a media ecology, a complex of material, meaningful relations among researchers and their objects of study, relationships structured by techniques of perception and communication.[1] I will learn, too, that the message from the mud depends not only on the media through which it is transmitted and translated but on who is reading and with what sorts of interpretative habits. MBARI researchers' exegeses of marine microbial texts are animated by environmental and ethical imaginations that have the ocean oscillating between the immersively immediate and the disorientatingly different, an alien medium intimately yet opaquely implicated in human affairs. For some of these secular scientists the microbial sea, made up of out-of-the-ordinary life forms with lessons for humans, even materializes as a quasi-spiritual medium.

THE RESEARCH VESSEL AND THE ROBOT

As we slowly swerve away from shore, the seven crewmembers of *Lobos* hustle across well-worn paths, securing swinging doors and tying necessary knots down the length of our 110-foot ship. Though *Lobos* usually carries some six people in the science party, today we are only four. Pete, coordinating the cruise, works as a postdoctoral fellow under MBARI microbial biologist Ed DeLong, whose work in environmental marine genomics, the sequencing of DNA from ocean water, is getting wide notice. In his early thirties, Pete is an amiable bear of a man who sets everyone at ease with his generous humor and quiet ability to cajole heterogeneous collections of oceangoing apparatus into coordinated action. A graduate student from nearby UC Santa Cruz whom I will call Adam wears a wool cap, windbreaker, and khaki pants that mark him as an outdoorsy type; he has signed on to participate in sampling and will be interested in extracting genetic material from the mud we collect. Nadine, dressed in a ship-smart slicker, is an assistant to an MBARI marine geologist and has been charged with learning about the ship's global positioning system.

We are just getting out to sea when the drugs begin to take hold. Nadine tells me she has taken meclizine. Adam has downed six Dramamine and four Vivarin. Pete promises that he, too, is well prepared, though he jokes that he has steered clear of scopolamine, which makes people hallucinate at sea. I, foolishly, have taken nothing against seasickness. Having ascertained that everyone else is dosed up, I nervously consult my stomach and listen with half an ear as Adam tells me about his fascination with genetics: "It's all the same. You can't tell bacterial and human DNA apart at first glance."

Perry Shoemake, deckhand on *Lobos*, approaches and asks us to sign waiver forms. We all hiccup over the phrase "including death." But, steeled by an odd conviviality prompted by this amusingly bureaucratic reminder of our mortality, we sign. After we ponder our deaths at sea, Perry, relaxed and reassuring, gives us a briefing on life—or, more exactly, on life jackets and lifeboats. Life, apparently, is about floating. And floating at sea, Perry says, is often about throwing up. He shows us the "place to yak," an area on the port side of the boat, a location, he instructs us, not monitored by the ship's cameras.

The boat is studded with surveillance for two reasons. One is safety—to make sure scientists do not wash into the sea. The other is to transmit images to the Monterey Bay Aquarium, a major tourist destination just down the coast dedicated to making the oceans ever more visible to ever more publics. The fish tanks at the Aquarium invite patrons to "come closer and see." The ship's cameras allow patrons to come closer and see us.

But the most important cameras on *Point Lobos* are not attached directly to the ship itself. They are built into the massive, 2.5-ton robot sitting on deck. This is *Ventana*, a remotely operated vehicle (ROV) that can be dispatched into deep water off the side of the ship (figure 4). True to its name, *Ventana*, Spanish for "window," offers a framed glimpse into the deep, a once-removed promise of transparency, an encounter with the sea alternating between immediacy and distance. *Ventana* receives commands through a stream of fiber optic cables running from the ship through what researchers call an umbilical cord connecting the robot to the *Lobos* onboard command center. This tether allows the ROV to travel down to

FIGURE 4. ROV *Ventana*, launching. Photo by Kim Fulton-Bennett ©2004 MBARI. Reproduced with permission.

1,500 meters, a zone of enormous pressure and heavy darkness, which *Ventana* can illuminate with full-spectrum and incandescent lamps. The images *Ventana* captures from the depths are transmitted up to the ship's control room, where they can be monitored in real time and, if desired, telemetered via microwave to shore and uploaded to the Internet, where a curious, clicking public can surf into a digitized deep.[2] The robot is outfitted, too, with remotely operated manipulator arms and a suction sampler, for collecting things like clams and tubeworms.

    Ventana encapsulates a technological history of deep-sea sensing. During the first oceanographic voyages, in the nineteenth century, ships such as Britain's HMS *Challenger* drew their knowledge of the abyss from dredging—bringing up objects from the bottom of the sea using buckets attached to piano wire. *Ventana*'s metal manipulators will later today deliver up coarse mud, sediment not too different from that curiously caressed (and sometimes tasted) by Victorian naturalists on *Challenger*. Sonar (SOund NAvigation Ranging), an invention of the early twentieth century created to detect submarines, also finds its place on *Ventana*, an aid to steering the robot. Sonar affords a dimensional portrait of the deep unavailable through the patchwork deployment of sounding lines. In her history of wire and acoustic sounding, Sabine Höhler argues that sounding with sound marked an arc toward visual representations of the deep: "Oceanographic research commencing in the mid-19th century could not rely on the direct observation of its object, but had to create its images of ocean depth through remote investigation. Depth became a matter of scientific definitions, systematic measurements, and graphic representations. In the course of a century, the *opaque* ocean of the 1850s was densely depicted in physical terms and transformed into a technically and scientifically *sound* oceanic volume."[3]

Piggybacking on innovations in underwater photography pioneered by the likes of Jacques Cousteau, ROVs now afford optical access to bits of this oceanic volume. The sensory trajectory through which the deep sea has been scientifically apprehended has traveled from the tactile, to the auditory, to the visual, with the submarine world becoming at once more intelligible and more fantastic. As I look into *Ventana*'s prominent zoom lens, attached to a Sony HDC-750A high-definition camera, I am reminded of the earliest exploits of William Beebe, inventor of the bathysphere. Crouched in his metal ball in the 1930s off the coast of Bermuda, Beebe looked out of his tiny portal and, speaking through a telephone wire run to the surface, delivered some of the first vivid descriptions of the abyss, later translated into watercolors by his associate Else Bostelmann. Rhapsodizing about the realm below, Beebe wrote: "The only other place comparable to these marvelous nether regions, must surely be naked space itself, out far beyond atmosphere, between the stars . . . where the blackness of space, the shining planets, comets, suns, and stars must really be closely akin to the world of life as it appears to the eyes of an awed human being, in the open sea half a mile down."[4]

This analogy to the cosmos has gathered gravity with *Ventana,* which, with its adaptations to extreme pressure and temperature, is sharply sculpted by space-age technology.[5] In trading telephones for telepresence and watercolors for the World Wide Web, the half-million-dollar *Ventana* also allows the ocean to be plugged directly into that peculiarly contemporary technoscientific medium, cyberspace.

The use of computer technology to sound the sea is a central goal of MBARI. Computer entrepreneur David Packard, of Hewlett-Packard, nurtured an interest in marine technology and founded the institute in 1987. The $60 million Aquarium, inaugurated in 1984, was inspired by Packard's daughters, both marine biologists; Julie Packard has long been executive director. MBARI was chartered to satisfy David Packard's more technical, research-oriented leanings. Known for his skill in winning military contracts for his company, Packard served in the 1970s as deputy secretary of defense under President Nixon, managing far-flung U.S. naval forces. By the time he left this position, he had gathered a store of knowledge about classified national marine technology. When the cold war ended, this intelligence, coupled with his wealth, put Packard in a unique position to move military technologies into the private sector as well as to commission custom-built devices like *Ventana*. MBARI has become a key player in developing robotic and telepresence techniques to research ocean ecosystems. In recent years, MBARI scientists have also been drawn to genomics, sequencing DNA from sea creatures like those we hope to gather today.

Ventana has been outfitted by operations technician Mark Talkovic with a bank of plastic cylinders that will be pushed into the seafloor to collect cross sections of methane-infused ooze. Looking for signs of life at the bottom of the ocean turns out to be a relatively recent possibility, not just technologically but epistemologically. In the early nineteenth century, naturalists thought the deep to be devoid of life, in part because of a prevailing belief that seawater was compressible—that, as James Hamilton-Paterson puts it in his remarkable book, *The Great Deep*, "seawater grew more and more solid until a point was reached beyond which a sinking object would sink no farther. Thus, somewhere in the middle regions of the great abyss, there existed 'floors' on which objects gathered according to their weight. Cannon, anchors and barrels of nails would sink lower than wooden ships, which in turn would lie beneath drowned sailors, who themselves lay at slightly different levels from one another, depending on their relative stoutness, the clothes they were wearing, and, quite possibly, the weight of their sins."[6] The ocean in which *Ventana* descends is no longer the thickening realm of the eerily lifeless. Nowadays, it is a medium layered with living things all the way down.

As I look at *Ventana*'s plastic cylinders, called pushcores, Adam tells me about marine mud. Among the DeLong lab's many interests, he explains, is a process called anaerobic methane oxidation, the metabolic breakdown of methane without the aid of oxygen. Microbes called methanogens—so named because they generate methane as a byproduct of their energy metabolism—are well known; they live in such comfortable environs as cow stomachs. Less understood are close relatives of these creatures that use methane as a source of carbon and energy. This process represents a kind of chemosynthesis, the production of organic materials without, as in photosynthesis, the need for light as an energy supply. Adam and Pete want to get hold of genetic material from microorganisms flourishing in the sunless settings of cold seeps to see if these creatures are capable of making enzymes that could address the question of whether methane-eating microbes are involved in Monterey anaerobic methane oxidation, and whether methanotrophy is a form of methanogenesis in reverse.[7] This is something the DeLong lab has been investigating since 1999. A key piece of the metabolic puzzle comes from work by outgoing MBARI postdoc Victoria Orphan, who provided evidence that anaerobic methane oxidation might unfold in multispecies microbial collectives, with methanotrophs oxidizing methane to carbon dioxide in coordination with microbes reducing sulfate.[8] Adam points out that methane oxidation is environmentally important because almost 75 percent of methane emanating from seep and other methane systems may be consumed by these "bugs," prevented from entering the atmosphere as a greenhouse gas.[9] If it were not, he laughs, we humans would be in "deep shit," smothered in methane.

Stopping himself abruptly, Adam points to starboard: "Look, there's a whale!"

I miss it.

We head inside to study a wall map of Monterey Canyon, on which serpentine lines representing the canyon's branches snake out from a shore-based origin point at Moss Landing (figure 5). This sonar atlas of a sucked-up sea delivers a parched blueprint of a topography invisible to an airborne eye. It is an apt symbol of the transparency promised by remote sensing and digital imaging. Still frustrated by my missed whale, my eye wanders up to a bend where, I am informed, a remarkable biotic feature sits at the bottom of the bay, 3,000 meters down. This is a "whale fall," the sunken carcass of a giant cetacean. Whale flesh and skeletons can become hosts to riotous ecologies. A dead whale delivers a dose of nutrients to the bottom of the sea more potent than the steady fall of "marine snow," bits of organic detritus that continually drift down to the deep. A whale fall is given its first going-over by scavengers such as sharks, after which invertebrate worms settle on the bones. The decomposition of whalebone lipids produces sulfides off which symbiotic microbes residing within these worms thrive. Some scientists think cold-water enzymes breaking down fats within whalebones might be good ingredients for cleaning detergents—a twenty-first-century descendant of the whale oil of *Moby-Dick*'s day.

We arrive at our dive site, just 12 nautical miles out of Moss Landing, at 36.78° N, 122.08° W. Monterey Canyon, deeper than the Grand Canyon, escorts deep water close to shore, one reason MBARI is sited where it is. When I step out onto deck, Perry and Mark are readying *Ventana* for deployment. Suddenly the full heft of the robot is over the rails. It begins to sink. Adam, Nadine, and I head back inside to watch a television tuned to *Ventana*'s channel. Marine snow rushes up from the bottom to the top of the screen, telling us that *Ventana* is descending. We begin our hour-long

FIGURE 5.   X marks the spot of the *Point Lobos* destination, March 7, 2003. Map by Norm Maher. ©1998 MBARI. Reproduced with permission.

wait until the robot reaches the methane-saturated ecologies in which Pete is interested. The color of the water on the monitor begins to change as wavelengths of light taper off, from light blue, to deep azure, and finally to black. Descriptions of the deep as dark and *therefore* mysterious, full of secrets, unknown, draw on a reservoir of meanings that associates sight and light with knowledge; indeed, the word *theory* derives from the ancient Greek for "to look on" and "to contemplate." No surprise, then, that seeing through the opaque ocean has become the governing goal of oceanography, the grail of techniques of remote sensing.

The emplacement of the medium of ROV TV in the medium of water suggests a maxim for media studies. Like seawater, media such as radio, television, and computers are not merely material, not merely delivery systems for messages, but in our very conceptions of their substance and form come suffused with meaning. Just as the opacity of seawater has been historically associated with mystery (and just as the once-assumed compressibility of seawater suggested irretrievability), so, for example, the fractured presence promised by television partakes of prior associations of sight with both distance and immediacy. Marshall McLuhan once suggested that "the medium is the message"[10]—that is, that media extend and modulate our sensorium, that "the effects of technology do not occur at the level of opinions or concepts, but alter sense ratios or patterns of perception."[11] But, though sense ratios or patterns of perception do indeed morph with media, opinions and concepts cannot be so easily centrifuged out of the swirl of materiality and meaning that make media. Franz Boas, a founder of American anthropology, famously wrote his dissertation—in physics and geography—on the color of seawater, which he later argued depended not only on scientific measurements but on angles of vision, frames of reference, modes of assigning meaning, and the whole suite of habits and beliefs he would later call *culture*.[12] In the wake of Boas, cultures came to be understood as social media saturating action and perception: forms of life.

I head outside and squint to see if I can make out Moss Landing, my most recent frame of reference.

MOSS LANDING

Moss Landing draws together various histories, and a maritime anthropologist will detect the outlines of many seas in this town from which I embarked upon my research. Never much larger than its present population of three hundred, Moss Landing was founded in the 1860s by a Captain Charles Moss, from New England, in partnership with Cato Vierra, a Portuguese whaler. Together, the two built a port for shipping barley from nearby Salinas. Centuries before, this had been Ohlone land, and then, with the establishment of Jesuit and Franciscan missions, an outpost of New Spain. By the time the Boston Brahmin author of *Two Years Before the Mast*, Richard Henry Dana Jr., anchored at Monterey in 1835 during a stint slumming it as a sailor, this area was part of Mexico. When Moss and Vierra arrived, twelve years after the United States bought California from Mexico, they found the harbor occupied by a visionary Frenchman, Paul Lezere, who nurtured a failed dream for a futuristic seaside metropolis to be named the City of St. Paul. Moss's Landing (as it was first called) operated until 1874, when the transcontinental railroad reached nearby Watsonville and barge traffic ceased. In the 1920s the town saw resurgence as a whaling station in the twilight days of the industrial hunting of these mammals, and from the 1930s it was a sardine cannery site, until the fishery collapsed. During World War II, Moss Landing was home to an African American battalion stationed to defend the coast against Japanese submarines—a reminder of the wartime sea, a zone not just of myth and mystery but also of strategy and state secrets.

Moss Landing has been a haven for commercial fishers—though in the age of corporate fishing their small-crew endeavors have suffered tremendously.[13] The town shifted toward nautical nostalgia in the 1970s, with antique stores moving up from Monterey's Cannery Row.[14] Ecotourism has arrived, and whale-watching boats now share the harbor with fishing vessels. Moss Landing has become a destination for kayakers and surfers, eager to immerse themselves in the wilderness attraction of the Monterey Bay National Marine Sanctuary. With the founding of California State University's Moss Landing Marine Laboratories in 1965 and, later, of MBARI, Moss Landing has become a center of marine research.

My settling as an anthropologist into Moss Landing began during afternoons at the Haute Enchilada, a Mexican folk art café stuffed with pottery collected by owner Kim Solano. Here, I held my first conversations with marine scientists and chatted with local fishermen, antique dealers, and migrant workers employed in nearby artichoke fields. Most of the farm hands I met were Mexican, living in the area temporarily, wiring money to Jalisco and Michoacán. Advertisements on public buses I took around Monterey County announced cheap calls to Mexico, promising a more quotidian telepresence than that offered by deep-diving robots like the somewhat romantically named *Ventana*. Fishermen, mostly Anglo, held forth at the café about their diminishing catch—and sometimes about MBARI, which they associated with government agencies dishing out regulations, although MBARI does nothing of the sort. A few complained about the cost of cleaning their boat hulls, keeping their crafts clear of hitchhiking pests from other ports—a problem to which biologists called attention with pamphlets profiling Monterey Bay's "Least Wanted Aquatic Invaders," the area's "alien species." Tensions between fishermen and scientists became apparent to me when I strayed into Moss Landing's fishermen's bar, The Bear Flag, with Adam, who was teased about his pristine surfstore-logoed sweatshirt and short pants.[15]

Commercial fishermen feel themselves squeezed out not just by scientists but also by a growing number of modest pleasure craft and houseboats claiming places on the piers. I learned about the live-aboard community when I ended up boat sitting for a retired construction worker I met at the Haute Enchilada, J. P., a 73-year-old man whose glass eye and prosthetic leg bespoke an adventurous, if accident-ridden, past. J. P. had been drawn to life on a boat because he could live inexpensively in a harbor where monthly slip fees were low. I lived for a few months aboard J. P.'s Chris Craft, decked out inside like a groovy 1970s Volkswagen bus, complete with a shag carpet and a jumble of Native American curios celebrating J. P.'s one-eighth Yukot ancestry. My stay on the boat tuned me into a different sea, that of people on the margins of middle-class life, living at an ocean's edge they associated with self-sufficiency. Some crewmembers from *Lobos* lived in this harbor. J. P.'s generosity in allowing me to live rent free in his boat suited my circumstances; I had just moved to California from Manhattan and was living on New York state unemployment while I waited to hear about grant and job applications. Shuttling back and forth between the harbor and MBARI helped me see how thick Moss Landing was with various ways of encountering the sea.

SENSING AN EXTREME MARINE WORLD

Ventana arrives at the seafloor and now floats just off the bottom, a thousand meters below *Lobos*. The science party congregates in the ROV control room, a wedge-shaped chamber squeezed into the front of the ship, on the lowest deck. In the dim light we can discern the inside-out shape of the bow. Once upon a time, when this ship was an oilfield supply vessel named *Lolita Chouest*, this room, the forecastle, housed sleeping quarters. Where sailors once sunk into seesaw sleep, scientists now remotely ride *Ventana* into the deep, revealed to them in a stream of images delivered to the banks of video monitors that crowd this dark room.

This array of computer and video screens is overwhelming, a torrent of sense data that feels like a direct feed from what Kant once called the *mathematical sublime*, that domain of difficult-to-get-your-head-around measures and magnitudes.[16] Arranged into columns, shoulder to shoulder, are Hewlett-Packard computer monitors showing sonar from *Ventana*, real-time diagrams of the ROV's thrusters and fuel tanks, and scrolling profiles of oxygen, temperature, and salinity levels at the depths through which *Ventana* has traveled. Video screens range amid these monitors. Some present exterior views of the ship. The starboard-side video, taken from a camera on the bow, is unsettling; if the ship is pitching (an up-and-down movement from bow to stern), this screen presents us with the inverse of what our guts feel. One screen offers a glimpse into the auditorium at the Monterey Bay Aquarium, where we see a podium from which docents lecturing to visitors can contact *Lobos* for the latest cruise news.

The central images in the control room, the ones we are here to see, are those affording looks through *Ventana*'s lenses. Though *Ventana* carries several cameras, we are most interested in the forward view, presented on several screens, at many sizes. The most prominent is adjacent to a VCR, next to a monitor hosting a video annotation system called VICKI, for Video Information Capture with Knowledge Inferencing. Next to this is a monitor displaying "frame-grabs" from ROV video; frames are captured when a researcher clicks an icon in the annotation display. On this trip, I am that researcher.

Pete has stationed me in a chair in front of VICKI. He hands me a timer to remind me when to change videotapes. As I look around the room, each screen sways to its own rhythm, providing a view of, well, everything at once. Pete sits to my left, in front of a keyboard from which he controls *Ventana*'s cameras and lights. The two ROV pilots (members of the ship's crew) sit in the leftmost two chairs, outfitted with joysticks for steering, or, as they prefer to say, flying the ROV.[17] The pilots also control two robot arms with which *Ventana* can grasp instruments and pick stuff up. Nadine sits in back, studying the navigation console. We all wear headsets, allowing us to speak with one another over the hum of the boat. Pete addresses me: "Stefan, go ahead and snap as many frame grabs as you want of the pushcores going in, coming out, whatever looks good." A CD player percolates a gentle reggae. The room is a sensory scramble, a layering of ocular, auditory, and corporeal disorientations: a multimedia experience, a dip into the *media sublime*, an overpowering flow of mediated sense data (figure 6).[18] To get our bearings in this milieu, we project our presences into *Ventana*, whose cameras are to be our steady eyes.

We see a clam bed, a sign that *Ventana* is near a methane seep. Pete moves his fingers over a mouse pad to control the camera. We are searching for peculiar colorations of mud, signs of distinctive metabolism. Pete instructs ROV pilot D. J. Osborne on where to position *Ventana*. D. J., slowly transposing his bodily motion, a fragment of his presence, into the robot deep below, tries to be mindful of the clams.

D. J.: If I sit down here, I can try not to disturb them.

PETE: They're pretty happy. They're gigantic! [*Looking at camera controls, realizing things are not what they seem*] Oh! We're zoomed in.

FIGURE 6. Oceanic sublime meets mathematical, media sublime. ROV control room on the *Point Lobos.* ©MBARI. Reproduced with permission.

In "Seeing in Depth," anthropologist Charles Goodwin argues that on oceanographic research vessels "perception is something that is instantiated in situated social practices, rather than in the individual brain," and he maintains that an "architecture for perception" is built into such ships, with oceangoing scientists and crew enacting "not simply a division of labour, but a division of perception."[19] Just so, Pete and the ROV pilots behave as different parts of a distributed body, adjusting to each other's perceptions and prosthetics. One pilot is hands, another, wings. Pete serves as the guiding gray matter. I act as a bit of memory, though, like everyone else, I am also part of a compound eye, grafted to screens of data which, as Goodwin would have it, "provide not just a window into the sea, but the resources required to move other inscription devices within it, including some of the machines that are producing these very representations."[20]

Classical sociologies of relations between scientists and ship's crew comment upon the different expertise these parties have, on how class hierarchies are often reinforced through divisions between "mental" and "manual" labor.[21] Though there are mechanical aspects of managing *Ventana* scientists would be ill prepared to undertake, operations in the ROV control room put scientists and crew into tighter relations of interdependence than typify much sea science. Scientists have to know how the robot works in order to think about sampling, and pilots have to be able to intuit what scientists might find interesting. Together, they enact what Edwin Hutchins, in his ethnography of navigation on board a navy ship, called "cognition in the wild," converting and communicating information through a cascade of representations channeled through analog and digital media and propagated through different people's embodied expertise.[22] Still, scientists are understood to direct the action:

PETE: Tubeworms! We landed right on them. Can I zoom in a little bit? Look at all of those! Let's do a flyby. Is it easy to get some of those tubeworms?

D. J: We can't guarantee that we'll be back to the tubeworms, so let's do it now.

PETE: Stop here. Come wide. Zoom in. There's one right next to your wrist there.

Pete means the wrist of *Ventana's* claw, not the pilot's closer-to-hand hand. The substitution is telling, a sign of presence transported. The ROV control system uses software similar to that of the Sony PlayStation, making joystick controls transparent to those, like younger ROV pilots, who have played first-person video games.

In her sociological account of deep-sea research, Chandra Mukerji argues that with such familiar techniques, benthic environments "begin to become less natural environments than areas that have characteristics of both the natural world and the social world of science. The expression of signatory techniques through the manipulation of equipment gives scientists a way to assert their culture, and not become overwhelmed by the scale of the ocean."[23]

In the language of flying that accompanies the operation of the ROV, we might hear a further assertion of the cultural command Mukerji describes. To be sure, this way of speaking, along with the notion that the ROV offers an extension of self, signals that people on *Lobos* construe their activity as exploring. But to imagine scientists on *Lobos* hungering after some exterior, transcendent position would be to miss the more intimate relations they develop to their subjects of study, relations for which the medium—water, video—is vital to the message. On *Lobos,* the sensation is not of detachment from nature but of a pleasurable, technological immersion *in* it—an experience of being "in the field" at once immediate and hypermediated. Media theorists Jay David Bolter and Richard Grusin offer that "our culture wants both to multiply its media and to erase all traces of mediation; ideally, it wants to erase its media in the very act of multiplying them."[24] Yes, but the world of *Lobos* also overflows this ideal. On board, awareness of visual mediation—of the always-negotiated social and technical boundaries of the interface—adds to and creates the sublime shudder

FIGURE 7. VICKI frame grab of *Ventana*'s vantage point as the ROV arm grasps a tubeworm. ©2003 MBARI. Reproduced with permission.

of encountering an alien world. The experience of a simultaneously immediate and hypermediated ocean emerges from a grappling dynamic between surfaces of flesh and machines that Haraway, following phenomenologist Maurice Merleau-Ponty, calls "infolding."[25] The technologically enabled origami of infolding sees elements of culture and nature, ROV and ocean, trading places, delivering an alternation between remote and intimate sensing, a rising and falling immersion in a multimedia ocean.[26]

After a few foiled attempts at wresting tubeworms from their moorings—"Did we lose that little guy?"—D. J. fastens the ROV arm onto a likely candidate. I am still next to Pete, in front of VICKI, and he instructs me to take frame grabs of the tubeworms we have now found near a methane seep (figure 7). These tubeworms are noteworthy because living inside their tissues are microbes that metabolize sulfide. This indicates that we might be near sulfate-reducing microbes, which, as Adam told me, some microbiologists suspect join with methanotrophs to accomplish anaerobic methane oxidation.

We are far from the sunny, humanly familiar ocean here. In fact, one biologist at MBARI argued to me that, rather than demonstrating that the sea is a zone of ecological harmony, sites like methane seeps should lead us to see it as a giant refuse heap; life exists here not because the medium is friendly to life but because life is so adaptable, can make its way in the most noxious environments. I am not sure this attitude is fair to tubeworms and their symbionts, though it is true that they have been persistently associated with the creepy, unearthly, even extraterrestrial—a chain of association more sensational than logical (one could argue that anaerobic methane oxidation is the most "earthy" process there is). But the alien connection is prevalent, particularly among scientists who see these critters as analogs for extraterrestrials. Pete tells me some enthusiastic artists have plopped tubeworms into fanciful seascapes of alien planets, a graft that makes little sense, since if metazoans evolved on other worlds it is unlikely they would resemble earth creatures.

Science fiction turns out to have been one of Pete's inspirations for becoming a deep-sea biologist. He tells me of having been a kid riveted by the television show *Star Trek*, by the idea of traveling and doing science in three dimensions (fittingly, the starship *Enterprise* and *Lobos* each employ viewscreens to convey images of an unfamiliar, exterior realm). As a teenager, Pete was also fascinated by deep-sea documentaries and took a keen interest in the migration of whales, reinforced by family trips to Sea World. He particularly enjoyed *Star Trek IV*, in which the crew of the *Enterprise* travels back in time to twentieth-century Earth to save the whales (specimens of which they find at a fictional "Cetacean Institute," portrayed in the film by the Monterey Bay Aquarium). Just as seventeenth-century European travelers to seaside Holland were trained by Dutch seascape paintings in how to see the sea, so Pete's eyes have been prepared for the deep by ecologically themed media entertainment and science fiction.[27] Not surprisingly, the environmentalist message of *Star Trek IV* is of a piece with MBARI's mission to understand the local waters of the central California coast. The message from the mud, I am coming to understand, is likely to be an ecological message, perhaps a message of warning to humans about fossil fuel use, agricultural intensification, and waste disposal. As at the Aquarium, which features primarily local sealife, "the old tropes of 'nature' as outside culture" are, as Eva Hayward puts it in her study of this site, "transplanted with ecological epistemologies of conservation."[28] In this ecological linkage between life forms and forms of life—famously articulated in the 1950s by Rachel Carson, whose books on Earth's endangered seas are easy to find at the Aquarium—we humans are nestled within ecosystems to which we have ethical responsibilities.

Pete says he was drawn to *Star Trek* because the scientists were not always perfect; they made mistakes.[29] Having been brought up in an Egyptian American Coptic family that urged caution about adopting the tenets of Darwinian evolution, he was aware of the need to be prudent about the limits of science. At the same time, Pete was taught respect for empirical ways of knowing; his mother had been a support engineer for some Apollo missions to the moon.

By now, a whole bush of tubeworms has been plucked free of its moorings—a process I have been uneasily documenting with frame grabs (does yanking up benthic life under incandescent light really constitute environmentally friendly science?). Later, I will visit the MBARI video lab, where such images are annotated. Thousands of tapes have been made using *Ventana* since 1988, in assorted formats. The three women who work in the video lab—a windowless, temperature-controlled room on shore—review footage and mark which creatures show up when. One of the video librarians, who holds a master's degree in invertebrate zoology, tells me their database has become unruly as organisms have changed names in response to genetically inspired rewritings of old classifications. Another calls bemused attention to the gendered division of land-and-sea labor that sees women like herself running between kids and nine-to-five work while a male ship's crew delivers ROV videos peppered with sometimes off-color jokes about the shapes of underwater formations.

We travel farther with the ROV. We see bamboo coral. Another push-core is pressed into the mud, capturing a clam. "This guy comes back with us!" says Pete.[30] I glance at my monitor and see a digital count of Greenwich Mean Time, tweaked to display numerals melting, Dalíesque, into their successors. Surrealist invention is allowed to flower in the corners of our computer screens, but an aesthetic of realism is sternly enforced for the screens delivering images from *Ventana*. We are meant to be watching a sort of real-time documentary about extraordinary things, not, say, a high-definition version of the bizarre works of Jean Painlevé and Geneviève Hamon, twentieth-century French filmmakers famous for their far-out movies of sea creatures, in which the viewer is constantly reminded of how much cinematic prodding it takes to make human eyes get their bearings in the refracting realm of the sea.[31] Still, looking at, through, and into water requires some tangling with *theory underwater*, recognizing that ways of seeing, schemes of explanation, are always informed, performed, and deformed by their medium. Pete continues in movie director mode, panning and tilting amid the ship's pitch, rock, and roll.

I realize that my stomach is rolling. I get up to find some air and no sooner do I step outside, onto deck, than vomit leaps out of my stomach into my ballooned mouth. The video timer goes off in my pocket. I race to the yak zone. Other people in the science party have gotten sick, too, in spite of the drugs. Adam remarks that it is hard to do science under these conditions; you're either sleepy or sick. *Lobos* is notorious for making people ill. A poem about the ship penned by an MBARIan ends,

> But I am experiencing technical difficulties:
> My heave compensator has failed,
> Compiler's in error, signals drowning in noise,
> Standard output is over the rail.[32]

When I reenter the control room, Adam has kindly changed videotapes and taken my place at VICKI. After a while, operations are finished and the ROV begins its ascent to the surface.

I ask ROV pilot Knute Brekke how he maneuvers *Ventana*.[33] He demonstrates the robot arms. The left one has several joints, named with words suggesting an amalgam of the mechanical (swing), bodily (shoulder, elbow, wrist, grip), and nautical (pitch and yaw). The right arm boasts hydraulics capable of lifting 500 pounds. I play with a *Ventana* arm as the robot rises to the surface. It is an experience of disconnection; there is no force feedback through the joystick. This reinforces a sense of weightless, outer-space-like travel. This is telecommunication without teletactility, a gap that makes explicit the work required to realize telepresence, immediacy from afar (other ROVs *do* have force feedback, narrowing the sensory gap).

Knute tells me he is sometimes up till midnight before a dive day, fixing hardware on the robot. I ask about an anecdote I heard back on shore from an MBARI software engineer. Normally, *Lobos* finds out where *Ventana* is through pinging a sonar signal off of it. Location information is fed to shore, where programmers receiving data points use a formula to construct a best-fit line for a day's ROV path. The pilots, though, have not found the sonar as fine tuned as they like; they want to know more about speed. They reoutfitted the ROV with a Doppler velocity log, measuring movement relative to seafloor (rather than in "absolute" space). Though this has been accurate about how fast the ROV is going, it has been less so about where it is. When Doppler data were first fed to shore, software engineers became confused about the ROV's line of flight, not knowing of the pilots' modifications. Knute says that computer folk sometimes confuse people at sea, too—as when they upgrade ship software from shore. Cursors move around on ship screens as though the boat is haunted.

It is a difference of perspective, of who is authorized to exercise remote power. My interview with shoreside software engineer John Graybeal a week earlier centered on how he thought about the sea as he put together a database for *Lobos*. The ocean on the screen was for him about projection, conceptualization. When he considered the ocean through the medium of his database, he said, "I don't think of it as a wet thing. It's a construct that places constraints on what we do. It's not the same ocean that we go to when we step outside and go to the beach." He said he "couldn't imagine someone solving a chess problem if they were up to their neck in water. *The ocean is not of us.* You can't live there." But the problem of giving form to information, which Lev Manovich in *The Language of New Media* describes as the task of software, is a problem which, for the ROV pilots at least, *does* require being "up to their neck in water."[34] It cannot sidestep the wetness of the ocean medium; deconvolving Doppler data, for example, requires practical knowledge of the materiality of seawater. On ship, oceanic data management is less of a chess problem than, perhaps, an exercise in underwater basket weaving, a material practice of manipulating form in a saturating medium, an activity more challenging than its folkloric association with easy college classes would suggest.

A call from the Aquarium interrupts my ROV tutorial. Through our shipboard monitor of the Aquarium auditorium, Knute and I see stock footage of the very control room in which we sit. I squint at the image of the docent on our screen and hear her tell her audience, in a tinny television voice, "Nothing is larger or more important than the ocean. We get food, minerals, and pharmaceuticals from it. And with the ROVs, we can enter this alien world. None of this would be possible without MBARI's research vessels. The scientists are on the boat right now! They can beam images to us live. If I go live to the boat, we see . . . not much. *Point Lobos,* are you there?"

Knute says yes.

"What did you do today?"

Knute explains about clams and tubeworms and cold seeps. The docent thanks him and proceeds, on the fly, with the help of a vast video menu, to explain what these are. She tells her audience that hydrogen sulfide smells like rotten eggs, that methane smells like what comes out of a cow when it farts.

Ventana is up. It does smell. Adam labels the mud-filled pushcores and transfers them into a holding box. Still scouting for whales on the horizon, I make out the shore and recall a whale-monitoring program run by the Moss Landing Marine Laboratories (MLML), the other research center in Moss Landing. If MBARI is high-tech and focused on the future, MLML is more modestly budgeted and attaches itself to more traditionally conservationist undertakings. One of MLML's projects has been the attempt to monitor the migration of whales using sea lions outfitted with cameras. During a conversation with the director of this program, whose offices are in a small trailer, I caught sight of a somewhat forlorn fiberglass model of a whale, to be used for training sea lion cinematographers. The sea lions, though they are much more at home in the water than robots, get nothing like the funding of the more technologically spectacular *Ventana*.[35]

We all end up on the *Lobos* bridge. Pete asks the captain about the ship's future. There is talk of decommissioning *Lobos*, transforming it into a craft that can maintain elements of MBARI's next big technological undertaking—a distributed ocean observatory, a network of remote sensing buoys providing continual Internet access to data from the sea. I later hear a talk at MBARI about how such networks would allow scientists to sit at home gathering oceanographic data. No need for seasickness pills, yak sites, or, indeed, fieldwork; what counts as presence in the field—presence upon which representations will be based—is transforming. The *Lobos* might do maintenance on nodes and cables in such a system (and *Ventana* might shift from a glamorous tool for immersive exploration to an everyday repair device). A crewmember jokes that without scientists on board they could do some proper salmon fishing.

Weeks later, on shore, I speak with Chris Scholin, an MBARI biologist working on a remote sensing module called an environmental sample processor, or ESP, an apparatus that might become part of an ocean observatory. The device is designed to test seawater for genetic signatures associated with ecologically interesting microbes or unwanted organisms. These might be invasive alien species or blooms of harmful algae, phytoplankton which, in too great quantities, he tells me, can have negative impacts on humans and wildlife as well as on local economies. Scholin says that genetic probes for such organisms would need to be augmented by further information in order to assess properly the hazards any samples may suggest; after all, one could find lots of genes from dead organisms. Closing a beach on the basis of false positives could be costly. But he is optimistic about "putting the oceans online," getting data from in situ, real-time sensors, using these to do long-term ecological monitoring, watching for changing abundances of dinoflagellates and other organisms like bacteria and viruses, which could offer clues about environmental change.[36] Scholin says that the ESP might someday be like *Star Trek*'s tricorder, a hand-held device that sniffs out signs of life. ESP, of course, is a perfect name for a tool that promises to jump beyond human sensing, to an expanded remote sensing that allows people to assay the ocean without going there. Extending associations between oceans and outer space, projected marine observatories have acronyms like MARS (Monterey Accelerated Research System), NEPTUNE (NorthEast Pacific Time-series

Undersea Networked Experiments; see chapter 6), and VENUS (Victoria Experimental Network Under the Sea). The vision feels science-fictional, floating in the future.

In these computerized days, the oceans are coming to be viewed through the robot cameras of entities like *Ventana* and their online interfaces. Understanding today's scientific sea means engaging with this media ecology. It also requires following how this system interacts with the medium of seawater, a churn of already in-place meaning and materiality. If "the medium is the message," or, as McLuhan later put it, "the medium is the *massage*"[37] (to draw attention to the sensual effects of media), media are already massaged by prior meanings. The media-mathematical sublime delivered by the online ocean at once tames, amplifies, and scrambles the old-fashioned seagoing sublime.

As Pete's message from the mud moves from sea to shore, media remain central to its fashioning. With the application of gene sequencing to consortia of marine microbes, the ocean is becoming something to be textually scanned, deciphered, read, at ever-higher resolutions. If there is a sensory itinerary toward accessing the previously unseen in marine biology, it should be no surprise to see a growing focus on harder-to-find, harder-to-discern life forms. Such a trajectory not only tracks the miniaturization new techniques have enabled but also sketches a story about the dwindling scale of life forms in the sea in the age of overfishing.[38]

READING THE GENETIC SEA

Just off the boat, I join Pete and Adam in archiving mud, which will be kept on ice until Adam can extract nucleic acids from it. Donning plastic gloves, Pete treats a few samples with RNA*later*, a product that prevents the degradation of messenger RNA, the substance that mediates—"translates"— between DNA and the proteins it specifies. Pete is curious to know whether he and Adam might be able to retrieve RNA from the mud, though DNA is the easier substance to work with and the favored material for the DeLong lab. We joke that RNA*later*, manufactured by Ambion, "The RNA Company," is ill named, suggesting not preservation but eradication— which Adam glosses, surfer style, as "RNA! Later, dude!" Pete remarks that when he touches mud with his ungloved hands—not permitted at the moment, since we do not want to contaminate samples—he finds it makes his skin smooth. He jokes that he should start a cosmetics company with the slogan, "It comes from the bottom of the sea!"—a phrase he repeats the next day at a port-tasting party held at a postdoc's apartment in Santa Cruz. The ensuing laughter prompts discussion of the chemical qualities of the wine, upon which people studying chemosynthesis are well qualified to comment. As if to confirm this observation, weeks later I meet a marine chemist from UC Santa Cruz, Phil Crews, who has started his own boutique winery.

The port tasting is a going-away party for one of the postdocs—an occasion for celebration but also a moment suffused with a certain exhausted melancholy. Present at the gathering are three marine biologist couples in the thick of trying to square their committed relationships with job offers from geographically distant universities. It is a problem that visits more and more academics in this age of scholarly temp work, and it is something these oceanographers have in common with this anthropologist. My fiancée lives in Pasadena and we see each other only occasionally, though her one-year job teaching anthropology in Claremont has made my fieldwork geographically—and, to begin, financially—feasible. Familiar to me, too, is the way marine biologists' negotiations between the personal and professional demand that they constantly articulate their research to a variety of possible employers—that they have flexible identities while building recognizable expertise. Pete and company's decoding of the message from the mud must be carefully managed, parceled out in strategic publications, tactically divided between postdocs on the job market, graduate students needing a push into a new project, and, of course, lab leaders like DeLong. Such dynamics are replicated across MBARI, which hosts some fifteen laboratories—in geology, chemistry, biology, and ocean engineering—each with its own principal investigator and a few postdoctoral fellows.

What will be the aim of extracting nucleic acids from the mud? Pete tells me the best outcome will be the construction of a genetic library representing the diversity of genes found in microbes in the cold seep environment. Such libraries allow comparison with microbial life in other ecosystems. Researchers in DeLong's lab concern themselves, we might say, not so much with mapping the sea as with mapping the DNA of sea creatures.

The lab is at the forefront of attempts to characterize the genetic profile not just of single organisms but also of whole environments.[39] Environmental marine microbial genomics seeks to describe sea-going microbes by sequencing the genes of entire communities of these creatures—from such diverse environments as deep-sea mud, hydrothermal vent systems, and surface waters.[40] It constitutes a radically new approach not just in marine biology but also in microbiology. The earliest days of microbiology depended almost exclusively upon microscopes, first turned to analytic advantage by Antonie van Leeuwenhoek in seventeenth-century Holland to peer at "animalcules" swarming in a drop of water. Two hundred years later, Louis Pasteur and Robert Koch isolated individual microbial strains for study by growing them in petri dish "cultures," a technique modified for use at sea in the 1940s by Claude ZoBell to cultivate microbes capable of metabolizing at high pressure.[41] "Culturing" entails concocting a nutrient medium congenial to organisms' dietary needs, simulating their environment so they can grow in the lab.[42] But it has proven notoriously difficult to manufacture laboratory cultures that replicate the conditions encountered by most marine microbes in their natural habitats, typified by complex ecological networks; the vast majority of microbes remain uncultured.[43] Biologists like DeLong have found ways around this impasse, even turning this obstacle into an asset. Nowadays, gene sequencing permits marine microbiologists to dispense entirely with the need to zero in on individual microbes—or even on populations of discrete cells. Volumes of water can be filtered directly for the genes they harbor, which in turn can be sequenced and read as signatures of microbial marine life.[44]

This is not merely a technologically innovative genre of genomics; it represents a theoretically novel mode of parsing the biotic world. This is a genomics beyond organisms, a practice that implicitly queries whether individuals are the only evolutionarily meaningful units. As DeLong explained in a 2003 interview in the *New York Times,* "A milliliter of seawater, in a genetic sense, has more complexity than the human genome."[45] In the view of DeLong and others, the microbial ocean can profitably be seen as a sea of genes.[46] Scientists at MBARI store, summarize, and analyze fragments of this sea in gene libraries.

How are libraries compiled? A few days after our cruise, Pete escorts me upstairs at MBARI, where he hands me to Steven Hallam, another postdoc in the DeLong lab, who promises to walk me through the process. Steven, a spry, jocular fellow in his early thirties, catches on to my anthropological attitude from the moment we meet, joking about the fetish objects of molecular biology: the centrifuges, which spin biological substances into component parts; the gene amplifiers, which "xerox" DNA; and the sequencers, which read the nucleotide stutters of the genetic text—strings of adenine, thymine, cytosine, and guanine, abbreviated A, T, C, and G. All the machines have pet names, he tells me, rattling off monikers like Harvey, Lola, and Eve—this last the designation of a device purchased from a company whose sales reps had recently dressed up as Raelians, the religious group that claimed the offshore cloning of a human in late 2002, whom they named Eve. Steven tells me that the most important machines in the lab do not carry names; these are machines so commonplace as to be almost invisible: freezers. These are where DNA libraries are stored. Steven opens a refrigerator-sized vault maintained at a subarctic −80°C, housing shelves stocked with small rectangular, clear plastic trays about the size of CD cases. Each tray, frosted with ice, is stickered on its side with the location and depth from which its contents hail. I see trays labeled "Monterey Bay," "Hawaii," and "Antarctica." Steven pulls out a typical tray, dotted with a twelve by eight grid of holes, each about half a centimeter in diameter. This is a 96-well plate, the basic storage format in this lab (more recent work uses a 384-well configuration). A collection of trays from one location at several depths (e.g., a site in Monterey Bay) constitutes a genetic library for that environment.[47]

What, exactly, is *in* a DNA library, and how does it get there? Steven leads me from the freezer to the lab bench. The first step, he says, is extracting DNA from organisms, a process that dissolves intact cells to release DNA and then eliminates unwanted proteins, lipids, and carbohydrates. This results in the purification of DNA (DNA extraction kits often include an agent for separating DNA that is derived from the fossils of diatoms, microscopic plankton that grow silicon cell walls). After extraction, one must decide what kind of library to make. One might make a *ribosomal library*, using genetic material involved in the production of ribosomes (cellular organelles, made of RNA, where protein is synthesized); such a library, based on genes many scientists consider genealogically stable, provides a possible inventory of taxonomic groups in a sample. One might also make a *functional library*, targeting a specific enzyme in a metabolic pathway to learn about links between cellular dynamics and ecological or biogeochemical processes. A *genomic* or *metagenomic library* represents a snapshot of the full genomic complement of a sample.

For a ribosomal or functional library, the next step is DNA amplification, making copies of a genetic material of interest. One needs DNA in bulk to get experiments going. Amplification is accomplished using the polymerase chain reaction (PCR), which enables the exponential copying of DNA segments. To copy segments from particular organisms, researchers stir into the genetic sample a few off-the-shelf DNA fragments called primers—sequences of genes that match only with specified target sequences. Such mixtures are then placed into a PCR machine like the lab's GeneAmp PCR System 9700 by Applied Biosystems. The system heats up and cools down the genetic mixture, in a repeating cycle, allowing sections of DNA double helices to uncouple—"denature"—and then reassemble—"renature"—into copies of the DNA element specified by the primer. To help this process along, PCR makes use of enzymes that survive at high temperatures. Some of these derive from extremophilic microbes at hydrothermal vents. DeepVent, "isolated from a submarine thermal vent at 2010 meters and able to grow at temperatures as high as 104°C,"[48] is one such extremozyme, and a product about which I learn more later. At the end of an environmental PCR screen, researchers have a heterogeneous mixture of DNA representing a sampled community.

After amplifying a sample, one needs to separate different genetic segments for closer examination. The first step toward this aim is cloning—cutting and pasting—genes into a circular piece of DNA known as a plasmid. The plasmid acts as a "vector" for inserting genes into a bacterial cell, such as *E. coli*, able to host genetic material from other organisms. *E. coli*'s incorporation of external DNA is aided by subjecting the bacterium to high heat and placing it in an incubator such as Forma Scientific's Orbital Shaker, inside which it is brought back down to 37°C, human body temperature. Zipping through the lab, Adam hazards that, as a "shit bug," *E. coli* (which lives in intestines) finds its comfort zone here. Transfer of genetic material into *E. coli* allows researchers to grow substantial quantities of such substances. (Making a genomic or metagenomic library is different; here, the lab uses a fosmid, a plasmid that controls copy number in cells).

What Steven and company are creating, then, are "mermicrobes"—organisms part sea creature and part everyday terrestrial bacterium.[49] When I offer this word to Steven, he grins and says the concern at this stage is to avoid producing chimeras, DNA clones containing gene regions artifactually combined from different organisms during amplification. Like the same-named fire-breathing monster of Greek mythology, such chimeras do not exist in nature; like mermicrobes, they are tangles of the natural and cultural. But, Steven cautions, the creation of such hybrids does not make of this kind of biology a Frankensteinian endeavor. Cloning microbial genes does not immediately scale up to the science-fiction-style cloning controversies that make the cover of *Time*—though, as I puzzle over plasmids, Pete joins Steven, shakes his head, and remarks that he shies away from telling more religious family members that his job entails cloning; they have been known to scold him for toying with creation. This is a sharp acknowledgement of ideological rifts in today's North American public culture about the proper relation between the objects of life science and the subjects of social life, between life forms like stem cells and forms of life like scientific research. If Pete and Steven see genes as manipulable

Dr. Edward F. DeLong of the Monterey Bay Aquarium Research Institute takes microbial plankton found in the bay and reads the community's DNA.

Peter DaSilva for The New York Times

A New Kind of Genomics, With an Eye on Ecosystems

FIGURE 8. Ed DeLong, looking through a dish of cloned colonies. Reproduced with permission of Peter DaSilva Photography.

bits of life, they do not therefore think genes in the lab *stand for* life as a moral whole; that status is reserved for organisms (and, for Steven, as becomes clear, for ecosystems as well).

Once genes have been cloned into *E. coli,* the bacteria are placed into a bed of agar, a gelatinous substance derived from seaweed (originally named *agar-agar* in Malay; *Gracilaria lichenoides* in the scientific lexicon). The bed sits a few millimeters deep in a circular petri dish, permitting clumps of *E. coli*—"colonies," in lab language—to form on the agar, dotting the plate with tiny, pale flesh-colored circles. Figure 8 shows DeLong looking through a circular plate specked with colonies of clones. The plate is labeled HF 70, which stands for Hawaii Ocean Time-series, Fosmid library, 70 meters down. I cannot help but compare this picture to DeLong's slide of Leonardo's Man against planet Earth; here again, a human has a whole world in his hands.

This petri dish of cloned colonies is the raw form of a genetic library, a form of what science studies scholar Eugene Thacker calls *biomedia,* "the technical recontextualization of biological components and processes."[50] The biological components at issue here, lengths of marine microbial DNA, have been placed in the context of *E. coli* to make them available en masse, as colonies, for scientific inspection. Once colonies are grown in media, they are transferred into slots on a 96-well plate. They are then dosed with antibiotics that permit only those clones that have taken up the plasmid vector to replicate. Steven tells me this is artificial selection at the molecular level. With a nod to Darwin, he offers, "We're doing this because this is how nature does it, and we haven't found a way to do it better." Commenting ironically on the conversion and regeneration of organic material in play here, he adds a religious joke: "Only the born-again will grow."

The joke points toward a key concern among these marine microbiologists: they are seeking to find meaning in the microbial life they study. Talk of the message from the mud—the extraction of information from sludge—should make this clear. The recontextualization at the center of Thacker's definition of biomedia alerts us to how meaning does not preexist interpretation; rather, the readjustment of context is that which *makes* meaning. Laboratories, after all, are devices for creating significance, for separating figure from ground, for adjusting what counts as text and context.[51] Steven's irreverent one-liner about the born-again is an ironic commentary on how difficult—and sometimes facile—the making of meaning might be. Scientists' quips about methane, *E. coli,* and shit—important associates and associations with mud—bespeak anxieties about how meaning might be made of messy things.[52] At the same time, such comments express a wonder and confidence that even the most chaotic substance can be made to signify. The language of *message* also suggests a communicative relationship between sender and receiver—a relation that unfolds in time, in which methanotrophic microbes are rhetorically positioned as delivering to scientists an ancient and alien wisdom from the deep history of Earth. While reinforcing scientists' sense of wonder, such messages are not, in the main, imagined through earnest spiritual rhetoric. The majority of the microbiologists I met were firm secularists, even agnostics and atheists. In searching for meaning in microbes, they trust rational approaches to reveal surprising relations—like those between biodiversity and biogeochemistry—though they also consider such connections to have social consequence, even ethical import.

With a range of associations from sullied to morally significant, methanogenic and methanotrophic microbes exhibit something of the flavor that anthropologists have assigned to the taboo. In Edmund Leach's 1964 "Anthropological Aspects of Language: Animal Categories and Verbal Abuse," we read: "Quite apart from the fact that all scientific enquiry is devoted to 'discovering' those parts of the environment that lie on the borders of what is 'already known,' we have the phenomenon, which is variously described by anthropologists and psychologists, in which whatever is taboo is a focus not only of special interest but also of anxiety. Whatever is taboo is sacred, valuable, important, powerful, dangerous, untouchable, filthy, unmentionable."[53]

So far, microbes have been described as bearers of important messages, as in need of protection from contamination, as versatile, as possibly chimerical, as invasive, as smelly, and as shit bugs. If not strictly taboo, microbes are certainly objects of interest and anxiety; their relations to humans matter to these scientists. And they come to matter precisely through their manifestation as media—as symbolic intermediaries between human selves and an oceanic other, as material things whose functions can be investigated as biomedia.

Microbes, then, become media for thinking about the world beyond the lab. For many of these ecologically minded and politically liberal scientists, deep-sea life forms are important for environmentalist forms of life. One of the microbiologists at the port party, a graduate student from a nearby university, e-mailed me an extended reflection on her search for meaning in marine microbes. Her words illustrate the connections at stake for many of these scientists:

> When I began to learn about marine ecology, ecosystems, cells, and genes, I was enraptured by the bizarreness of hydrothermal vents, by the mysteries and promises those still-freshly-discovered habitats embodied—and I began to learn about the roles microbial actors played. Marine microbes are the linchpins not only of marine food webs, and thus ocean life, but indeed the linchpins of global biogeochemistry, global ecology. They are the "spark of life"; not only do they hold clues to the first sparks of life, and the accretion of those sparks into the mighty flame we now behold around us, but they remain the great Spark. They fuel the biosphere, they transform waste into nutrients, they adapt. I've never been religious, but my heart just sings to contemplate the miracle of this planet's complexity, how much sense it makes, how beneath every layer there's another and yet we can piece together the workings and learn new emergent principles. And yet, what dire immediacy in this pursuit, today. So many beautiful creatures, wrought by millennia of evolution, being snuffed out in today's cataclysmically changing world. I feel the calling to help mediate the gulf of what we don't know about the workings of the oceans, and what the public doesn't know they should even care about. I want to learn how these communities control our biosphere, in hopes that we can grasp, as a species, what the likely paths can be before us, so we can make an informed decision among them and walk into our future with our eyes open.

This student's words reveal sentiments of wonder and a respect for nature, a view of humans as reliant upon (though often ignorant of) this order, and a rational optimism that scientific explanations of life forms can broker forms of life responsible to the future well-being of all life. This position follows a long tradition in American science of considering biology to hold civic lessons for researchers and their publics.[54] The marine microbiologist becomes a medium for propagating information about the subvisible realm of microbes to the human-eyed world.

When I walk into the lab the next day, Chris Preston, a research technician in Ed's lab, greets me. She is "picking colonies," plucking small clumps of *E. coli* out of gummy agar with a toothpick and transferring each colony into a hole on a 96-well plate. Each tiny well contains a liquid growth medium called Luria-Bertani broth. The work, Chris sighs, is numbingly dull, and in bigger labs robots do it. She sits me down at the bench and gives me pointers for taking over, and for the next hour I busy myself picking colonies. As I poke the little dots of *E. coli,* placing each in a well of its own, I am not sure if I am transfixed or completely bored out of my brain. My gaze wanders to a haiku written on the lab whiteboard:

> picking colonies
> meditation or boring?
> just buy the robot

Steven sprints by and informs me that he favors listening to minimalist composer Philip Glass while doing this work. Chris has put on Norwegian techno. People in the lab shuttle around, in attitudes of distracted concentration, waiting for various processes to finish, staring into the middle distance, considering what they must do next. I, too, have invisible materials in my head, such as fragments of argument from sociologist of science Karin Knorr-Cetina's *Epistemic Cultures.* Of lab work in molecular biology, Knorr-Cetina writes, "Scientists act like ensembles of sense and memory organs and manipulation routines into which intelligence has been inscribed; they tend to treat themselves more like intelligent materials than silent thinking machines."[55] The techno music gives such intelligence an enabling soundtrack, immersing researchers in a mundane media sublime, a flow. Such activity may support Knorr-Cetina's claim that, "if anything is indeed irrelevant to the conduct of research in molecular biology, it is the sensory body *as a primary research tool.*"[56] But while messaging mud and swaying to the sounds of the Orbital Shaker are, yes, not primary research protocols, they *do* allow researchers to fasten their bodies to their work. Nonetheless, it is true that the parameters within which these scientists must embody repetitive routines can become confining and tiring. Occasionally, someone groans that they need a break and wishes it were warm enough to use MBARI's beachside volleyball court.

Steven, now installed next to me, transfers small volumes between wells on different plates and goes through lots of disposable pipettes. He complains about the trash molecular biology generates; it is scandalous, he says, that here we are doing science about ecology while generating this toxic material that will head straight to landfills. I turn back to the *E. coli* I am stabbing and, reading the legend on the petri dish, realize it contains gene fragments from Hawaiian waters. I try to connect a visual of blue water with the gel before me and find it impossible. I see each well as a Russian doll, containing a unique colony containing a unique plasmid containing unique DNA. At the end of my toothpicks, I have completed one plate in a genomic DNA library.

DNA libraries are understood as collections of information encoded in catalogued packages. If *Ventana* promises oceanic transparency, genetic libraries promise legibility. In play here is not so much the uncanny oscillation between immediacy and hypermediacy that produces the intimate and alien ocean for ROV users but rather a sense that the unfamiliar will yield to decipherment, which may lead to unexpected knowledge. The library analogy is persuasive to scientists in part because of the linguistic metaphor of the genetic code, a set of correspondences between nucleotide bases and amino acids. The textual metaphor might also suggest that different plates are volumes in series that treat of microbes from different places. Each plate, after all, has a call number, like a library book. But DNA is not a language—it has neither grammar nor semantics—and so the analogy of these plates with books containing text is inexact at that level.[57] The libraries are anyway read more hypertextually, with meaning read out in relation, pragmatically, not written in beforehand. The libraries have much in common with *Gould's Book of Fish*, that volume described in Richard Flanagan's novel of the same name, a liquid book that spawns new chapters every time it is opened.[58]

These wet databases of fragmented life, of life as media technique, are different from the jars of dead fish MBARI maintains in an air-conditioned room downstairs, a place humorously known as the Necropiscatorium. This assortment of squids, jellyfish, and other creatures has some of the romance of a nineteenth-century natural history museum.[59] The collection, however, is falling into disuse. "Only a few people care anymore," says Steven. "We are losing sight of the organism." People today are more interested in gene libraries, because of the multiple stories they can tell in concert with bioinformatics programs—because of their potency as biomedia, organic stuff that can be reformatted into novel configurations. In "Unpacking My Library," Walter Benjamin writes that "not only books but also copies of books have their fates."[60] So, too, with DNA libraries.

Back upstairs, Steven instructs me that when one completes a 96-well plate labeled something like MB10—a legend that stands for "Monterey Bay, 10 meters down"—one might compare it to other plates or libraries, looking for similarities across environments. With a genome library one could head back to the PCR machine and, using a specific DNA primer, see if already known genes appear in an unanalyzed library. An example: In 2000, the DeLong lab discovered a complex of genes in Monterey Bay near-surface marine bacteria specifying the production of a light-absorbing pigment, rhodopsin, not previously known to exist in oceanic microbes and of signal importance for comprehending how energy from the sun is cycled by the sea.[61] Creating primers complementary to genes involved in rhodopsin production enabled DeLong's team to use PCR to see if rhodopsins were present in genomes from other environments, like Hawaiian waters. DeLong discovered rhodopsins at different ocean depths, spectrally tuned to distinct frequencies of light. This finding has enormous significance for understanding the rates at which ocean plankton can turn light into life— processes which, in parallel with such activities as microbial methane munching, are instrumental in keeping Earth habitable for aerobic light lovers like us.

I run into Adam, pipetting small volumes of microbes from one plate to another. He laughs, "It's an act of faith that there's anything in these tubes!" In some ways, there is nothing particular about the ocean driving the way this lab science is done. It is, much of it, standard molecular biology.

Chris disagrees, though clarifying that the oceanic particularity requires expertise to discern. She tells me that in trying to figure out what might be unique about a library from Hawai'i, for example, she consults the log of the ship on which samples were collected, looking through records of salinity, temperature, and the like. She shows me a website from which she can retrieve such data and comments on the constant need to think of context. Sometimes, she tells me, if she is looking at a microbe living near the surface, she will think back to surfing or scuba diving, trying to remember temperature and light shifts. Meditating on further connections between the lab and the sea, she tells me a chilling fact about ocean samples: when trying to determine whether particular marine microbes gather their carbon from near the bottom or higher up in the water column, some labs look at the concentration of carbon-14 in their bodies. Because of twentieth-century nuclear testing in the South Pacific, surface waters around the world are enriched in "bomb carbon," which because of the ocean's slow circulation has not yet reached its deepest regions. Steven interrupts us to say, with a mixture of outrage, world weariness, and scientific detachment, that such information can be crucial for "doing forensic science on the living ocean."[62]

The frequencies of war hum in the background of our work now, too. Everyone I talk to hates the Iraq war, but no one knows what to do, short of showing up at protests. Most people in the lab dislike President Bush, who, distrusting climate science, by implication dismisses the value of these aspiring scientists' vocations and probably the relevance of any message from the mud the lab might decrypt. Whereas the technologies of the cold war generated much of MBARI's instrumentarium, marine scientists here are not much preoccupied with a vision of science in the national interest. They are committed to a moral economy of following questions where they lead—a position enabled in part by the ample funding of the Packard Foundation, which gives them steady salaries that shield them from constant grant writing (even as MBARI operates with an industrial model that keeps inquiries from wandering too far from the objectives of the institute, which demand that science always be done with an eye toward using new oceanographic technologies).

Another echo of war arrives in the form of an e-mail from MBARI's long-range research vessel, the *Western Flyer*, now in Mexico's Gulf of California.[63] Everyone on board is reading *The Log of the Sea of Cortez*, a book written by John Steinbeck about a voyage he took to the Gulf with Ed Ricketts, during World War II, to collect invertebrates. Remarking on the scrambled sensibility that came from being at sea while life on land slid into war, Steinbeck wrote, "Our radio was full of static and the world was going to hell."[64] As if to cement MBARI scientists' feeling that they are stuck in a perverse repetition of history—and politically paralyzed at sea, like Steinbeck—the institute's ship takes its name from the very vessel Steinbeck and Ricketts chartered to Mexico. What has changed between the eras of Ed Ricketts and Ed DeLong, of course, is the speed at which warfare and science now unfold, when information is produced far faster than meaning, when intelligence about a suspicious warehouse in the desert or an anomalous gene sequence in *E. coli* outpaces the ability of experts to decipher plausible functions.

If making gene libraries provides a first step to reconstructing genetically enabled microbial processes, a proper reading requires gene sequencing—spelling out the bases constituting DNA. The DeLong lab has a modest sequencer the postdocs have named Lola, after the German movie *Run Lola Run,* a film that features three alternative endings for the same story. I ask Steven whether "Lola" was chosen because sequences never come out the same way twice. He chuckles enigmatically. More pertinent is that Lola runs too slowly—it takes six hours to sequence a 96-well plate. MBARI sends most materials out to be sequenced at the Institute for Genomic Research, the entity behind the shotgun sequencing that propelled the Human Genome Project toward completion. Shotgun sequencing works by shredding multiple copies of the same organismic DNA, which are then sequenced by armies of speedy robots. Using bioinformatic software, computers churn through all these overlapping sequences and reassemble them into something researchers hope looks a lot like the actual, correct, sequence. When such sequences are sent back to MBARI, people like Pete, Steven, Adam, and Chris move into the realm of computational comparison, matching their sequences with those at other labs and building relationship trees for consortia of microbes, trying to find out how they are related. At this juncture, marine biology goes digital—of which much more in the next chapter.

LOVELOCK MEETS LOVECRAFT

We have still not heard the message from the mud. Adam continues to toil to get genetic material out of sediment we collected with *Lobos,* sorting it into neatly named types, a difficult task since, as Pete points out, "whatever message we read will come from *everybody* hanging around in the mud— microbes, protists, even worms, should we have accidentally ground one up." But Pete remains enthusiastic about using nucleic acids to learn about methane metabolism, though he tells me he is wary of thinking about microbial matters only bioinformatically; organic stuff often drops out, he says, in the digital domain. He compares digital gene sequences to CDs and organisms to full-bodied analog LPs and, warming to his comparison, says, "Hey, I still like my old vinyl records."

Pete points me to a poster detailing results the DeLong lab will present later in the year. I scan the poster, titled "Construction and Analysis of Shotgun and Fosmid Genomic Libraries of Anaerobic Methane-oxidizing Archaea from Deep-sea Methane Seeps," reviewing, in technical language, some of what I have seen at MBARI. The poster reports on work undertaken with DNA libraries made well before my *Lobos* trip, though on similar samples. What, then, is the message of the mud? What have these scientists read from such libraries as they have compiled? Steven says there is a parable in the poster, on which he is first author. He will tell me over lunch.

When we meet up at the Haute Enchilada, Steven reports that during his undergraduate studies at Sarah Lawrence he was impressed with the work of Teilhard de Chardin (1881–1955), a Jesuit paleontologist keen to riddle through the relation of religion to science. I learn that Steven's father is a Methodist minister and that Steven has long been a student of spirituality and alternative epistemologies, convinced that science is just one—albeit, compelling—way of knowing among others. Steven has been ruminating on an account of anaerobic methane oxidation that proffers a moral about the dangers of seeing the sea as infinitely flexible, endlessly able to absorb human insults like toxic waste or the sinking of carbon dioxide into the deep to keep it out of the atmosphere (a practice in which the Bush administration expressed interest a few years previous). He begins with a quick sketch of atmospheric chemist James Lovelock's Gaia hypothesis, the notion that Earth is a homeostatic system, delicately tuned to and by the life forms it supports. Gaia, he points out, for all its associations with romantic visions of natural balance, could care less about the well-being of humans; indeed, we may produce the very feedback systems that lead to our own demise. Inspired by our recent visit to the unforgettably odd Necropiscatorium, Steven brings up H. P. Lovecraft, an early twentieth-century fantasist famous for his ornamented horror stories about aliens and his panic about the biological degeneration of Western civilization. Fusing Lovelock and Lovecraft, Steven offers the following fever dream—in the spirit of Lovecraft's weird accounts of "the unsampled secrets of an elder and utterly alien earth"[65]—a sort of psychedelic rereading of the anaerobic methane oxidation paper:

> Once upon a time, when the Earth was young, there was very little oxygen in the atmosphere. Instead, the atmosphere was mostly composed of methane and carbon dioxide and the oceans were warm and shallow. Life evolved to thrive under these greenhouse conditions. Methanogenic microbes feeding on carbon dioxide and other simple carbon compounds produced vast quantities of methane and this methane was in turn consumed by methane-oxidizing microbes found primarily beneath the ocean's surface. In cooperation with sulfate-reducing organisms, the methane-oxidizing microbes built towering reef cities formed from mineralized carbonate and filled them over countless generations with their collective brood.
>
> And affairs continued in this tranquil equilibrium for one and a half billion years, until the genesis of oxygenic-phototrophic metabolisms and the oxidation of the atmosphere. Life forms able to adapt to elevated

oxygen levels thrived and radiated. Meanwhile, those content with living in anoxic places were pushed to marginal zones, to extreme environments—subterranean worlds and still waters, mud flats, and seafloor spreading centers. The great reef cities fell into ruin and were subsumed into submarine strata, a cryptic but lingering record of the lives of these ancient organisms. Despite this catastrophic reversal of fortune, these ancient ones held onto the edges of their once great empire and there they waited.

And here's the moral of this conjectural tale: They knew, these ancient ones knew, to the very core of their genomic fiber, that it would all be okay, because through their DNA they had bequeathed the knowledge and the drive to return and rebuild. Because it turns out that all of the anthropogenic processes connected to climate change—fuel emissions, deforestation, cattle grazing—may well have the result of bringing back the ancient atmosphere. You see, these ancient organisms are patient. And here are the ironies—a good story always has ironies—they have no imperial ambitions, they have adapted to live and lurk in the marginal zones. But when the madness of humanity resurrects the ancient atmosphere they will be ready and willing to return, to rebuild their ancient dwellings beneath the sea and continue their eldritch cycling of methane. And the primordial balance will return. Until the next big catastrophe.

Steven's generative fusion of Lovelock and Lovecraft discloses an ocean alien to human purposes, purposes themselves perhaps unwittingly alien to their own scaffoldings.[66] Steven tells me he wishes we humans had something like an ocean ethic, an ethos that could detour us away from funneling everything through the marketplace or technocratic politics. A deep-sea biologist from MBARI, Kurt Buck, walks by. He asks Kim, behind the counter of the Haute Enchilada, if he can get some used vegetable oil to experiment with as a biofriendly fuel. Steven says to me, "I would argue that the lesson to be learned is how to compose our lives and our societies based on the properties of Earth systems under study and the properties and interactions of organisms living in balance within these systems."

It is a good start that MBARI funds vanpools from Santa Cruz, I observe, though mordantly funny that the institute is built on the beach, which after a century of climate change could all be under water. Steven wants a more spiritual sensibility to animate our engagement with the ocean, though he also says, "But I am not a priest for Gaia. Gaia doesn't need priests. Gaia needs scientists." It seems to me that what Steven desires is a sophisticated "gaiasociality," a way of weaving biogeochemical knowledge and political commitments responsibly together. The task is huge. Steven has to get back to publishing, to doing the science he trusts.

Steven's fantastic fable will not soon appear in the pages of a scientific journal. If MBARI scientists were to deliver a moral from the message from the mud, it would more likely come in the shape of a sober policy recommendation, from an elder statesperson like DeLong. As a postdoc, Steven is being professionalized, and his story is infused with questions about the meaning of his career and the creatures to which he has devoted his work. The paper that Steven eventually publishes in *Science*, "Reverse Methanogenesis: Testing the Hypothesis with Environmental Genomics," begins by pointing out that "microbial methane consumption in anoxic sediments significantly impacts the global environment by reducing the flux of greenhouse gases from ocean to atmosphere"[67] and, after arguing that a better characterization of microbial metabolism is on the way, leaves it at that, as well the paper might, since the study does not reach into methods and theory from climate science.

Although Steven's Lovecraft-Lovelock account is wryly satirical about human shortsightedness, it has a serious strand. Alienated from the ocean, humanity may be destined to damage it. And ancient life forms, with ways alien to contemporary oxygen-addicted humans, may inherit the earth.[68] This story of evolution, devolution, and revolution coils together narratives of the conjoined origins of self and alien in the sea. The ocean becomes a spiritual medium, delivering messages from apparitions of an ancient world to listeners living today—just as the media of telephone, radio, and television were once believed to carry transmissions from ghostly denizens of alien realms.[69]

Mixing the horror of Lovecraft with the romantic rationalism of Lovelock produces a volatile concoction. I hear in the tale the twin impulses that Haraway locates in the American millenarian imagination: "Apocalypse, in the sense of the final destruction of man's home world, and comedy, in the sense both of the humorous and of the ultimate harmonious resolution of all conflict through progress, are bedfellows in the soap opera of technoscience."[70] In the froth of Steven's science fiction is a struggle to make meaning out of microbiology, to tell a story that will compel humans to slog out of their own mud, our own mess. It is a call to tune in to the alien medium of the ocean, revealed through a media ecology of ROVs, DNA libraries, and bioinformatics programs. If the alien ocean is a medium refractory to human apprehension and presence—dense in its darkness, crushing in its pressure, suffocating in its substance—it is also host to creatures whose very otherness is crucial to human life support. Steven's is a story about time and limits—about deep time and contemporary emergency. The relation between humans and microbes is structured by their shifting positions on this timeline. Anthropologist Johannes Fabian, in *Time and the Other,* writes that "time, much like language or money, is a carrier of significance, a form through which we define the content of relations between the Self and the Other."[71] The ocean Steven describes is ancient, enduring, and often alien to human origin stories and destinies. He is not alone in describing this sea.

2   Dissolving the Tree of Life

Alien Kinship at Hydrothermal Vents

Movie theaters invite out-of-body experiences. Sinking comfortably into my seat at the New England Aquarium's IMAX theater in Boston, I am sitting next to Pete Girguis, chief scientist on the *Point Lobos* cruise I joined a year earlier in Monterey. We are readying to watch *Volcanoes of the Deep Sea*, a king-sized documentary about hydrothermal vents produced in 2003 by *Titanic* director James Cameron.[1] Vents are undersea locations where superheated chemicals emerge from Earth's crust, nourishing a variety of creatures capable of living without sunlight. Vents are celebrated for hosting heat-loving microbes—hyperthermophiles—that engage, not in photosynthesis, but in chemosynthesis, the production of organic materials using energy from chemicals, such as hydrogen sulfide (or methane; see chapter 1), poisonous to most other organisms. Pete is in town to give a seminar at Harvard about research he has done at such vents while working inside *Alvin*, the famous three-person submersible operated by the nearby Woods Hole Oceanographic Institution.

Some marine microbiologists maintain that vent hyperthermophiles are the most conserved life forms on the planet, direct lines back to the origin of life, to the root of what evolutionary biologists call "the tree of life," that diagrammatic structure linking together all living things on Earth in a great genealogical branching. This is the tale told by the film. Other biologists believe that the genetic relations of such microbes are so tangled as to threaten to obscure the root of the tree, requiring a radical reformatting of the origin-of-life framework offered by evolutionary biology. This is a tale I tell in this chapter, a story I compose from readings of the microbiological literature and from conversations with key protagonists in evolutionary and marine biological debates.

Eager to hear Pete's view on how origins and ocean science are portrayed to a popular audience, I have taken the subway over from MIT, where I now teach anthropology. The film is an ideal conversation piece—even as the magnitude of the movie often suspends our murmured speech, demanding with its 12,000-watt sound our full submersion in this twenty-first-century spin on *20,000 Leagues under the Sea*.

The gargantuan screen before us—65 feet tall by 85 feet wide, described by IMAX as "the world's most immersive theatre entertainment"—lifts our vision into a helicopter-eye view of the open ocean. We race toward a distant ship. Actor Ed Harris, who played a rugged diver in Cameron's *The Abyss*, narrates: "The research vessel *Atlantis*. Destination is the Mid-Atlantic Ridge system, 500 miles ahead and 12,000 feet down." We are introduced to *Alvin*, deployed from *Atlantis* and its predecessors over the past forty years to survey the submarine world, including the network of fault lines, the undersea ridge system, that circles the planet. A voiceover primer on hydrothermal vents follows: "Water, descending into fissures in the seafloor, was apparently interacting with the hot rocks beneath to reemerge as a black cocktail of poison chemicals. . . . The sub's temperature probes indicated water hot enough to melt lead. And it was heavily laden with hydrogen sulfide. There should have been nothing alive here at all."

With our vision now routed through the vantage point of *Alvin*, "outfitted with high-resolution cameras and 4,000 watts of illumination," we see towering volcanic chimneys that sit thousands of feet below the sea surface, spewing black superheated minerals into the ambient water, feeding ecologies of tubeworms and chemosynthetic microbes (figure 9). Harris continues: "As sunlight creatures, we didn't think to look in the dark for life. Nor did we think to look in water hot enough to boil us alive." Pete laughs that, technically speaking, we are not seeing a boiling vent; high pressure in the deep only allows water to superheat, not to bubble into gas. Pop science often offers a different mapping of the world than research science. Still, Pete knows that popular representations sometimes reach into the framing of the most staid research. He tells me about an instance when popular and scientific mappings literally traded places. It was Pete's first *Alvin* dive, and the chief scientist insisted on using a map from *National Geographic*, even though the charts from which the map had been assembled were in the *Atlantis* library. The sub got lost, because the magazine contained a typo.

The film lingers over a vent covered with tubeworms sprouting bright red petals. Harris's voiceover tells us that "scientists were astonished to discover that tubeworms' red color comes from blood containing hemoglobin,

FIGURE 9. Hydrothermal vent system. Reproduced with permission of Verena Tunnicliffe.

very much like our own." Inviting our reverent amazement, Harris asks: "So, who *are* these fantastic creatures of the dark who share with us the very blood that flows through our veins?" This is just the first moment that *Volcanoes* seeks to connect vent creatures to humans through a shared biotic substance—in this instance blood, a potent symbol of kinship, especially for those of us for whom the notion that "blood is thicker than water" signals a belief that the "natural ties" of blood relations secure enduring, indissoluble connections of family solidarity and obligation. Pete whispers that on one trip he took to a tubeworm site a colleague became enamored of the idea of bleeding sulfide-metabolizing tubeworms to make synthetic blood for people suffering from sulfide poisoning. Dreams of blood transfusions between humans and animal others are hardly new; in the late 1800s, tragic transfers between sheep and people prompted research leading to the identification of the ABO system, which underwrote for a time the belief that blood groups corresponded to narrow racial types, even as the collateral notion of the "universal donor," group O, made a hash out of such supposed natural kinds (with the discovery of A and B in nonhuman primates later reopening the question of cross-species relationships). Tubeworms, living in symbiosis with sulfide-oxidizing bacteria residing in their tissues, may be symbolically easy to import into such vampire tales, but their bodies apparently are not.[2] Unfortunately, Pete reports, there is no coagulant in tubeworm blood—a major problem in any transfusion, and especially for hemophiliacs. So much for the simplicity of these blood ties.

But *Volcanoes* is bent on connecting vent creatures to humans through a grand history of our watery linkage, an attempt I find anthropologically intriguing, for stories of ultimate and shared origins reveal much about the values and anxieties of the movie's American target audience. The film takes us deeper, into a scientific-cinematic-mythological vision quest into the underworld. Our apprehended eyes zip past shrimp grazing on vent towers, and we plunge into a digitally rendered rocky crevasse: "When scientists began to . . . probe deeper into the chimney walls themselves, they made the most astonishing discovery. In total darkness, bathed in the poison breath of the inner earth, at 3,500 pounds of pressure per square inch, at temperatures exceeding 230 degrees Fahrenheit, lives the microscopic hyperthermophile."

In the film, these heat-loving microbes appear as computer-animated creatures resembling translucent gray jellybeans. They sport red tendrils that sway languidly out of their bodies, suggesting antennae but also umbilical cords severed from the walls of their womblike thermal environment. Trying earnestly to analyze the film's heart-of-darkness journey into the watery, mineral uterus of Mother Earth, I keep thinking of vent biologist Colleen Cavanaugh's funnier preview of this scene in a conversation we had the previous week, in her Harvard office. Apparently, this segment inspired much comic "microbe dancing" around her lab.

As the blobs do their digitally designed ballet, Harris goes on: "We didn't even think to look here for life. There is no harsher environment on earth. There is no creature more alien to us." After allowing us a moment to gaze at the outlandish ovoids, the story twists, flipping the script of strangeness back into familiarity, guiding our vision into the body of one of these extreme-loving microbes, zooming in on its naked genes: "But as we journey down deeper among the molecules of its DNA, we reach the four base chemicals of life's universal alphabet. This is the language of human DNA, and in *this* we are most certainly related." A ladder-like double helix resolves itself in the heart of our hyperthermophile and the arrival of genes on the scene now brokers a claim of common origin at vents: "There is a good chance that *this* is where life began on Earth and here among the embers of that long-ago dead star [early Earth] is where we began *our* journey, five billion years ago." Through the blood of tube-worms to the genes of hyperthermophilic microbes, we have traveled back in time, taking a detour into the nearly alien, to arrive at a genetically imagined kinship with the deep sea, spelled out in the metaphorical "language of DNA."

The unexpected genetic kinship this story seeks to reveal makes an implicit call to recognize the unity of life on Earth. The film's message is a restatement of Darwin's reassurance to his religious readers, in *On the Origin of Species*, that "there is grandeur in this view of life,"[3] that aesthetic and spiritual meaning might be gleaned from learning how evolution makes kin of us all. To an audience familiar with paternity tests and adopted children's search for their biological parents, *Volcanoes* advances the claim that genetic classification will reveal our true ancestors and press us to acknowledge our long-lost cousins, today's hyperthermophiles. It delivers a moral demand right out of Dickens to accept our connections to the lowliest branches on our family tree, and it summons us to this responsibility though painting the deep sea as a motherly matrix and nursery for life on Earth. In *Volcanoes of the Deep Sea*, the tree of life and the tree of scientific knowledge are planted firmly in the life-giving sea.

I look in this chapter at the historical emergence of this narrative, for the sea was not always such a lively, family-friendly place. More central, I examine how a key scientific support of this story may now be disintegrating. Some microbiologists argue that this tale of deep, direct lineage cannot be so neat. Microbes have the capacity to exchange genes with their contemporaries continually, particularly in the fluid space of the ocean, and this may muddle any attempt to root the tree of life. I describe the techniques used to make that case and inquire into its ramifications for understanding a possible hyperthermophilic origin for life in the ancient ocean. There are intense debates among biologists about whether it is possible to shore up the tree representation and about whether the category of *species* makes any sense for gene-juggling microbes. To the vexation of scientific participants, creationists of the "intelligent design" school have lately jumped into these waters, interpreting the dispute as exposing flaws in evolutionary theories of descent, seeing room for a divine designer to slide into biological narratives of genesis. Secular scientific unease about the embattled tree, meanwhile, manifests in troubled talk of "alien genes." But microbiologists also seek to get the story back in line, back into a frame of naturalistic meaning, reimagining the tree of life as a net—a primeval example of a worldwide web—refiguring Nature not as a unitary parent but as the original genetic engineer. They script gene-trafficking life forms as anticipating, anchoring, their own biotechnological form of life.

Cultural associations of the deep sea with origins have infused scientific accounts of hydrothermal vents and their resident hyperthermophiles but are lately, with the potential liquefaction of the root of the tree of life, in transformation. I argue that the uprooting of the tree has followed, para-doxically, from taking literally the notion that DNA contains information that can be employed to trace lineages. Less technically and more anthro-pologically, I maintain that novel modes of conceptualizing biogenetic kin-ship are in the making, along with new ways of understanding the flow of genes as information, substance, and property—and, indeed, as tokens of biopolitics, those relations of power organized through the management of biological systems, including those of which humans are a part. Lateral connection rather than relation by descent is emerging as a new idiom of biogenetic relatedness—perhaps no surprise in this age of genetic modifi-cation, reproductive technology, and organ transplant. To learn about one way such microbial, molecular, and mobile life forms and biopolitics are gathering substance, we must go back to the bubbling, shifting terrain of the deep. I begin by dipping back into the ocean as it appeared just before Darwin.

A SHALLOW HISTORY OF THE DEEP

Intimate entanglements of human vitality with an alien ocean have not always featured in scientific or popular narratives of the deep. What was feared for many centuries as a mysterious realm of sea serpents became, in the early nineteenth century, a space imagined as static and barren—empty of currents, temperature changes, nutrient exchanges, life. Naturalist Edward Forbes held in the 1840s that below about 300 fathoms (1,800 feet) there existed a "probable zero of life," which he termed an *azoic zone*.[4] After Darwin, however, this netherworld came back to life. Darwin postu-lated that stable environmental conditions in the deep might actually sup-port organisms that had changed little over Earth's history; the deep sea could harbor "living fossils, evolutionary throwbacks, and missing links."[5] The Victorian imagination, in step with then nascent scientific archaeology, came to associate the deep with the early history of Earth, "as if there were a correlation between going deep and going back. Thus the deeper one went, the more primitive would be the life-forms encountered, the more prehistoric and inchoate."[6] As literary theorist Gillian Beer argues,

"Evolutionary theory implied a new myth of the past: instead of the garden at the beginning, there was the sea and the swamp."[7]

Such visions informed late nineteenth-century naturalist Thomas Huxley's belief that the seafloor was covered in a primordial ooze, a substance he called *Bathybius hackellii*, in honor of Ernst Haeckel, the German biologist who had postulated the existence of a protoplasm from which all life had descended, which he named *Urschleim* (from the German *Ur*, "primitive, original, earliest," and *schleim*, "slime").[8] When naturalists on *Challenger* demonstrated that Huxley's *Bathybius* was simply a precipitate of calcium sulfate—an artifact of storing deep-sea mud samples in bottles filled with alcohol—the goo vanished as an object of study. But associations of the deep with the ancient endured, and the figure of the sea monster resurfaced, often dressed in dinosaurial garb. There now came a shift from understanding the sea through the frame of biblical chronology (and via such imagery as Jonah and the whale) to comprehending the ocean through secular geological and evolutionary attitudes. By the end of the nineteenth century, naturalists had substantially revised their accounts of the deep, calling into question views of the abyss as a timeless lair to living fossils when creatures dredged from the deeps during the laying of undersea telegraph cables were revealed to be both familiar and novel life forms and not, say, trilobites.[9]

In the twentieth century, marine science found the recesses of the deep sea to be full of creatures—anemones, squat lobsters, sea cucumbers, and much else—living in environments as dynamic as anything terrestrial. Much of this discovery unfolded in connection with U.S. military work during the cold war. Antisubmarine warfare research, organized around technologies of deep listening, disclosed layers of sea life migrating between depths and picked up underwater rumblings that turned out to be volcanic eruptions and seaquakes, often epicentered at subduction zones where one tectonic plate moves beneath another or at mid-ocean ridges where seafloor spreading originates. Deep volcanism provided a key piece of evidence in the theory of continental drift, an account of geological change generally accepted by geologists only in the 1960s.[10] Some of the more fine-grained characteristics of undersea volcanic sites could not be fully investigated until the U.S. Navy declassified its maps of the seafloor after the cold war. One reason locations of deep-sea fissures were not made public previously was that they held deposits of nickel, copper, and manganese—minerals with industrial and military applications. The United States sought to appropriate knowledge of and access to these resources, casting the deep as a site for national secrets.[11]

In 1977, geologists diving in *Alvin* on the undersea Galápagos Ridge in the eastern Pacific Ocean first located hydrothermal vents.[12] Seawater, seeping into the mid-ocean ridge, heats to great temperatures and reacts with magma, exiting vents as plumes of "black smoke." Biologists on the next *Alvin* dive at Galápagos in 1979 found a bevy of organisms adapted to life in the vicinity of mineral plumes. Deep-sea biologist Cindy Van Dover writes: "Vent water is enriched in reduced chemical compounds, especially hydrogen sulfide. . . . A variety of bacteria thrive on the sulfide, using its chemical energy through chemosynthesis in much the same way that plants use energy from light to produce organic carbon through photosynthesis."[13] Chemosynthesis by microbial life forms is the basis of hydrothermal ecosystems and their patterns of symbiosis and chains of predation.[14]

In the wake of the discovery of vents, as *Volcanoes of the Deep Sea* implied, the idea that the ocean might be home to "living fossils" was revived. As geologists Raymond Binns and David Decker write in a 1998 issue of *Scientific American*, "On the cooler fringes of the hot springs, there are mussels, several newly recognized kinds of anemones and long-necked barnacles, which until recently were thought to have died out with the dinosaurs at the end of the Mesozoic era, 65 million years ago."[15] The lost-world narrative, the story of the land that time forgot, so beloved of the Victorian science fiction of Arthur Conan Doyle and Jules Verne as well as of the degeneration-obsessed horror of H. P. Lovecraft, is reactivated here; indeed, vent sites have been given such tongue-in-cheek biblical and archeological names as Garden of Eden and Lost City. And, most important, as we have seen, the microbial inhabitants of these realms, heat-loving microbes, have been portrayed as analogs for the most ancient life on the planet. Hyperthermophiles push not only at the metabolic limits of life but at the very threshold of its beginning.

Given the enduring pull of visions of the sea as a primitive world, it might be no wonder that popular descriptions of vent creatures and their habitats reach toward tropes of dimly sensed primeval throngs. *New York Times* science writer William Broad, for example, writes, "It was a major revelation to learn that highly complex ecosystems were powered by [chemosynthesis]—that we and all the other light-eaters of Earth shared our planet with an alien horde that thrived in total darkness."[16] Here, as in *Volcanoes*, with its "creatures of the dark," a journey into the deep reads not just as science fiction but also as a voyage into the heart of darkness to discover the secrets of a lost world inhabited by archaic, obscure others. Broad's alien "horde" is reminiscent of the terms nineteenth-century social theorists used to describe the putatively primeval kinship groups of colonized peoples—notably social evolutionist Lewis Henry Morgan, who in 1866 named the "promiscuous horde" as characteristic of the "savage" stage of human development, a phase for which, he suggested, family trees could not be easily made because lines of descent would be confounded by the absence of legitimate patriarchal, patrilineal monogamy—that Victorian ideal believed to secure unilineal flows of property through tracing paternity. The alien of the alien ocean is sometimes figured as a primitive other, marked as such through the trope of darkness, a figure that both suggests an absence of enlightenment and calls up the fearsome and fascinating dark bodies of racialist discourse. As Fabian argues, "There is no knowledge of the Other which is not also a temporal, historical, a political act."[17] Siting an ultimate origin for human selves in a hydrothermal other—the view offered by *Volcanoes*—often mobilizes narratives about primitivity, lineage, and legitimacy, narratives that gather meaning from the racial and colonial histories through which those concepts have traveled.

Any origin story has multiple—even contradictory and mutually exclusive—versions. Not all scientists agree on how or where to root the tree of life—or even on whether a stable rooting is possible. To understand why, we must climb down the branches of this tree, through thickets of tricky scientific theory, and into hot water.

THE TREE OF LIFE, UNDERWATER

In *On the Origin of Species*, Darwin named genealogy—relations of ancestry and descent—as the ideal principle from which to generate what he called "natural classification."[18] "I believe," he wrote, "this element of descent is the hidden bond of connexion which naturalists have sought under the term of the Natural System."[19] Darwin hoped to improve upon such efforts as Linnaeus's *Systema Naturae*, which grouped creatures not by lines of ancestry but by physical similarities, believed to be markers of immutable essences. One problem with arraying organisms by similarity was that resemblances resided in the eye of the beholder. More, it was not obvious which similarities—of, say, reproductive organs, locomotion, or habits—should claim priority in classification.[20] A "natural system," "genealogical in its arrangement,"[21] Darwin held, would make the matter clear, cleanly tracking the divergence of species, which he felt were distinct and coherent, if mutable, biological forms.

Such a genealogical system, Darwin offered, could be represented as a branching tree, a structure he borrowed from Victorian practices of family

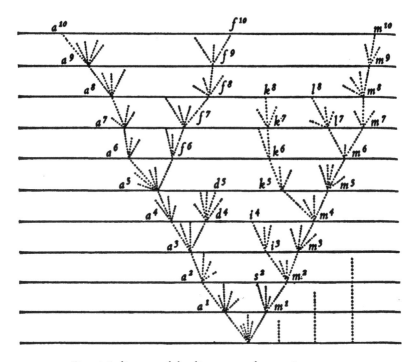

FIGURE 10. Darwin's diagram of the divergence of taxa, 1859.

recordkeeping, a form that had in turn developed from earlier elite European traditions of legitimating property inheritance through diagrammatic appeals to illustrious ancestors (figure 10).[22] Gillian Beer has suggested that, though Darwin's forking figure in the final pages of *Origin* "could as well be interpreted by the eye as shrub, branching coral, or seaweed . . . Darwin saw not only the explanatory but the mythic potentiality of this diagram [and] its congruence with past orders of descent."[23] Committing himself to the tree image and reading such kinship representations onto the organic world, Darwin naturalized and universalized such structures, suggesting through a now commonplace figure/ground reversal that his culture's method of kin reckoning was itself an emanation of a natural logic organizing all relatedness.

For Darwin, trees traced heredity, a concept figured at the time through the metaphor of blood and bloodlines. In the twentieth century, after the rediscovery of Mendel's work, the idiom of genetics gradually replaced that of blood—a conceptual trajectory repeated in *Volcanoes of the Deep Sea*, with its travel from blood-tipped tubeworms to DNA-programmed hyperthermophiles. As the universal coin of heredity, genes were enlisted to do something blood was never asked to do: provide the substance linking together *all* living organisms on earth, through the fivefold ramification of creatures into the kingdoms of animals, plants, fungi, protists, and bacteria and across the twofold division between prokaryotes and eukaryotes—that dichotomy based on the absence or presence of a nuclear membrane around an organism's DNA. Genes became the currency of genealogy, the substance traced by the tree of life.

It should come as no surprise, then, that when microbiologists in the late twentieth century sought to place deep-sea creatures like hyperthermophiles on the tree of life, they turned to these organisms' genetic properties and relations (a practice prompted, additionally, by the lack of a fossil record for most microbes). Using the gene sequence data that became increasingly available toward the end of the twentieth century, microbiologists sought to construct molecular phylogenies—genealogical trees based on DNA and RNA—hoping to confirm, revise, and sharpen organismic phylogeny.[24] Examining creatures like vent hyperthermophiles in this light revealed a curious fact: they appeared to possess genes from both eukaryotes and prokaryotes (genes involved in replication looked eukaryotic, those for metabolism looked bacterial), a finding that some biologists thought called for a reconsideration of branching events on the tree of life and, also, perhaps, a deeper rooting of the tree itself (a response not unlike that prompted by the platypus in the nineteenth century: was it a mammal? a bird? should the system be rethought?). By the early 1990s, such hyperthermophilic microbes, along with their recently discovered cold- and salt-loving and methane-metabolizing similars (from other environments), had been gathered into a newly created taxonomic category *above* the kingdom level: the domain.[25] In a 1990 article titled "Towards a Natural System of Organisms"—an explicit nod to Linnaeus and Darwin—biologist Carl Woese argued that genealogies guided by molecular genetics led the way toward a view of life on earth as divided into three domains: Bacteria, Eucarya, and—the new microbial group—Archaea (meaning "ancient ones," from the Greek for "ancient"), so named "to denote their apparent primitive nature"[26] (see figure 11). Woese's nomenclatural move, widely accepted by the mid-1990s, expanded the armature of evolutionary taxonomy by creating a new level, a step taken in part because of Archaea's apparent genetic uniqueness.[27]

Molecular biologists believe that finding genes that point cleanly back to ancestral destinations requires particular kinds of genes, genes that accumulate random mutations at a constant rate, that are sheltered from the

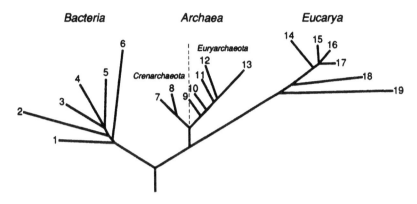

FIGURE 11. Three domains of life. From Woese, Kandler, and Wheelis (1990: figure 1). Reproduced with permission.

vicissitudes of natural selection and can therefore act as "molecular clocks." Woese argued that small-subunit ribosomal RNA—genetic material from ribosomes (cellular organelles where protein is synthesized)—was a good candidate for such a chronometer because it is present in all genomes and, since its function is so fundamental, it is not likely to be subject to the unpredictable pressure of natural selection.[28] More important, Woese considered ribosomal RNA the center of what he termed the cell's "information processing systems"[29]—a complex of genes coding for replication, transcription, and translation; in other words, the part coding for coding itself. As one researcher ventriloquized this view, "Transcription and translation genes are central to the 'essence of the organism.' They encode the hardware that reads the exchangeable genes for the cellular software."[30] Family trees tracking this "essence" initially revealed Archaea as distinct from Bacteria and Eucarya. The traits Archaea shared with other prokaryotes and with eukaryotes looked early on to be explained by the possibility that a hyperthermophile akin to contemporary Archaean thermophiles might be ancestral to all earthly life. As deep-sea oceanographer John Baross put it, "The rooted global phylogenetic model of Woese and colleagues, based on 16S rRNA sequences, predicts that the oldest of extant organisms are thermophilic and that the common ancestor to all extant organisms . . . was also thermophilic."[31]

Science writer Richard Ellis reports in *Aquagenesis: The Origin and Evolution of Life in the Sea* that the idea of a hot origin for life was reinforced for many researchers by the hypothesis that hyperthermophiles today inhabit the kinds of extreme environments that may have characterized

Earth during the Archaean geologic era, 3.5 billion years ago.[32] John Corliss, John Baross, and Sarah Hoffman argued in 1981 that lipids and amino acids could have originated in chemically rich water only if temperatures soared above boiling and if high pressure prevented this heat from denaturing such chemical configurations.[33] More, according to Baross, during a time when Earth was frequently bombarded by asteroids, "the only safe havens for microbial communities . . . would have been the deep ocean and the subseafloor. The subseafloor would have had the advantages of retaining significant levels of water even after an ocean-evaporating impact while continuously producing, from hydrothermal activity, the carbon and energy sources necessary to sustain life."[34]

The cosmic context offered here prompted some to speculate that hydrothermal ecosystems could be "models for sites where life might have originated on this planet and where extraterrestrial life is speculated to exist on Mars and Europa."[35] Popular books reporting on these life forms capitalize on the association between the deep sea and outer space, offering such titles as *Dark Life: Martian Nanobacteria, Rock-Eating Cave Bugs, and Other Extreme Organisms of Inner Earth and Outer Space*[36] and *Evolution of Hydrothermal Ecosystems on Earth (and Mars?)*.[37] Van Dover leans on the extraterrestrial analogy: "For those of us lucky enough to be involved in this research, it is like discovering life on another planet and having the privilege of being among the first to study that life."[38] Some scientists take cosmic visions of hydrothermal life a step beyond the alien analog, arguing that such life may have originated on Mars and then been conveyed to Earth in meteorites (see chapter 7).[39] Whether Martians or Earthlings, however, hyperthermophilic microbes, according to such models, found sustenance in vents. The family tree for life on Earth, very much as *Volcanoes of the Deep Sea* would suggest, looked as though it would be rooted in deep superheated ocean water.[40]

THE SOLVENT OF INFORMATION

During my early research at MBARI, I tested out my understanding of molecular phylogenetics on Ed DeLong, telling him that it sounded to me as though the information instantiated in genes (in the form of Woese's small-subunit ribosomal RNA) was pointing the way to the origin of life. To adapt an old cliché about family relations, genetic information as a kind of kinship substance seemed to be thicker than water, durable through the wet worlds inside and outside the organism through which it had to be transmitted.

But DeLong drew a more complex picture. He told me that in the late 1990s several papers in microbiology had begun to point out that assumptions about stable genealogy might be all wrong for organisms like Archaea. When microbiologists tried to corroborate ribosomal RNA phylogenies of Archaea, Bacteria, and Eucarya by choosing genes other than those related to "information processing systems," the trees that resulted were often quite contradictory, with genetic trajectories pointing every which way. The tree of life was in a brambled state.

Crafting phylogenies relies on a varied set of materials and tools. In the first instance, one must have to hand DNA libraries like those described in chapter 1, collections of genetic fragments from organisms one is interested in classifying. Information from such libraries is encoded into DNA sequence databases, which offer transcriptions (in the more translatory than genetic sense of this term) of genetic material into As, Ts, Cs, and Gs.[41] Researchers can then compare sequences from databases using computer software designed to generate possible relationships between genes. One of the most intriguing relationships for origins-of-life research is that of shared descent. Similarities in sequences can be arranged to map possible common ancestry, conjecturally diagrammed on a tree.[42] Software packages like PHYLIP, MacClade, and PAUP aid in the making of family trees by allowing researchers to play around with different parameters for generation lengths, molecular clock rates, or parsimony (economy of explanation), to take just a few examples.[43] Computer-aided phylogeny is one kind of bioinformatics. As MBARI's Steven Hallam informed me as he was putting together trees connecting various Archaea to one another, phylogenetic trees based on the same data can have radically different structures depending on the algorithms and assumptions used to construct them.

In 1996, molecular biologist Kim Borges signaled the prospect of classificatory systems that might embrace polyglot categories (much like that more famous Borges, the Argentinean fabulist known for his tales of self-contradicting catalogs). Borges explained, "It is possible to support four possible rooted universal tree topologies, depending on the protein encoding genes chosen."[44] It depends, in other words, on which genes one chooses to read from a DNA library. A variety of theory-driven decisions must be made on the bioinformatic path to a tree, and these choices take on a life of their own once plugged into programmed procedures. Rooting a tree requires grounding one's logic in better or worse assumptions, as one team of microbiologists studying Archaea has noted: "There is in fact in principle no way . . . to root . . . a universal tree based only on a collection of homologous sequences. We can root any sequence-based tree relating a restricted group of organisms (all animals, say) by determining which point on it is closest to an 'outgroup' (plants, for example). But *there can be no such organismal outgroup for a tree relating all organisms*, and the designation of an outgroup for any less-embracing tree involves an assumption, justifiable only by other unrelated data or argument."[45]

The problem here is one of bootstrapping and of the radically singular[46] character of "all organisms." A universal, rooted tree relating all organisms cannot feasibly be placed in a corroborating exterior context, since this would entail specifying the initial object of inquiry: all organisms and their relationships.

Given the cascade of logical decisions endemic to building family trees, to figuring out the relations of unfamiliar life forms, it is no surprise that consensus phylogenies have been elusive. As Adrian Mackenzie argues, "Bioinformatics, when it works on comparing sequences, cuts across genealogical descent and historical unidirectionality."[47] The genetic similarities possible to discover in a database do not necessarily map onto similarities in the organic world, let alone onto accounts of how such similarities come into being.

However, DeLong told me, some researchers were beginning to believe that discordance between family trees might result from more than methodological heterogeneity. A few microbiologists suspected that the genetic information contained in microbes like the Archaea—and Bacteria, too—was not itself behaving in a predictable generation-to-generation way. It may be difficult to build trees, not only because of the tools scientists use, but because genes may often be transmitted laterally, within generations, in addition to cascading vertically "down" generations.[48] Microbes shuffle genes back and forth with their contemporaries, an activity mixing up their own and others' genealogies. Such lateral gene transfer could make it extremely difficult to arrive at a root for the tree of life.[49]

Discerning that I was unusually interested in this scrambling of the genealogical logic underwriting evolutionary notions of relatedness, DeLong handed me a photocopy of a 1999 article from *Science* by microbiologist W. Ford Doolittle, "Phylogenetic Classification and the Universal Tree." Doolittle argued that molecular phylogenetics might in fact reveal that one cannot construct stable phylogenies for microbial life and might therefore not be able to follow a singular genetic trail back to the first earthly life. The relationships of early organisms may not be treelike in nature. As Doolittle envisioned it, the tree of life might turn out to be a "net"—here extending the 1993 position of Hilario and Gogarten, who argued that with horizontal gene transfer "The Tree of Life Becomes a Net

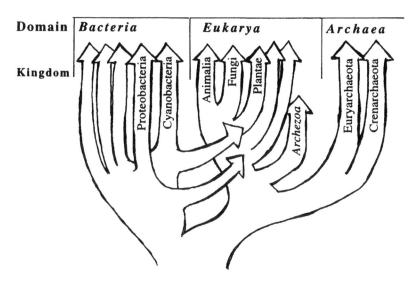

FIGURE 12. "Current consensus or standard model" of the "universal tree of life." Reprinted with permission from *Science* 284 (1999): 2124–28. ©1999 AAAS.

of Life."[50] Doolittle explained: "If . . . different genes give different trees, and there is no fair way to suppress this disagreement, then a species (or phylum) can 'belong' to many genera (or kingdoms) at the same time: There really can be no universal phylogenetic tree of organisms based on such a reduction to genes."[51] A later paper by John Paul, a professor in the Department of Marine Sciences at the University of South Florida, put it this way: "The sequencing of complete genomes of archaea, bacteria, and eukaryotes has clearly 'shaken the tree of life'"; it "has shown that microbes are mosaics of acquired genes. In fact, the rooting of the universal tree is highly problematic because of multiple transfer events between the kingdoms." Doolittle's paper offered a striking, hand-drawn contrast between a traditional "universal tree of life" graphic and a new, netlike picture of the history of life (see figures 12 and 13).[52] Calls for rhizomatic, reticulated representation as an alternative to the linearity of the tree diagram—calls that will strike a familiar chord with readers of the maniac philosophy of Gilles Deleuze and Félix Guattari—find an evolutionary biological analog here.[53]

So much for the genealogical origin of all species.

In addition to questioning the universal applicability of the tree diagram for representing the history of life on Earth, Doolittle was challenging a commitment at the center of Woese's ribosomal RNA reasoning, the idea

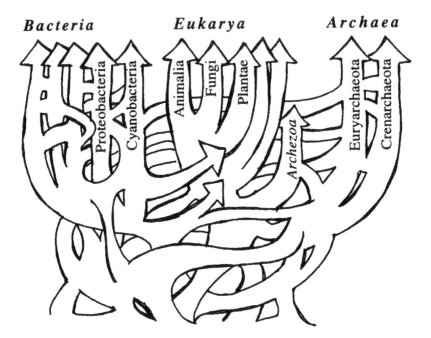

FIGURE 13. "Reticulated tree, or net, which might more appropriately represent life's history." Reprinted with permission from *Science* 284 (1999): 2124–28. ©1999 AAAS.

that transcription and translation genes are central to the "essence of the organism"—that those parts of the genome bound up with "information processing systems" constitute a stable genealogical core around which other genes might come and go. Doolittle pointed out that "cells do not actually know what is fundamental to them, which of their genes encode hardware rather than software."[54]

Doolittle's indictment of the language of computation points to the slipperiness of a key term at the heart of contemporary genetic thinking: *information.* Colloquially, "information" has signified *meaning,* the interpretative, contextual content of communication. In information theory and computer science, meanwhile, the word has designated a quantitative measure of the formal, syntactic complexity of a message (like a telegraphic transmission) and has nothing to do with what a message might mean. When "genetic information" in the 1950s became a metaphor for the biological specificity of material relations between biochemical structures like nucleic acids and amino acids, genes became increasingly described in terms of formal properties and relations (e.g., "coding" functions). Such description sidelined the physical properties of nucleic acids and the bio-chemical milieu within which they made sense.[55]

Woese's designation of "information processing systems" as constituting a hard genomic core renders genes in these systems abstract even in the act of concretizing them, positing information as a preexisting formal feature rather than a relational, emergent quality.[56] But "genetic information" might not be so neatly abstractable from the substance in which it is expressed. Eugene Thacker is not quite correct when he argues, in *Biomedia*, that "a bioinformatics researcher performing a multiple sequence alignment on an unknown DNA sample is interacting not just with a computer, but with a 'bio-logic' that has been conserved in the transition from the wet lab to the dry lab, from DNA in the cell to DNA in the database."[57] On Doolittle's account of the ontological indiscernability of hardware and software, something is elided in such translations: the material conditions in which the genome exists, a physicality that *allows for* lateral gene transfer. This observation is of a piece with recent recognitions that genes are labile, mobile, and malleable beasts. In *The Century of the Gene*, Evelyn Fox Keller writes: "For almost fifty years, we lulled ourselves into believing that, in discovering the molecular basis of genetic information, we had found 'the secret of life'; we were confident that if we could only decode the message in DNA's sequence of nucleotides, we would understand the 'program' that makes an organism what it is. . . . But now, in the call for a functional genomics, we can read at least a tacit acknowledgement of how large the gap between genetic 'information' and biological meaning really is."[58]

I add that with the coming into legibility of lateral gene transfer, molecular biologists and microbiologists call new attention to how genes are enmeshed in a swarm of material relations, including those that may fracture bits of the genealogical, informatic backbone meant to undergird "natural classification."

In *The Forest of Symbols*, anthropologist Victor Turner argues that symbols are things "representing or recalling something by possession of analogous qualities or by association in fact or thought."[59] On this definition, genes are not just tokens of particulate inheritance but symbols which, through association with heredity, have come to recall relatedness by descent. Thus many of us talk about our biological inheritance from our parents as being "in our genes," but we do not describe viral infections as in our genes, even though they are. Lateral gene transfer places the simple symbolic association between genes and descent at risk. In "Prokaryotic Evolution in Light of Gene Transfer," Gogarten, Doolittle, and Lawrence put this somewhat apocalyptically: "Gene transfer will obliterate patterns of vertical descent within groups that exchange genes at high frequency, producing discordant relationship among genes with different ancestries within the same cells."[60]

Any anthropologist interested, as I am, in kinship—in how people understand relatedness—will be captivated by Doolittle's tale of genealogy-jumbling genes. In 1968, David Schneider wrote that "in American cultural conception, kinship is defined as biogenetic. This definition says that kinship is whatever the biogenetic relationship is. If science discovers new facts about biogenetic relationship, then this is what kinship is and was all along."[61] One expectation of Schneider's pronouncement—which extends well beyond the American middle-class case to include people the world over who subscribe to scientific models of heredity—was that, as the disciplines of biology got a firmer fix on genes, Western folk notions of genealogy would find tighter confirmation. In other words, the arrangement of life forms would correspond to and underwrite forms of life. Doolittle's "Fun with Genealogy," a response in the *Proceedings of the National Academy of Sciences* to a colleague's paper on the tree of life, illustrates the normative formulation. The commentary opens with a playful disclosure about Doolittle's biogenetic relation to that paper's senior author: "Russell F. Doolittle and I recently have ascertained that we descend from a remote common ancestral couple—Ebeneezer and Hannah (nee Hall) Doolittle—via eight intermediate nodes on Russell's side and seven, including two more Ebeneezers, on mine. We share Y chromosomes, if there have been no adoptions or other irregularities in either lineage."[62] Here, biogenetics *is* kinship. As anthropologists studying reproductive and genetic technologies have found, however, "new facts about biogenetic relationship" do not always have such straightforward or consolidating meanings for kinship ideologies or practice.[63] Janet Carsten argues that, in the age of in vitro fertilization, surrogacy, cloning, and the laboratory transfer of genes across species—when cultural practice must be mobilized to guide biological systems toward social goals like parenthood, patentability, or profitability—the notion that "nature" grounds social relations in a straightforward way "appears to have been 'destabilised.'"[64]

"New facts about biogenetic relationship" in accounts of the polyvalent relations of microbial life clearly trouble Darwin's faith that "natural classification" will always follow from sorting out lines of inheritance. The new facts of gene transfer may, in some instances, unfasten links between genealogy and classification, between trees and the organisms they purport to organize. Lateral gene transfer in microbes places in jeopardy the vertical inheritance needed to root the tree of life.[65]

From an anthropological angle, what I see happening is this: A culturally reinforced biogenetic understanding of relatedness pressed researchers into phylogenies based on genes, splicing a genetic logic into the arborescent image of the family tree. Biologists believed they could discern the relations of species by tracking genes through the objects—individuals and populations—of neo-Darwinian theory. Rather than pointing toward a common ancestor, however, this focus fragmented the very object of inquiry. "Natural classification," rendered into the information processing systems of genealogy, turns back on itself to take apart its own constitution as naturally classificatory. In the bargain, the stability of the category of *species* for microbes has been called into question: "A potential result of rampant interspecific recombination is the blurring of species boundaries, and the failure of any one gene to reflect the evolutionary history of the organism as a whole."[66] This may not be surprising, since the particulate, discrete logic that has inflected biology since the rise of gene thinking has long been at work disaggregating the genealogical ontologies of such subjectively named categories as *variety* and *race*. As Harriet Ritvo demonstrates in *The Platypus and the Mermaid*, taxonomic systems have long embedded philosophical commitments containing the seeds of their own undoing.[67]

Biologists' adherence to the tree image has been aided in part by their treating genetic information as an abstract property. As Doolittle's attention to lateral gene transfer suggests, there is no part of the genome—no self-evident information-processing backbone—sitting outside the material circumstances of organisms. Commitments to the notion that genes are best considered abstractable *information* explicitly dissolve when they run up against the contingencies of lateral gene transfer, a process that refers us to such physical mechanisms as conjugation (contact between microbes), transformation (uptake of DNA from ambient environments), and transduction (transfer of genes via viruses) as well as to such material items as plasmids. What N. Katherine Hayles has called "the materiality of informatics"[68] takes shape with respect to the emergence of genetic information in the circumstances of particular cellular, organismic, and environmental media. Bioinformaticians who build phylogenies based on envisioning information as a sheerly formal entity risk floating free of the contingencies of the physical world. Thacker suggests that "the biological and the digital domains are no longer rendered ontologically distinct, but instead are seen to inhere in each other; the biological 'informs' the digital, just as the digital 'corporealizes' the biological."[69] But although this may be true in some cases, in the dynamics I describe here biological and digital ontologies destabilize, deform, as they are diffracted through one another.

The story, of course, is not over. Lateral gene transfer may not always "obliterate patterns of vertical descent," as Gogarten, Doolittle, and Lawrence warn.[70] University of Arizona microbiologist Howard Ochman, for example, argues that gene transfer introduces only a little waviness around the limbs of the tree of life.[71] In his diagrams, this looks something like the vibrations around the tines of a tuning fork. Lateral gene transfer, Ochman says, mostly unfolds among already physiologically similar groups and is therefore a force and resource for the coherence of types. Not all genes transfer equally, or easily. This argument rests on renewed concern about the materiality of genes, now as durable devices that cannot be transferred across organisms without physical work. But although Ochman's position is optimistic on the matter of whether tree representations can be salvaged, his work continues to invite the question of how species should be defined, and, indeed, what should count as relatedness at all.

REARRANGING GENESIS

A year after taking up my position at MIT, I was able to join a graduate seminar treating "advanced topics in fields of aquatic chemistry, aquatic biology, molecular ecology and oceanography." Ford Doolittle, visiting for fall 2004 from his home institution in Canada, was one participant. DeLong was also in attendance, having relocated (conveniently for me) from MBARI to MIT. I had met Doolittle in 2002 in New York, at the American Museum of Natural History at a meeting on "Assembling the Tree of Life," where he had complained to me that "people have deep paradigmatic commitments to trees; they are wedded to genes and trees." Dr. Doolittle—like his fictional near-namesake, drawing our attention to menageries of creatures that confound species boundaries—reiterated his argument that genes could not stabilize trees for these creatures because "as few as five percent of the genes may be faithful to one another," a phrasing that both retains and refuses the link between relatedness and heredity. In "Fun with Genealogy," he extends such metaphors of legitimacy, suggesting that lateral gene transfer may reveal dynamics "akin to adoption in their confusing effects" and asking, "To what extent is our desire to look at early evolution in terms of cellular lineages preventing us from seeing that it is about genes and their promiscuous spread across taxonomic boundaries, which then have no permanent significance?"[72] In the MIT seminar,

Doolittle continued to argue that gene transfer was gnawing at the tree representation of relatedness for microbial life. He pointed out that

> the mere fact that you can construct a tree doesn't mean that you've got a phylogeny. For example, a tree of people in this room based on the color of their clothing could be constructed, but no one would think that was a phylogeny. What's happening now is that "phylogeny" is being extended to describe any pattern of similarity that can be mapped onto a tree. But that's not what Darwin meant by natural, genealogical classification. Extreme people like me wouldn't deny that there *is* a pattern, just whether it is equivalent to a tree of life. The question is: Are you discovering what an organism is, or are you simply deciding what to call it?

As Doolittle spoke, I recalled a lecture in which DeLong had presented a treelike map of genetic distances between various microbes and immediately cautioned his audience against seeing it as a family tree, declaring "this is not a phylogeny"—a phrase that might remind readers of the Surrealist painter René Magritte's famous canvas depicting a pipe floating above a calligraphic legend that reads, in French, "This is not a pipe." Representation does not easily map onto reality, to be sure, but more: sometimes a tree is just a tree. Or just a diagram we imagine as treelike.[73]

According to Doolittle, the historical and cultural power of the tree image—solidified in the persistence of family tree imagery and, one seminar participant added, now reinforced in digital genealogy programs—has kept people from seeing how ill the fit is between the tree diagram and lines of microbial connectedness and genesis. It has even kept us, Doolittle argued, from recognizing the artificiality of the tree for organizing human systems of kin reckoning. Doolittle offered the incoherence of *race* as a genealogical category as an analogy to what he saw happening to *species* in microbial biology: "It will be interesting to pay attention to population geneticists' human phylogenies in the coming years. They will produce trees that will trace particular parts of the human genome and these trees will probably be mapped onto particular geographies—but this will *not* thereby mean that these trees have isolated human 'races.' The trees may only trace a few traits. That will not mean that trees are tracking 'essences.' We in microbiology are trying to define something analogous to 'race' in human populations, and we won't find it."

But if the tree diagram is at risk from lateral gene transfer, it is not clear that biogenetically rendered kinship is thereby drifting off the radar (nor, for that matter, *race*—though, as I suggest below, it is persistently reimagined vertically, not laterally). We may find ourselves in a novel realm of

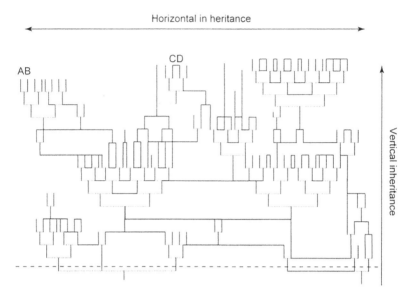

FIGURE 14. "Synthesis of Life." Reprinted from *Trends in Microbiology* 12 (2004), with permission from Elsevier.

what anthropologist Roy Wagner once called "analogic kinship,"[74] with genes functioning as vehicles of analogic relation that *enable* natural classification: witness Doolittle and colleagues' recent coining of "horizontal kin" to designate organisms linked by lateral gene transfer.[75] As Doolittle put it in the seminar, "Both shared descent and shared exchange can create coherent connection," a phrasing that recalls the two primary paradigms in classical anthropological kinship theory, descent and alliance, here articulated in a unifying language of genetics rather than of, say, patrilineality or exogamy.[76] Doolittle and his colleagues have recently proposed a "Synthesis of Life" that makes use of both tree and net linkages to represent phylogenesis, defined, they record, by the OED as "the evolutionary development of a species or other group of organisms through a succession of forms," which, they write, "in no way requires that species or other groups be produced solely through divergence, nor that diagrammatic representation of the evolutionary development of species must be a bifurcating tree" (see figure 14).[77] What is preserved in this new map, however, is the figure of the gene, continuing to serve as a token, a black box, representing the flow of "life."

There is an instructive difference between this netlike synthesis and what Darwin in *Origin* termed life's "inextricable web of affinities." Darwin's web was not a map of shifting allegiances but rather, as Beer has suggested, a set of "fixed patterns and achieved limits"; indeed, it was synonym for his tree.[78] For the Victorians, webs called to mind woven fabric, made of steady, stable, unchanging threads. For double-clickers like today's scientists, webworks are made up of threads that move, that can be routed to link anything to anything else. There are no necessary boundaries to such networks.

Little wonder, then, that the genetically rendered "life" mapped in such formulations as the "Synthesis of Life" is becoming partially agenealogical. Strict correspondences between life, genes, genealogy, genesis, and classification have been denatured. It would be easy to say that this dissolves one of Darwin's dearest premises, if one imagines he wanted the "hidden bond of connexion" of descent to lead to a unitary ancestor. But the final sentence of *Origin* reads as open to multiplicity: "There is grandeur in this view of life, with its several powers, having been originally breathed into a *few forms* or into one."[79] Some polygenists—those creationists of Darwin's day who believed that there were distinct Adams and Eves for different races—may have found Darwin's phrasing congenial to their views, but it is almost certain that they were not a readership he meant to please.

Mention of creationists brings up the intriguing fact that some of the most avid readers of Doolittle are religious thinkers of the intelligent-design school who interpret his questioning of the genealogical model of relatedness as undermining evolutionary theory itself. Molecular biologist and doctor of theology Jonathan Wells of the Discovery Institute, a non-profit association in Seattle that promotes "belief in God-given reason and the permanency of human nature,"[80] writes:

> Ten years ago molecular biologists were hopeful that DNA studies would confirm Darwin's "tree of life." But as they compared more and more genes it became clear that relationships among the major kingdoms of life are a tangled thicket rather than a branching tree. According to Canadian molecular biologist W. Ford Doolittle, scientists should stop trying to force the data "into the mold provided by Darwin." Clearly, the so-called "fact" of evolution is nothing of the sort. Darwinists simply assume that all organisms are related through common descent, because to admit otherwise opens the door to the possibility of divine creation and design.[81]

Doolittle's challenge, however, is not to such staples of Darwinian theory as mutation, selection, or, indeed, descent with modification but simply to whether phylogenesis always and only happens through descent. Intelligent-design advocates have tweaked Doolittle's account to support the typological thinking characteristic of creationism. The twist in intelligent design's essentialism is this: advocates now lean on the discordance of organismal and molecular phylogenies to claim that organisms and molecules have independent histories, unmodified by descent. Reproducing Doolittle's revised tree of life figure from *Science*, intelligent-design proponent Casey Luskin writes, "The three major 'domains' of life—Bacteria, Archaea, and Eukarya have a distribution of characteristics which does not allow a tree to be constructed to describe their alleged ancestral relationships."[82] Instead of imputing this distribution to the phylogenesis that might emerge from lateral gene transfer, Luskin suggests that both organisms and their molecular components have designed, if distinct, origins (i.e., an intelligent designer installed similar molecular structures in diverse lineages of creatures—an argument similar to that which would have chimp and human hands as modular parts added to independently created lineages). We are witnessing here the rise of a molecular creationism—a creationism still organized by pure types all the way down.

But just because it may be impossible to root the tree of life in a unitary ancestor does not mean molecules have typological permanence. More, trouble with the tree representation on its own does not make it impossible to make a case for an ancient origin for life in hydrothermal environments—or in any other environment for that matter, given appropriate evidence.

Such evidence, of course, will remain unconvincing to creationists. Indeed, as I was at work on this chapter, I read in the *New York Times* that about a dozen science museums, primarily in the American South, had been turning down *Volcanoes of the Deep Sea* because of its references to evolution and, according to Richard Lutz, chief science advisor for the film, the suggestion that life may have begun in hydrothermal vents. "Carol Murray, director of marketing for the Fort Worth Museum of Science and History, said the museum decided not to offer the movie after showing it to a sample audience, a practice often followed by managers of Imax theaters. Ms. Murray said 137 people participated in the survey, and while some thought it was well done, 'some people said it was blasphemous.'"[83]

And here I thought Pete and I had seen a family film. In the *Times* article, James Cameron weighed in: "It seems to be a new phenomenon . . . obviously symptomatic of our shift away from empiricism in science to faith-based science."[84] The Fort Worth Museum of Science and History, though, after initially taking to heart the anxieties of fundamentalist patrons, eventually reversed its decision and showed the film.

Intelligent-design adherents and creationists are correct to see something in descent with modification that competes with a view of divinely ordained organic stasis. In her analysis of the literary structure of *Origin*, Beer suggests that, in employing the tree to represent not just the genealogy of life on Earth but also the promise of science to reveal this pedigree, Darwin effectively "makes the tree of life and the tree of knowledge one," not only supplanting the biblical story of creation with evolution but also declaring the forbidden fruit of knowledge capable of nourishing testable trees.[85] The controversy over *Volcanoes* reveals tensions that many people in the United States feel between religion and science, frictions that center on who is authorized to make sense of the existential, material, and potential meanings of biology as flesh and discipline, to connect life forms to forms of life.[86] Creationists want a single meaning, a single connection. Others remain in suspense about multiplicities, even as some scientists resist the dissolution of the tree of life not only because of the skeptical stance that normatively animates any scientific attitude but also because such a transmogrification may uproot their own tree of knowledge—of species names, of models of descent all the way down, and even of one-way evolutionary time. Biologist E. O. Wilson, famous for his entomology and infamous for his sociobiology, in a lecture at the Assembling the Tree of Life workshop admitted the lateral logic of hypertext into his view of relatedness only in his suggestion that every species in the world have its own website.

Darwin's frame "a few forms" offers a path into imagining multiple organismic relationships, forms of biogenetic relatedness at once unitary and multiple, at once familiar and strange. Bapteste et al.'s diagram captures the ambivalence—the dual attachment—here, suggesting trees rhizomatically growing into and out of each other. This image in mind, DeLong's "microscopic forests of the sea" (and a related naming by biologist Paul Falkowski of microbial marine life as "the ocean's invisible forest")[87] can be read not just as reaching toward analogies with rain forests but also as invoking the multidirectional vitality of trees' roots, entangled with one another. The upside-down, inside-out image of "forests of the sea" recalls the material and sensory confusion described by Baudelaire in "Correspondences," the poem from which Turner took the title of his *Forest of Symbols*. The "vertical and horizontal correspondences" that animate Baudelaire's poetry of "ambivalent analogy"[88] reappear in the siting of "life" in such networks as the "Synthesis of Life," which operates as a system for redistributing that itinerant and unstable scientific symbol of life itself, the gene.

What scientists on either side of the tree debate have in common is the idea that "relations" structure both kinship and knowledge systems, with the two providing what Marilyn Strathern describes as "figurative or metaphorical resources for one another, borrowed back and forth."[89] "The term 'relation,'" Strathern writes, "already denoted intellectual practice—narration, referring back to something, making a comparison—before it became applied to social ties and specifically to ties of blood and marriage."[90] In other words, diagrams of life and of knowledge have long structured one another—a mutual constitution that is, if anything, stronger today, when scientists imagine both life and knowledge as created by bringing bits of information into relation.

HOT WATER

Where do the newly described relations of gene transfer leave marine hyperthermophiles as possible universal ancestors? When I first began following this debate, it seemed to me that the movement of genes across organisms asked us to consider the physical conditions under which transfer could happen. I became curious as to whether attention to lateral gene transfer might shift emphasis back from genetic information to genetic material, and I wondered whether it might turn out that for many of the organisms that trade genes—like marine microbes in hydrothermal vents—the particular environments of the water they inhabited might be important for gene transfer. I had read that "terrestrially-derived models of gene transfer may have no validity in marine or other environments."[91] And, in an article on gene transfer in bacteria, I found this pronouncement about hyperthermophiles: "Comparisons of completely sequenced genomes verify that bacteria have experienced significant amounts of lateral gene transfer, resulting in chromosomes that are mosaics of ancestral and horizontally acquired sequences. The hyperthermophilic Eubacteria *Aquifex aeolicus* and *Thermotoga maritima* each contain a large number of genes that are most similar in their protein sequences and, in some cases, in their arrangements, to homologues in thermophilic Archaea."[92]

Genes seemed to be moving across the domains of bacteria and archaea in hydrothermal environments. Did these aqueous settings shape processes of transfer in significant ways? The declaration that models of gene transfer on land might have "no validity" in the sea sounded overdrawn—but what if water, hot water, might actually act as a medium through which genetic "information" might be broken down to flow across boundaries? Perhaps information was not always thicker than water. Certainly, the flow of genes in marine environments might be more intense than elsewhere: "Given the very high concentrations of bacteria and viruses in seawater, and the tremendous volume of water in the ocean, it follows that gene transfer between organisms takes place about 20 million billion times *per second* in the oceans."[93] Might there be something about water impinging upon and affecting lateral gene transfer?

This was a question I posed to Doolittle in an early conversation. He was suspicious of my curiosity about water: "What are you, a Pisces?" he asked, teasingly. He did not see that water as such had shown "genetic informa- tion" to be a metaphysical conceit, directing attention to the material cir- cumstances in which genes actually exist. He said: "It's not so much that recent work on extremophilic microbes and lateral gene transfer has *mate- rialized* the genome. We [microbial biologists] knew genes were material all along," but that "bioinformatics has *dematerialized* the genome." But it seemed to me another dynamic was in place here, too: bioinformatics seemed to be *allowing* lateral gene transfer to become visible. It was not water, but, ironically, information—computationally rendered—that was being used to dissolve abstract, disembodied, notions of genetic informa- tion; with bioinformatics, it was becoming clear that the "information" manipulated *in silico* was not genetic information but rather information *about* genes. In 1970, molecular biologist François Jacob reflected, "Today the world is message, codes and information. Tomorrow, what analysis will break down our objects to reconstitute them in a new space?"[94] *Information*, like *gene*, has been a productive concept not because it is so exact but because its multiple meanings have reconstituted the space within which scientists break down their objects and build up their ques- tions, within which they see the material world.

Still, what *were* the physical effects on genomes of being located in hot water? And had the assumption of a durable information-processing core directed attention away from such effects? Perhaps. The placement of hyperthermophiles at the bottom of the tree of life had, by the late 1990s, been called into question—in part because of lateral gene transfer and in part because the environment of superheated, high-pressure water might influence the molecular constitution of hyperthermophilic genomes, with resulting dramatic effects on the rate at which relevant molecular clocks might tick. By 1999, according to the life science editor for *Science*, Elizabeth Pennisi, some microbiologists, like the University of Maryland's Frank Robb, "suspected that the placement of the hyperthermophiles, microbes that live at extremely high temperatures, toward the bottom of the tree might be an artifact resulting from assumptions about how fast these microbes evolve and on their initial discovery in extreme environments reminiscent of what early life might have experienced."[95] Archaea's placement near the bottom of the tree of life "was based in part on the finding that the DNA in the ribosomal genes of the organisms has lots of guanine and cytosine bases, an indication that their genomes had been around long enough for certain bases to become overrepresented." But it turns out that these bases were probably not randomly accumulated but selected for in the genomes of organisms living in extreme environments, for they help stabilize DNA under thermal stress.[96] This possibility could mean that earlier estimates of antiquity were based on an erroneous reading of the thermophilic molecular clock.[97]

I concluded this: a commitment to Archaea as "ancient ones" living at the bottom of the sea had come together with a treatment of "genetic information" as abstracted from its local substantiation in particular organisms and circumstances, such as variously hot, cold, differently pressured, diversely salty water. Like Haeckel's *Urschleim*, hyperthermophilic archaeota as the originary genetic stuff of earthly life may turn out to be an artifact—in this case not of storage in alcohol (as with the gunk called *Urschleim*) but of the storage of their sequences in an informatic medium that factors out the contingencies of their coming into being.[98]

Information talk continues to saturate discussions of the tree of life. Thus, a marine biologist at Penn State who works on hydrothermal vents put the phylogenetic issue to me in terms directly borrowed from information theory: the "signal" of origin may not be easily audible through the "noise" of gene transfer. "Signal" and "noise" are not properties in nature but rather terms that refer to scientists' own attempts to capture information they think relevant to their classification systems. Anthropologist Gregory Bateson once wrote that "the elementary unit of information . . . is a *difference which makes a difference*."[99] From the point of view of many microbiologists, the difference represented by hyperthermophiles now resides not so much in a putative primordial status as in the possibility that their genetic relations may undermine their classification.

Although the image of a scrambled communication from a remote source may destabilize any originary nature for hyperthermophilic Archaea, it does not eliminate the figure of the alien. As ultimate paths back to the origin of life have been obscured, rhetorics of alien origins have shifted sideways. As agents of lateral gene transfer, microbes are now described as installing alienness at many sites in the tree of life. Discussing lateral sources of genome diversity, Bapteste et al. write of "the substitution of a resident gene by an alien version,"[100] and Daubin, Moran, and Ochman name such dynamics "alien-gene acquisition."[101] Genes become alien visitors, inhabiting and rewriting the logics of life from within. Recent discussions of mobile elements in the human genome, like transposons, retrotransposons, and materials having origins in retroviruses incorporated into early eukaryotic and mammalian lines, suggest that lateral gene transfer shaped nonmicrobial life forms as well.[102]

The "alien gene" describes an alterity that reproduces itself only at the moment that it becomes familiar, incorporated. Once installed in novel "nets of life," however, it continues to do defamiliarizing work, reformatting the time and space of evolution itself. The alien gene reshuffles time by scrambling spatial representations of history (e.g., the tree of life), leaving in its wake a twirl of possible chronotopes, materializations and maps of time in space.[103] The alien gene as a kind of time traveler is an emissary of that ocean that washes over memory, an entity revealed, fittingly, by relational bioinformatics databases, forms of nonnarrative memory through which a plurality of possible pasts and futures can be imagined.[104] The alien gene is a sign of an unresolved, unfamiliar logic that now inhabits life forms, that warps the very form of life itself.

As rhetorician Richard Doyle puts it, "The very essence of the alien presence . . . [is] its characteristic ability to proliferate and mutate, disturbing the various taxonomical categories that we bring to bear."[105] "Life" becomes uncanny, alien to itself, often most keenly when we seek to replicate or reproduce it in our classifications—through such vehicles of representation as the gene. It is through this process that the hyperthermophiles of *Volcanoes of the Deep Sea* can oscillate between being "alien to us" and "most certainly related."

FROM TREES TO TYPES

With the possibility that much that is solid about the tree of life will melt into water or ramify at the root, microbiologists have been scrambling to find new classifications. To be sure, there are those who welcome the new complexity. Steven Hallam of MBARI told me: "Even though in a professional setting I use all these tools that group things together, ultimately in my personal life, I don't think of the world that way, in fundamental species concepts. If anything, I like the idea of that picture of the tree that's all knotted and tangled. It fits with my view of the world, of everything being connected, as parts in a body, of a Gaian synthesis."

Even Steven, however, like most microbiologists, is hardly ready to give up the classificatory ghost. Biologists want tools for sorting things out. Dissatisfaction about the match of species to microbial affiliations in the wild has prompted some to propose the *phylotype,* a genetic but not necessarily genealogical classification. In this formulation, two microbes might belong to the same phylotype if they show more than 70 percent genetic similarity.[106] This framing is tricky since there are some categories of microbes which, though genetically similar, have different niches in the sea.[107] To account for these differences, some microbiologists describe *ecotypes*—like those genetically similar marine bacteria discovered by the DeLong lab that float at different depths and use different frequencies of light. But the notion of *ecotypes* can be confusing for biologists because "organisms currently grouped together because of . . . shared [physiological and morphological] properties may be quite distantly related."[108]

This language of *types* echoes preevolutionary nineteenth-century arrangements of organisms into stable classes such as *race,* a framing that highlights stability over the quality often of most interest in evolution: variation. Unlike those older categories, these new types have obviously unsteady relations to genealogy, and, for ecotypes at least, reveal themselves *in relation* to environments that they and their human classifiers both have a part in calling into being. We are not looking exactly at the typological thinking of polygenist racists or classic creationists, who believe that God creates durable life forms once and for all. Ecotypes are not little microbial races, stable types that replicate themselves neatly down generations. Classifications like the phylotype, however, do have something in common with the morphed renascence of race in the genomic age, which seeks to label whole individuals based on percentages of putative ancestry. Recent projects to account for human diversity at the level of single nucleotide polymorphisms—single base substitutions that lead to such traits as Tay Sachs or sickle cell—are reactivating racial typologies like "Ashkenazi Jew" and "African American" as rough and ready categories for medical diagnosis, pharmaceutical marketing, and treatment, even if, as Doolittle pointed out, such classification efforts have nothing to say about whether these categories are natural kinds.[109] Within forms of life shaped by social structures and representations that reinforce investments in race as a category—what Michael Omi and Howard Winant term *racial formations*—even life forms with tangled lineages can be arranged to reinscribe typological categories from earlier orders of classification.[110] If racialized imagery can be glommed onto deep-dwelling microbes, it should be no surprise that wringing its biologized spirit out of humans will take more than breaking biology down into genomes; *race,* like *species* and *life,* is being broken down to be built up again in shapes both old and new.

What is most notable about the typological thinking growing out of the rotting roots and desiccating branches of the tree of life is a focus not on what organisms are but on what they do. Exchanging modern categories for medieval ones, DeLong and others have proposed organizing microbes by *guilds,* groups of organisms that resemble one another in the environmental conditions within which they metabolize—in other words, like "associations formed . . . for the prosecution of some common purpose" (OED).[111] In addition to the hyperthermophiles I have been working with here and the methanotrophs that starred in chapter 1, there also exist barophiles, capable of living at pressures greater than 400 atmospheres, reaching to the very bottom of the sea floor; halophiles, which withstand high salinity; and psychrophiles, cold-loving microbes with a minimum growth temperature of 0°C or less that live in polar places.

This categorization of life forms based on what they make and do (or, listening to the meaning of *-phile,* what they love) rather than on how they are related to one another is consonant with the logic of biotechnology, which aims to use the abilities of these creatures to create new products. We are close to the listing in *Moby-Dick* of products made from whales—combs, corsets, candles—a consideration of creatures through what they can provide humans. The dissolution of the tree of life does not jeopardize biotech enterprise; in many cases, the liquefaction of its Linnaean-named limbs permits practitioners to improvise.[112] As a unitary genetic origin story recedes, attention shifts from a mythic past to a utilitarian future, and from natural history to cultural enterprise.

If nineteenth-century visions of the sea as sublime store of vitality have the ocean as, in the words of Chris Connery, "not simply a nutritive Outside, but a reorganized, chaotic version of the body itself, where bones, blood, marrow, and human energy are simply divided up and redistributed among the various life elements of the sea,"[113] then twenty-first-century portraits of gene-trading microbes rewrite this vision in informatic language. If the sea was once a chaotic and cosmic amnion, an archive of life primeval and life to come, those pasts and futures nowadays read more like a mix-and-match database than a straightforward, stratigraphic archaeological record. Melville, in the simultaneously straight-talking and satirical "Cetology" chapter of *Moby-Dick,* classified cetaceans neither by family nor behavior but by books, ranging them according to proportion, on analogy to sheets of paper folded in two, eight, and twelve, commenting thereby on the folly of putting whales to paper ("According to magnitude

I divide the whales into three primary BOOKS . . . I. THE FOLIO WHALE; II. the OCTAVO WHALE; III. the DUODECIMO WHALE").[114] Today's microbiology has a less ironic attitude to bookish metaphors, arraying microbial life as genomic text to be read and rewritten from DNA libraries.

EXTREME BIOTECHNOLOGY

Alien genes and their dynamics of transfer have ramifications beyond the rarefied world of microbial taxonomy and origins-of-life research. The extreme temperatures at which vent microbes thrive are of interest in gene amplification, since enzymes from these creatures can be used to make biochemical reactions run hotter and faster. DNA polymerase derived from hyperthermophiles well withstands the temperature fluctuations required in PCR, the lab process used to copy DNA. Broad's *The Universe Below* provides an example of extreme biotechnology, DeepVent, a product we met in chapter 1:

> From a hot spring more than a mile deep in the Gulf of California, amid dense thickets of tube worms thriving in otherworldly darkness, Holger W. Jannasch, a microbiologist at Woods Hole, in 1988 isolated an Archaeal hyperthermophile of the *Pyrococcus* genus. The microbe was obviously special. It grew at temperatures of up to 104 degrees Celsius and could withstand much higher heats for short periods of time. New England Biolabs, Inc., of Beverly, Massachusetts, took the microbe, isolated its DNA polymerase, cloned it, and then sold the enzyme, beginning in December 1991. It was the first time a deep-sea microbe had been brought to market. Appropriately enough, the trade name of the DNA polymerase was Deep Vent. "Thermostability, Fidelity & Versatility from the Ocean Depths" read one of the company's ads.[115]

The transfer events here—a microbe from Mexico's Gulf of California transported to Massachusetts; a gene from an archaeon copied into another creature's genome (likely *E. coli*); a polymerase moved from lab to market—place Archaea at the center of assemblages of deep-sea submersibles, biotech labs, and corporations. These microbes are hyperlinked not just to other organisms through gene transfer but also to new kinds of biotechnological science, capital, politics.

Lateral gene transfer, then, has implications not only for how scientists think about genes, phylogeny, kinship, and nature but also for biopolitics. I have in mind here something of a remix of Michel Foucault's classical articulation of biopolitics, which he described as a nexus of government power and knowledge about life that focused state attention on the sexual doings and reproductive health of national citizenries, linking individuals to populations in regimes of management. The biopolitics of our contemporary world comes in more varied forms; genomics and transgenics are not eugenics, and they enable different biopolitical constellations, ones not so neatly organized around genealogy and birth, or, for that matter, through human bodies.

Philosopher Giorgio Agamben has argued that biological life as such—what he calls "bare life"—has come "to occupy the very center of the political scene of modernity."[116] Bare life is that vital minimum made to stand for "life itself" and fit to be entered into calculations of who and what will be permitted to grow and reproduce. The substance that stands for bare life in biological terms has taken many forms. If bare life was for social Darwinists and eugenicists the principle of heredity and the sanctity of national germplasms, today, for biotechnologists, genetic material and information (as substances that can be frozen and FedExed, as information that can be e-mailed) has become that minimum that can be made available to the calculations and interventions of such biopolitical practices as patenting genes or creating transgenic lab animals. Following Aristotle, Agamben glosses the distinction between what I am calling biological "life forms" and social "forms of life" as one between *zoë*, "the simple fact of living" (what he also calls "bare life") and *bios*, "the form or way of living proper to an individual or a group."[117]

In discussions of lateral gene transfer, I propose, we are seeing the rise of an informatically inflected bare life that is increasingly agenealogical, molecular, and modular. How this *zoë* will link to appropriate forms of social life is in the making. Nikolas Rose suggests that "biopolitics now addresses human existence at the molecular level: it is waged about molecules, among molecules, and where the molecules are themselves at stake."[118] Molecular biopolitics makes elements associated with living things—genes, proteins, tissues—mobile, transferable across locations and organisms. If *sex* was the pivot point of classical biopolitics, tying together individuals and populations as well as citizens and states, *transfer* may be the practice through which new biopolitical links—between persons and patents, polymorphisms and politics—will be forged.[119] The mobility and delocalization facilitated by the dynamics of transfer also mean that nation-states are no longer the exclusive bankers of biopower; corporations, universities, patient advocacy groups, and many others reshuffle not just the substance at stake in biopower but also relations between society and biology to begin with. Bare life is, then, as Haraway suggests, "never 'bare,' never just itself."[120] The microbiopolitics of inserting gene-circulating microbes into worlds of political and economic action always dovetails with questions of symbiopolitics.

I learned about some potential biopolitical consequences of dissolving the tree of life when I signed up for a biotechnology workshop titled "Extremophiles: Theory and Techniques" at the Center of Marine Biotechnology (COMB) in Maryland in summer 2001. Located in water-front Baltimore, COMB is housed in a glass-windowed office building covered by an undulating roof suggesting the mantle of a mollusc. Wandering around a public area on the ground floor, just before the con-ference, I read on a colorful poster aimed at children that "scientists work to improve the future using cutting-edge biotech tools to investigate the oceans—earth's last frontier. They strive to develop products to improve our lives—new food sources; cures for cancer, AIDS, Alzheimer's; and clean up the planet, from the bay outside to marine ecosystems world-wide." Painted on a wall upstairs was the legend "Give a man a fish, he eats for a day; teach a man to fish and he eats for a lifetime—ancient Chinese proverb." The biotechnological sea was to be a space of bounty—fitting in an age when fish populations are dwindling. COMB scientists do some-times work on biotech for aquaculture, and the quotation brought to mind the biblical miracle of loaves and fishes—though placed in the secular Western frame of ancient Eastern wisdom, realized through the high-tech promise of fish cloning.

Or gene cloning. In the workshop's first presentation, microbiologist Michael Madigan announced that "extremophiles represent valuable genetic resources for solutions to unique biological problems."[121] In a talk titled "Archaeal Molecular Biology," another scientist opined that "the hyperthermophiles are gold mines for biotechnologically important prod-ucts that demonstrate enhanced biological activity and stability at high temperatures." Several talks discussed archaeons' ability to break down hydrogen sulfide, a toxic waste of the mining and power industries—a capacity that made them of interest in bioremediation. The COMB website announced work on biotech "tools with which to process and degrade a wide variety of natural and man-made substances," "reversing the effects of environmental contamination."[122] This was a technical fix to a social problem and gathered its rhetorical force from its phrasing as a natural solution. Extremophiles, already denizens of the underworld, would be pressed into service in the mines; through biotechnology, their abilities to live in poisonous, often oxygen-scarce environments could be exploited to make them ideal machinic laborers.

If Earth-healing secrets were locked up in extremophiles, how might scientists go about finding them?[123] Through the hypertextual reading activities of bioinformatics. Identifying practical properties of Archaea depends on comparing gene sequences across creatures for signs of useful proteins. Using data from archaeons, scientists at COMB were looking to redesign macromolecules, using comparative genomic analysis to infer, reconstruct, and modulate such properties as protein thermostability. COMB's Frank Robb writes, "We have been able to 'install' increments of thermostability based on comparative structure and sequence studies. The potential therefore exists to modulate enzyme stability by rational design."[124]

Researchers go online to scout for interesting genes, where they can employ computational tools to compare genomes point by point. Using a technique called Cross BLAST (Basic Local Alignment Search Tool),[125] they compare microbial genomes base by base, codon by codon, or protein by protein, looking for spots of similarity. Searching for similarities, of course, might be aided by having a sense of which creatures are related, on having to hand a family tree. Given that trees looked to be unsteady, though, I was curious at the workshop about the extent to which biotechnologists consulted the embattled tree. When I asked Robb how he located organisms with promising properties, he said, "Well, I wouldn't want to be relying on phylogeny *these* days" and suggested that finding genes proceeded opportunistically, looking associatively in databases.

One conference speaker maintained that lateral gene transfer held an important lesson for scientists. The process, he said, was akin to human biotechnological enterprise. "Natural genetic engineering," he argued, "is very common."[126] The nature at the heart of this biotechnology was not one of biological constraints. Paul Rabinow has pronounced that, in the culture of late modernity, "nature will be modeled on culture understood as practice. Nature will be known and remade through technique and . . . become artificial, just as culture becomes natural."[127] The "natural genetic engineering" of Archaea gives an organic heritage to a future in which biology becomes technology. The microbial marine biologists at the workshop were locating in gene transfer a hypertextual logic tethered explicitly to "nature," to how these things really work. Life was becoming an informational vector of baroque connectivity, which in turn was coming to be seen as the nature of nature itself. The age-old dynamics of these life forms looked a lot like these researchers' forms of life. The tree of life was always a net. Nature was always a genetic engineer.

KINSHIP IN HYPERTEXT

Following Schneider's pronouncement that "if science discovers new facts about biogenetic relationship, then this is what kinship is and was all along,"[128] and putting this together with Bapteste et al.'s "horizontal kin," it might be said that this view naturalizes transgenics as an operative logic of kinship (or, indeed, folds Rabinow's *biosociality* back into that older genetic determinism, sociobiology, which put culture on nature's leash). But one could also ask why this hypertexty genetic connectivity should be called "kinship" at all. Why be so literal minded as to insist on following genes into all the networks they connect? One reason: When COMB promises to employ genetic engineering to do the good work of saving the environment, we hear a story that conjoins "the simple fact of being alive"—bare life, or *zoë*—with "the appropriate form given to a way of life of an individual or group"—*bios*.[129] Genetic materials are attached to social and political commitments—to kinship "substances" and "codes for conduct," to use Schneider's terminology—in ways that sit, however uneasily, at the boundary between nature and culture. And kinship, as Strathern reminds us, "connects the two domains."[130] Kinship is, as Haraway notes, "a technology for producing the material and semiotic effect of natural relationship, of shared kind."[131] The "nature" that can produce these special effects is changing. The familiar seascapes of these natures are now visited by "alien genes," saturated by the strange but newly naturalized flows of genetic engineering. I advocate a view of "new facts of biogenetic relationship" as kinship facts, because so seeing them allows for the deployment of other kinship tropes of shared responsibility and risk, solidarity and dispossession, as well as those classics of the kinship literature, joking and avoidance relationships—to say nothing of inheritance, incest, and violence in an age when symbiopolitical relations include pharmaceutical corporations as new kinds of in-laws. Gender, race, and sexuality—to take just three examples of biopolitically charged kinship topologies—are joined by novel entanglements of life forms and forms of life. I call this polyvalent, proliferating relationality *kinship in hypertext*.[132]

As portions of the tree of life dissolve, we are left with floating, mobile elements of a molecular biopolitics, fragments for a rewritable book of life. Life itself, relatedness, genetic engineering—even creationism—modulate into the molecular, the space of a flexible biology which, broken down to be built up again, may nonetheless be mobilized to naturalize such forms of life as biotechnology.[133] Translated into the text of the information ocean,

Mother Sea is becoming unsexed, as genome torques, unwinds, gender as a grounding sign of life.[134]

More traditional tales, of course, will still be told in this liquefied language. When Ed Harris in *Volcanoes of the Deep Sea* guides us into a computer-animated hyperthermophile, taking us on a "journey down deeper among the molecules of its DNA [to] . . . reach the four base chemicals of life's universal alphabet," finding that "in *this* we are most certainly related," we hear a simple translation of Darwin—even as the quick cuts of lateral transfer threaten/promise to turn this book of life into an overworked screenplay full of alien cameos.

In interpreting message-bearing methanotrophs and hyperlinked hyperthermophiles, scientists translate associations of the ocean with origins into a microbial, genetic idiom. But rendered into DNA libraries and bioinformatic data, these extremophilic microbes present puzzles to researchers who would read them for narratives of genealogical origin or biogeochemical history. The unfamiliar signals these organisms send through these new media are marked by the energy of the alien, a figure conceived as benignly oblivious to human purposes, as an unfamiliar interloper, and as a primordial version of ourselves. This alien ocean is animated by organisms whose metabolic difference from humans makes them radically unfamiliar and whose crosswise propagation calls linear genealogy and even evolutionary timekeeping into confusion. Fitting for an information age, these critters are imagined through and as high-tech instruments—as alien biotechnology, reprogramming the tree of life.

3   Blue-Green Capitalism

Marine Biotechnology in Hawai'i

The Fourth Asia-Pacific Marine Biotechnology Conference, convened at the University of Hawai'i, had been postponed from late 2001, when the attacks on the Twin Towers and Pentagon threw international planning into disarray. By April 2002, when I arrived in Honolulu, the fallout of those disasters had drifted into the general air—so much that a keynote speaker at the meeting declared that marine biotechnology might "solve problems for the nation, including homeland security." Undersea creatures, this American biologist suggested, might contain compounds active against anthrax—a microbe of no small worry at that time, when it had been making its way to U.S. newspaper and government offices in anonymously mailed envelopes. Marine microbes held promise for everyday applications, too. One project at the university fixed on extracting pigments from cyanobacteria, also known as blue-green algae. Carotenoid pigments, which protect such microbes from overdosing on light, could have utility as colorants for cosmetics and foods like farm-raised salmon, dyed pink to make them look like wild fish. Most commercial carotenoids are synthetic; versions derived from marine organisms could be marketed as natural.

In this chapter, I examine how marine microbes are made meaningful in the circuits of biotechnology, particularly in Hawai'i. Two channels of instrumentation become important in fashioning microbes as currency for biotechnology. First: laboratory techniques, including cell purification and cultivation, gene sequencing, freezing, the making of DNA libraries, and methods for arresting or amplifying the expression of genes for targeted compounds or enzymes. Second: institutional apparatuses, including government, university, economic, and legal instruments such as tax breaks, investment legislation, patents, and material transfer agreements, which specify the frames within which biotech exchange might unfold. These two instrumental regimes, organizing life forms and forms of life, respectively, are calibrated to one another, with the institutional calling on the biotechnical but also capable of floating free when biotech becomes speculative.

This combination of technology with regulation is generative of what Sarah Franklin and Margaret Lock call *biocapital*, a kind of wealth that depends upon a "form of extraction that involves isolating and mobilizing the primary reproductive agency of specific body parts, particularly cells"[1] and that also banks on promises about the commercial products and profits such mobilizations might deliver in the future. Marine biotechnology in Hawai'i also relies, I suggest, on the cultural force of images of Hawai'i as a tropical oceanic paradise full of natural promise for health and rejuvenation, a view held by many mainland Americans, a pool from which biotechnologists in the Islands are mainly drawn. A vision of the ocean as endlessly generative mimes and anchors a conception of biology as always overflowing with (re)productivity. Taking seriously this symbolic component allows me to look closely at the sentiments animating Hawaiian marine biotechnology—at a social form I term *blue-green capitalism,* where blue stands for speculative sky-high promise and green for a belief in biological fecundity. I track the materialization (as well as evaporation) of blue-green capital across such sites as the biotech conference, the offices of Hawaiian biotech companies, the World Wide Web, the University of Hawai'i's test-tubed cyanobacteria collection, a university ocean-sampling location known as Station ALOHA, and the Hawai'i state legislature. Attention to the local specifics of biotechnology also reveals why some of the legal instruments of biotech are fracturing in Hawai'i, as Native Hawaiians challenge the right of biologists to turn Hawaiian marine life into an alienable resource for the blue-sky dreams of biological enterprise.

CHANNELING BIODIVERSITY INTO BIOTECHNOLOGY

At the 2002 Asia-Pacific Marine Biotechnology Conference, an international workshop dedicated to exploring how marine life might be mined for pharmaceutical, industrial, and aquacultural biotechnology, the invocation of homeland security by our keynote speaker sounded strange. It was odd not only because the speaker's delivery was halfhearted—as though he were struggling at once to fathom a distant catastrophe and avoid coming across as a biotech vulture—but also because less than a third of the participants hailed from U.S. institutions; most were based in Asia. Of the 101 attendees, only thirty-two came from the United States (of these, twenty-five were based in Hawai'i, with a few credentialed at schools in Asia, primarily China and Taiwan).[2] But if the U.S. national appeal sounded off-key, it was taken from an identifiable tune, sampled from the then recent chartering of the University of Hawai'i's marine biotechnology center.

In 1998, the university became home to the Marine Bioproducts Engineering Center (MarBEC)—a novel institutional entity known as an engineering research center (ERC), a hybrid of academia and industry chartered by the National Science Foundation. ERCs like MarBEC would "focus on next-generation advances in complex engineered systems important for the Nation's future. Activity within ERCs lies at the inter-face between the discovery-driven culture of science and the innovation-driven culture of engineering. . . . ERCs provide the intellectual foundation for industry to collaborate with faculty and students on . . . [producing] the knowledge base for steady advances in technology and their speedy transi-tion to the marketplace."[3] MarBEC would train students in the discovery and manufacture of marine bioproducts, where "products" referred to items created using organic processes as well as to commercial goods.[4]

In this vision, biotic material becomes an object of commerce, trans-mutable into value, money, capital. While pursuing their degrees, students might work as interns in biotech companies, and companies might put money into MarBEC in exchange for rights to the licensing of products. For marine scientists, this signals a shift of funding since the end of the cold war and the days of grants from the U.S. Office of Naval Research. The University of Hawai'i would join several other schools with growing marine biotechnology concerns, including the University of California at Santa Barbara, the Scripps Institution of Oceanography, and the University of Maryland, whose Center of Marine Biotechnology we visited in chapter 2. That center was founded on the notion that "marine biotechnology is one of the greatest frontiers of scientific exploration and commercial endeavors for the next century. Compared with the terrestrial environment, the oceans of the world remain largely unexplored and represent a major por-tion of the Earth's genetic resources. Using the tools of biotechnology, this vast and diverse potential source of new foods, pharmaceuticals, minerals, and energy could be applied to help meet the needs of the world's expand-ing populations and economies."[5]

On the Pacific edge of this frontier, Hawai'i's MarBEC promised distinc-tive local biota as well as a unique financial and social environment.[6] In his opening address to the Asia-Pacific meeting, the mayor of Honolulu, Jeremy Harris, said that although the state had survived on a "tourist monoculture," he believed Hawai'i's "true destiny is as a center for high-tech, knowledge-based industry."[7] He invoked a standard image of Hawai'i as "a bridge between east and west," emphasizing that the state's time zone allows online investors to trade in U.S. and Asian stock markets in the space of a single day. This view configures the cyberspatialized Pacific Rim as an extension of an unfettered American frontier economy and as a site where capital meets its Western limits only to find openings into Eastern markets; capitalism becomes a sea serpent, ringing the world, eating its own regenerating tail.

Next on Harris's list of archipelagic assets for biotech was Hawai'i's migration history: "We also have in Hawai'i a very diverse human gene pool, good for developing new pharmaceuticals." Turning on its head 1990s biotechnological interest in Iceland as a genetic goldmine because of the purported homogeneity of that island's population, Harris conjured up a converse human heterogeneity for Hawai'i, one he suggested would make the state ideal for internationally recognized clinical trials. Global science and finance converged in a genetically imagined multiculturalism, with state citizens a reserve for bioeconomic experimentation. The mayor's pronouncement implicitly reframed as fortifying genetic fuel for biotech the late nineteenth-century history that saw Chinese, Japanese, and Filipino labor imported to the Islands for sugar plantation work. As he spoke, I imagined Iceland and Hawai'i as inverse islands floating on opposite sides of Earth, antipodal in their difference though capable of trading places with the flick of a framing. Measures of biodiversity, after all, are calibrated to the historical and geographic contexts employed to make them legible. Islands isolate as well as connect. The volcanic territories of Hawai'i and Iceland are what Mike Fortun, in his ethnography of the volatile politics of genomics in Iceland, terms "lavaXlands," where X marks the spot—the chiasmus, the fissure—where fluidity and stability crisscross.[8]

X is the unknown answer to the questions that crosshatch the matter of marine biotechnology in Hawai'i: Is this archipelago a biodiverse, bountiful paradise or a biodepauperate string of volcanoes in the middle of nowhere? Is the hunt for biotech treasure in the waters of the island chain a practice of virtuous bioprospecting (mining biodiversity to create products to save the state from tourist monoculture), of avaricious biopiracy (stealing the inheritance of indigenous Hawaiians), or a vain quest for sunken treasure? Who benefits when organic matter from the sea is transformed into property? Is the market an ethically robust mediator for human relations with the sea, a proper channel through which to deliver the healthy secrets of the ocean to the public, to make microbial life forms meaningful for forms of life?

The conference began in earnest when molecular biologist Eric Mathur took the stage to speak under the title "Tapping into Nature's Diversity for Novel Biomolecules Discovery." Mathur represented Diversa, a San Diego–based biotechnology firm. Wearing a red-and-black floral print shirt that stood out among the more sedate blue aloha shirts in his audience, Mathur was here to get us hyped. Setting out a grand grid for his talk, he showed a slide of a phylogeny of life on Earth, declaring that "the tree of life contains a broad array of metabolisms," and that Diversa, in search of new natural products, was "cultivating organisms from all over the tree." More, the company was extracting promising genes and recombining them to design new products. "Nature," he announced, "has produced diversity we can fine tune." He cited *Streptomyces diversa*™, a microbe used to produce antibiotics, immunosuppressants, antiparasitic agents, and herbicides. The presence of Diversa's name and trademark sign in the taxonomic designation recalls the naming of plants and animals after gentleman naturalists—with the twist that *Streptomyces diversa* is a proprietary strain, meaning that the company owns the exclusive right to produce this microbe (a genetically engineered derivative of *S. venezuelae*) and to employ it in research and discovery. Haraway's claim that genomic and transgenic sciences are brokering a shift from "kind" to "brand" in biological nomenclature could not be better substantiated (even as an attachment to Latinate names simultaneously suggests a desire for time-honored authority).[9] For Mathur, as for extremophile workshop participants in Baltimore, nature is a genetic engineer whose creations can be discovered, improved, capitalized—a vision Mathur illustrated iconically when he displayed a slide of Diversa's stock market logo pointillistically rendered in gene clones in a DNA library. "Biodiversity," he said, "is the basic building block for biotechnology."

Because the ocean constitutes the majority of Earth's biosphere, biotechnologists imagine marine biodiversity to be immense—and largely undiscovered. William Fenical, a Scripps advocate of marine biotech, articulated this view in a 1999 interview in *Discover* magazine. A full-page photo showing Fenical holding a sea fan against his cornflower-blue aloha shirt has him declaring, "The ocean's right there. It's diverse as hell, and it's waiting for us" (figure 15).[10]

We should pause over this enthusiasm for diversity, for it is a key support for biotech capitalism. *Biodiversity* began its life as *biological diversity*, a term advanced by conservationists to describe nature as a store of variety that might be measured and valued, in both ecological and economic terms. Since its coinage, *biodiversity* has become infectiously polyvalent.

FIGURE 15. Fenical in *Discover.* © Michael Sexton
Photography. Reprinted with permission.

Cori Hayden lists the meanings it has accreted: "an ecological workhorse, essential raw material for evolution, a sustainable economic resource, the source of aesthetic and ecological value, of option and existence value, a global heritage, genetic capital, the key to the survival of life itself."[11]

Propelled by the 1992 UN Convention on Biological Diversity into the language of policy, *biodiversity* has slid into synonymy with *nature*, turning the biological world into a site filled with variety in need of protection and even, perhaps (resonating with its bureaucratic cousin, ethnic diversity), political and economic representation. For marine biotechnologists, particularly in America, *marine* biodiversity represents a wide-ranging, frontier form of biodiversity: healing waters writ large, full of new genes awaiting amplification, delivering what marine microbiologist Rita Colwell (director of NSF, 1998–2004) has called "entirely new 'harvests' from the sea."[12] Insofar as humans make use of this new nature by capitalizing it, the prevailing sentiment goes, they must do so "sustainably"—protecting, husbanding, and curating diversity, always understood as a positive value, a life form with a moral valence for human forms of life. No wonder a biotech company named itself Diversa.

Biological oceanographer Paul Falkowski from Rutgers University followed Mathur with an impassioned lecture about why marine biotechnology was still awaiting a breakthrough product. Marine biotechnology, he told the audience, "is fundamentally idea-limited. We don't think in terms of an array of *products* and this is because most of us are in academia." More, marine biologists "always want to work with their favorite organisms, because they've learned to sentimentalize nature, especially the sea." We have to look closely, he said, at microbes, "the workhorses of the ocean." In order for marine biotechnology to take off, academia and industry must work together; practitioners must recognize that—he underscored the point by shouting it—"markets are not sentimental!" Falkowski then took his turn with the tree of life, slapping down an overhead slide of branching limbs of gene sequences, asking, "Where do you invest?" The audience was silent. "It's obvious," he announced, "Where the genomes have been sequenced!" *Not,* he reprimanded, where you find your favorite organism.

Falkowski's talk poured cold water on the usual PR for marine biotechnology, which emphasizes the unique bounty of the sea while also trading on a romantic, conservationist sentiment. Recall the Center of Marine Biotechnology promise to clone only genes, "leaving organisms where they belong—in the environment."[13] Contrary to Falkowski's intuition, such statements suggest that markets are actually created and sustained by sentiment. Anthropologist Sylvia Yanagisako has argued that economic actions, "including capital accumulation, firm expansion, and diversification," are "constituted by both deliberate, rational calculation and by sentiments and desires."[14] In other words, Falkowski is wrong. Still, it is worth analyzing his commitment to unsentimental markets. In his formulation, the market morphs into a new path toward that important aspiration of science: objectivity.

Scientific objectivity in the ideal sense celebrated by sociologist Robert Merton—secured through the organized skepticism and communal information sharing of disinterested knowledge seekers—is a scarce resource in the world of biotechnology, a zone in which information slides between public and private, and in which biological materials are often proprietary.[15] In 1989, Chandra Mukerji argued that deep-sea scientists during the cold war were able to see themselves as doing disinterested, apolitical work because of the breathing room given them by state funding that encouraged a range of exploratory projects adjacent to—but not instantly applicable to—state concerns.[16] Scientists' interests in plate tectonics, deep ocean currents, and the composition of seawater could piggyback on state needs to know about navigation, submarine movements, and deep-sea disposal sites for radioactive waste. Oceanographers' objectivity was safeguarded as a resource by the state as they became an elite reserve labor force that could be enlisted to legitimate policy decisions.

Times have changed since the end of the cold war, and new norms are in the making, partly because marine scientists must increasingly turn to biotech rather than state funding. In an era when many scientists announce at academic conferences that they are beholden to nondisclosure agreements, auditable objectivity is not always accessible. What is available instead is a faith in the "market" as a force that will secure a measure of effectiveness and guarantee a purchase on a "real world" at once scientific and economic. Such faith is part of a moral economy, a set of commitments to those proper reciprocal obligations that will secure the continuance of a form of life. The exchange of money and promises of useful products stand in for the pursuit of scientific truth.

Mathur and Falkowski represented a market enthusiasm hard to find in other conference talks—though market language got a show-stopping enunciation from a representative of the Marine Biotechnological Institute in Japan, who made explicit the less-than-communal features of academic-industry relations when he said, without irony (and to the visible discomfort of American biotech boosters), "It is important to have an 'old boys' network in place because this way we can get secrets from companies. We are a private institute that sends graduates to companies. Our students can be spies." The first three Asia-Pacific Marine Biotechnology Conferences were held in Asia—Japan in 1995, Thailand in 1997, and the Philippines in 1999—and I got the impression, corroborated by a few Thai and Filipino scientists over lunches, that many participants felt bowled over by the outsized market rhetorics delivered by Mathur and Falkowski, rhetorics that had biotechnologists made in the image of a genetically enterprising, engineering nature.

One Thai scientist, from Chulalongkorn University, was offended by Mathur's talk. Diversa, she told me, was "playing with evolution," making new things for their own sake. She protested that she was simply trying to come up with tools for learning how shrimp reproduced in aquaculture ponds. Biotech for investigation was good, but modification was not. Microbial biotechnology by recombination of the kind Diversa promoted, she said, would endanger us all as the organisms the company engineered evolved to be more robust. "If scientists make all these aliens," she warned, these new life forms would eventually attack us and "we could all die in fifty years." She was unnerved by what she judged to be cheery American assessments of the modifiability of nature. Her sentiments had fish aquaculture as a sensible multiplication of existing life (even though shrimp ponds are not exactly models of environmentally friendly practice).

National and institutional differences surfaced throughout the conference. The small delegation from the People's Republic of China concerned itself with seaweed aquaculture.[17]Questions around nutrition and disease prevention in aquacultural ponds dominated the remarks of scientists from the Philippines. Speakers from India's National Institute of Oceanography discussed microbes in mangrove ecologies as instruments for decoloring paper and textile factory effluents, as biotechnologies for remediating contamination on the Indian coastline. Several spoke of capitalizing on the Indian government's subsidizing of the nation's entry into international biotech, discussing how seeking patents in the United States would allow products to be approved by the FDA and thereby to move fluidly into global markets (a bioproduct from a green mussel aimed at diabetes treatment was one example). Like American marine biotechnological endeavors, such projects in what science studies scholar Wen-Hua Kuo calls "bioglobalization" were often articulated in terms of national initiatives, though no one naturalized their nation's reach into the world ocean in quite the same way as the Americans, who often framed the promises of marine biotechnology as coextensive with the taming of a frontier territory.[18]

Over lunch, some of the American contingent enlarged upon the impact of marine biotechnology—and well beyond the idiom of the economy. One man held forth on how marine biotechnology was important because "we came from the sea. The ancestor of all life will have a marine genome." Finding *that* out, using that knowledge to make new kinds of creatures, this MarBEC participant prophesied, "will be a real paradigm shift; biotechnology will transform how we think about life. The most important event was in 1980, with the patenting of the first organism. As we keep going, this will lead to religious wars. We're changing our understanding of life in the biggest way since Jesus Christ and the New Testament."

This Christian "we"—offered by someone who later clarified that he was an atheist—plugs into a discourse of suffering and redemption. The swing between promise and apocalypse characteristic of American millennial culture here colors the prospect of marine biotech. It participates in a dual imaginary in which the ocean is both in trouble and, through the enterprise of humans, the source of its own curative power. This is the ocean as an alien with a healing touch.

I listened to a great many lectures about growing massive colonies of blue-green algae in vats or greenhouses. I did not understand the significance of it all until later, when Patrick Takahashi, an emeritus professor of biochemical engineering, closed the conference with a dinner talk in which he imagined intensifying aquaculture to incredible dimensions. Here is how he put it in a lecture he gave to the Intergovernmental Oceanographic Commission of UNESCO, a copy of which he later gave me over a fancy fish lunch at an upmarket Waikīkī hotel:

> The next frontier is the open ocean. Largely not owned by any nation, nutrient-rich fluids at 4° Celsius are available 1,000 meters below the 20 degree latitude band surface. Just in this natural solar collector region, if only one part in ten thousand of the insolation can be converted to useful energy, the needs of society would be satisfied. . . . Picture, then, a grazing plantship . . . supporting a marine biomass plantation with next generation ocean ranches. . . . Then consider several hundred, no, thousands of these productive platforms. Current international law dictates that each, under certain circumstances, can legally become a nation. Imagine the United Nations in the 22nd century. . . . European seafaring nations might again consider colonization, this time in the open ocean, where there are no obvious downsides, such as the sociological problems that came with the era after Columbus. One cannot guess what Greenpeace might do, but there are no native populations, not even whales, as permanent residents in the middle of the ocean.[19]

Takahashi's expansive vision reaches into the extraterritorial sea to realize its apotheosis: an alien ocean brought within colonial range through humanity's planktonic emissaries. It is a chlorophyllic remix of the promise of the "Blue Revolution," the promotion of fish farms in the Third World as scaled-up food resources (named, forgetfully, it would seem, after the much criticized Green Revolution of the 1970s).[20]

Takahaski's dream is a perfect example of what *Harvard Business Review* authors W. Chan Kim and Renée Mauborgne in 2004 called a "blue ocean strategy," a set of tactics for tapping into and creating "uncontested market space." Kim and Mauborgne imagine this blue business ethos through the image of the uninhabited ocean and contrast it to a "red ocean strategy," which sees competitors battling bloodily, tooth and tentacle, for limited space.[21] "Blue oceans" are, of course, a riff on "blue skies," zones of research or investment with no immediate applications, which may or may not come down to earth in the future.[22] Blue skies are notional spaces for such blue ocean ideas as Takahashi's plantations without politics—aqua-farms populated by phytoplanktonic biomass—his invitation to Europe to restage its colonial past in a solar-powered sea of sociological emptiness.

In what follows—the story of MarBEC, its discontents, and contests in Hawai'i over the terms of blue-green capital—I show that such seas are anything but socially empty. Instead, they are infused with sentiment, saturated with desire for sky-high dreams and sunken treasure.

SEAS AND SENTIMENT IN MARINE BIOTECHNOLOGY

By the time I returned to Hawai'i in summer 2003 for an extended research visit, MarBEC was struggling, sinking. In 2002, representatives from the NSF arrived to assess MarBEC's progress since 1998 and judge whether funding should be renewed. The verdict was no. Chemical engineer Michael Cooney suggested to me that multiple directors, changing objectives, and overambitious guarantees were responsible. He said MarBEC had invested the majority of its funds into sequencing marine microbes—something that could have been done through collaborations. And though MarBEC had initiated many projects, it had yet to patent or commercialize any product.[23] To be fair, he advised, it takes a lot of time, resources, and luck to develop a marine bioproduct—from discovering a promising lead in an ocean organism, to extracting likely chemical compounds, to engineering these into new materials, to patenting useful modifications, to testing the effectiveness of medicines in clinical trials, to enlisting corporate partners in marketing products.

The complexity of what MarBEC had been attempting was made clear by the dizzying array of flowcharts I picked up at the Asia-Pacific conference, detailing MarBEC's promised pipelines of production. Nothing flowed the way the charts anticipated. Cooney compared MarBEC to a "dot-com," one of those hyper-hyped late 1990s Internet-based companies that primed investment with savvy Web presence. And indeed, the MarBEC website—part of what had drawn me to this fieldsite—had been expertly realized, giving the impression of a network of people who knew exactly what they were doing. Perhaps their logo—a swooping light blue spiral suggesting the profile of a tumbling wave—should have tipped me off, for it also resembled, with its Fibonacci curl, the curve of a snail shell. Maybe MarBEC had been an elaborate shell game (see figure 16).

In 2003, some participants in the experiment that was MarBEC were still hoping to revitalize academy-industry collaboration in Hawai'i. Speaking with these scientists and business folk gave me a sense of the

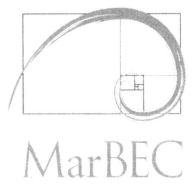

FIGURE 16. MarBEC logo.

sentiments from which they believed a meaningful marine biotech might emerge. It would be a form of life suffused with the logic of the free market, whose operating systems might be coordinated with those of lab biology. I met a former MarBEC participant at an office near the University of Hawai'i Mānoa campus. He retrieved a like-minded colleague. Both men, like myself, would be placed in the category of *haole*—the Hawaiian-language term describing European-descended white people.[24] Both wore floral aloha shirts. Having recently rearrived in the Islands, I was still adjusting my eyes to these garments, so emblematic for mainlanders of tropical souvenirs but de rigueur in cities like Honolulu as casual office wear, especially on days like today, "Aloha Friday."

The two explained their vision of a marine biotechnology center. Their goal was simple: "to transfer the knowledge we have about the ocean directly to the public through biotechnology." A biotechnology center, they said, would benefit the Hawaiian economy. They handed me a brochure, "Hawai'i's Biotech Industry," published by the State of Hawai'i's Department of Business and Economic Development and Tourism. Pitched at companies considering relocation to Hawai'i, the brochure described the state's attractions, most prominently a year-round growing season. But a more exciting enticement for biotech was an "extensive range of readily accessible biodiversity," from locales that included the open ocean. Before it went under, MarBEC, which had initially touted deep-sea vents,[25] had already begun to look to such surface waters. One report read, "Though not generally perceived as being an 'extreme' environment, the surface waters of the North Pacific Ocean are now recognized as a 'treasure chest' of microbial diversity and a sea of potential for new discoveries."[26]

Most important for any aspiring biotech company in Hawai'i were new regulatory attractions. In 2001, the state legislature offered investment incentives and tax credits to arriving high-tech companies. Once these benefits were taken into account, one of my interlocutors said, interested corporations "don't have to do much math!" It would be a "win-win situation" because companies could help diversify the Hawaiian economy and provide jobs for local youth, enticing them to stay in the state rather than head for the mainland (here, markets are nurtured not only through sentiment but also by government). Marine biotechnology would provide Hawai'i with a distinct market identity.

I began to fear that these scientists had taken me for a journalist, since our conversation increasingly felt like a Hollywood pitch meeting. Hawai'i! The ocean! Biodiversity! Biotechnology! Money! I noticed the two were fatigued—one, I gathered, had been up since 3 A.M., telephoning New York and surfing the Web for investors.

Genomics and biotechnology are shot through with the language of promise, as though biotech has inherited the potentiality associated with genes themselves. MarBEC's publicity announced "the promise for discovery of new antibacterial, anticancer and antifungal agents" in marine organisms.[27] Sociologist Charis Thompson has argued that the "biotech mode of (re)production" operates with *promissory capital*, "capital raised for speculative ventures on the strength of promised future returns."[28]

In Hawai'i, such promise is symbolized by a recurrent image in the logos of marine biotech companies, an image that any surfer would see as portending exciting things to come: the wave. Oceanit, a Hawai'i-based company conducting research on algal biotechnology for toxin testing, brands itself, for example, with a pair of cresting swells.[29] Such waves also suggest leisure and surfing—tourist images of Hawai'i. We will be speeding along, the logos seem to say, but relaxing while we do it. But if waves can buoy oceangoing ventures toward new shores, they can also crash. Surfing the Web in search of companies in Hawai'i, I found several signs of sunken marine biotech startups. There was Kona-based Aquasearch (its logo a stylized blue wave, splashing up green bubbles), one of MarBEC's industrial partners, devoted to investigating marine microalgae for new medicines, food supplements, and animal feed additives—"bringing the ocean to life" as their slogan had it. In June 2002, Aquasearch filed Chapter 11 bankruptcy.[30]

In 2005, a few University of Hawai'i scientists organized a new marine biotech center, CMMED (or "sea med"; Center for Marine Microbial

FIGURE 17. CMMED logo.

Ecology and Diversity), its logo a blue double helix morphing from the bio-genetic curve of DNA into the curl of a breaking wave (figure 17).[31] (When last I looked, in 2008, CMMED had itself morphed into a new entity called the "Pacific Research Center for Marine Biomedicine"—its logo three wavy blue lines—which now joins basic research into pharmaceuticals with study of the relation of harmful algal blooms to human health.)

During my time in O'ahu, one biotechnology company *was* riding energetically along: Hawaii Biotech, Inc. MarBEC had once hoped to part-ner with this thirty-person firm, and employees were cautiously opti-mistic about marine biotechnology, waiting to see if a university-based partnership might yet be consolidated. Located near Pearl Harbor, the company is headquartered in a four-story modernist concrete industrial building. Of the thirteen people on the management team and board of directors, at least eleven have degrees from prestigious mainland institu-tions (Harvard, Yale) and extensive experience at biotech companies including Genentech, Novartis, and Eli Lilly. Centering its efforts on bio-pharmaceuticals, Hawaii Biotech has had success with vaccines for dengue and West Nile viruses.[32] In 2003, Hawaii Biotech received a grant from the U.S. Department of Defense to develop compounds against anthrax, sift-ing through microbes for leads against bioterrorism. This was to be where MarBEC would come in. A portion of the grant was written to enable Hawaii Biotech to collaborate with MarBEC to identify molecules in, for example, the university's cyanobacteria collections.[33] As MarBEC stated in a 2003 annual report, "With the phase-down of support from NSF, MarBEC has teamed with its industrial partners to pursue new directions in research and product development. One major new activity, which promises significant funding for MarBEC's team, involves screening of aquatic and terrestrial organisms for developing new vaccines in support of homeland security."[34]

I met with Mark Goldman, early fifties, then director of drug discovery at Hawaii Biotech:

> MG: Here in Hawai'i, we need to collaborate. Hawaii Biotech brings molecular assays into the picture. But we depend on UH [University of Hawai'i] for samples and chemical expertise. We want to establish with UH a practice for screening natural products. There's value in those natural products collections. They have cultures locally, we test them locally. It's faster than FedEx!
>
> SH: Why would samples from the sea be particularly interesting?
>
> MG: It's all about survival. The organism's role is survival—and secondarily to that to procreate. And organisms have developed unique ways to defend themselves. They can say: back off, you don't want to eat me. It's these chemical processes that promote survival. Most of the planet is seawater, which is an untapped resource. In a milliliter of water, there are thousands of microorganisms struggling to survive. Maybe I can get lucky and one of these organisms synthesizes something that gets at the disease I'm studying.

Goldman's explanation offers a portrait of the ocean as a neo-Darwinian soup—something like the agonistic "red ocean" described by the authors of "Blue Ocean Strategy." This vision resonates with the "unsentimental" view Falkowski advocated, now located not just in the market but, more, in nature itself—and constitutive of biodiversity as such. On this view, marine biotechnology is not a sentimental enterprise—swept up in the romance of the sea—but a practical business that calls upon the work of natural selection to act as a screen for possible profits. Biological difference, sited in the sea, holds utility. The competitive red ocean can be "tapped" for lively promise.

As I left Hawaii Biotech, I noted that the building that housed the company was still engraved with the old name of its major tenant: the Hawaiian Sugar Planters Association (now the Hawai'i Agriculture Research Center), the organization that ran the plantations that brought Chinese, Japanese, and Filipino contract laborers to O'ahu. Plantations and their human labor might be seen as nineteenth-century engines of biocapital, especially since the capacities of workers to reproduce themselves at work from day to day and even to supply children to the plantation economy were key features of this system. As Sidney Mintz argues, plantations were "a synthesis of field and factory," sites where cultivation and capitalization worked hand in glove.[35] What is intriguing about biocapital in the making now is the promise that the bodily capacities pressed to create biowealth will increasingly belong to plants, like cyanobacteria, and not always or only to people (or animals).[36] Takahashi's aquaterritorial plantation is a symbol of the dream that the new biology will escape histories of human exploitation. That dream has a recent ancestry, in the aquaculture projects that began to pepper Hawai'i in the 1970s, predecessors to the more molecular-minded marine biotechnology of today.

Those forays into producing fish as capital were often undertaken by mainland haoles who moved to Hawai'i for sun and surf, employing local people as technicians in enterprises such as shrimp farming. Often these mainlanders were ex-military folk once stationed in the Islands. Military-men-turned-hippies bought Hawaiian land on which they contracted local labor to farm fish.[37] I met one such person at the Asia-Pacific Conference, a Southern Californian whose biography had taken him from antisubmarine warfare research to aquaculturing "hippie fish" with friends in Hawai'i to founding a marine biotech company to extract "drugs from the sea."[38]

What happened to the early marine biotech startups and to MarBEC? Michael Cooney compared the scramble to find marine bioproducts to the California gold rush. For mainland U.S. scientists who move to Hawai'i—often in mid-career and midlife (one even told me of his dream to "start over," taking the next step in the "westward migration")—this archipelago comes to look like the Promised Land. Indeed, European and American misapprehensions about the Islands as a source of wealth are commemorated in the name of O'ahu's signature landmark, Diamond Head, so-called when explorers erroneously believed this volcanic crater full of diamonds. The natural landscape, Cooney argued, is still playing tricks on the minds of mainland speculators. Popular sentiment has lately been captured by the idea that Hawai'i is overflowing with biodiversity. In a way this is simply a scientific rendering of the pristine nature that has been sold to tourists as one of Hawai'i's attractions.

As a vacation destination, Hawai'i has become "a site of white Edenic regeneration,"[39] a seascape domesticated for tourists in such postcard movies as Elvis Presley's *Blue Hawaii,* with its lush landscapes and sensual natives. Hawaiian jungle and seaside nature—long generating a gold rush on Island property values—nowadays appears as "biodiversity." But just because Hawai'i looks to the naïve eye like a rain forest, Cooney suggested, does not mean it has the same variety of, say, the Amazon. As the most isolated archipelago in the world, it is, by some measures, biologically rather spare, except for the assortment of introduced species. "Volcanic islands, in the middle of deep-sea oceans," Cooney said, "are not necessarily the best sources of marine biodiversity."

BIOTECH GOLD

But what about these oceans? What about those projects aimed at trans-forming cyanobacteria into bioproducts? Is there any sense in which the cyanobacteria floating in archipelagic waters *are* filled with biotech gold? Is it true that, as MarBEC's mission statement had it, island waters host a "'treasure chest' of microbial diversity and a sea of potential for new discoveries"? It would not be the first time someone tried to squeeze gold out of ocean water. In 1925, Germany mounted an expedition to research the feasibility of extracting gold from seawater in the South Atlantic to pay off the nation's war debt.[40] This enterprise—a watery permutation of the medieval dream of discovering the philosopher's stone, that magical material that would turn base metals into gold—represents one of the earliest attempts to extract wealth not just from beneath the sea but from its substance. As it happens, it is possible to get gold from ocean water, but it requires an almost supernatural effort, as the Germans determined. And as MarBEC learned. After having found that MarBEC's website concealed not proprietary secrets but a suspended promise of wealth, I was haunted by a question Mike Fortun asks in his study of the promises made in the 1990s by the corporation deCODE Genetics to "mine" a database of genomes from Icelanders: "How do you tell a real genomics company from a counterfeit one?"[41]

Perhaps, I thought, in this empire of floating signs—websites and screensavers promising to transmute water into wealth—it would be useful to be a literalist and find the referent, the treasury, the blue-green gold, the bedrock from which MarBEC researchers sought "to 'mine' marine organisms that produce novel carotenoids and terpenoids."[42] Maybe looking at the University of Hawai'i's store of cyanobacteria could give me clues about where it all began. I had read on the Web that, "through the University of Hawai'i, MarBEC has access to extensive collections of marine microbes, which are being screened for new bioproducts." I visited one of the "culture collections MarBEC [had] to offer" to such industrial partners as Hawaii Biotech.

When I saw the Patterson Culture Collection, the largest assortment of cyanobacteria in the world, I was astounded by its sheer physical presence. The collection, housed in a campus chemistry building, consists of hundreds of 5-inch test tubes, each containing a distinct strain of algae (figure 18). I saw walls of tubes filled with green specks, spirals, and smudges, a cyano-bacterial media-mathematical sublime.

Greg Patterson was the University of Hawai'i chemistry professor who collected these cyanobacteria between 1986 and 1999 and bequeathed the

FIGURE 18.   Cyanobacteria collection. Photo by the author.

samples to the university upon his retirement. The curator of the collection, Georgia Tien, inherited the task of organizing this archive three years before my visit, finding it in disarray because of underfunding. Patterson's record keeping was difficult to trace. He "didn't leave a wide paper trail," she told me, and some samples had not survived cryopreservation. What Tien had reconstructed, however, was impressive, and quite beautiful. The Excel database downloadable through MarBEC's website gave some idea of the cosmopolitan character of the 2,308 strains in the collection. Some were collected from the Mānoa campus itself—from such offbeat sites as ornamental fountains—or at Waikīkī beach. Others came from Palau, the U.S. Virgin Islands, and Guam. A series of samples illuminate Patterson's traveling trajectory: I found listings from Moon Beach Hotel, Okinawa; Kona Surf Hotel, Hawai'i; and Hotel Tahiti, French Polynesia.

Tien said she was happy curating the collection, a job that had that day entailed making subsamples of strains to be stored in liquid nitrogen as backups. I was back in the world of biomedia, the recontextualization of organic matter for ends beyond what life forms or elements might do on their own. It had been a long path for Tien to this job as librarian of bacteria. After studying zoology in Massachusetts, she made her way to Hawai'i, where she worked at Bernice Pauahi Bishop Museum, Hawai'i's most prominent natural and cultural history museum. There, she dealt with "jars and jars of dead animals. They were always collecting more, killing more." She grew tired of it: "There's an emotional component. I will no longer work with any living thing that is multicellular, just cell cultures and microbes." Through the University of Hawai'i, she worked for Aquasearch, screening for compounds active against cancer. The process involved "chemically extracting the slime and putting it on cancer cells to see if it is a yes or a no." She said she had a disposition for cataloging: "My ancestors built the Great Wall, I can deal with this. I do needlepoint, so I like minutiae. I take a lot of pride in this; if it goes down the tubes, I'll feel bad. This is a live collection."

And the life of this collection is multifarious. Famous cyanobacteria, like *Spirulina*, are popular as health food additives. Others are toxic. All are unicellular but can grow filamentous, sheetlike colonies. In aggregates, particularly in closed environments such as aquaria or test tubes primed to help them flourish, cyanobacteria exude slime.

Gesturing toward the wall of tubes, Tien told me, "There is a huge drug-producing potential here." Antimicrobial, antifungal, antiinflammatory, and anticancer compounds have higher biomass in cyanobacteria, she said, particularly in tropical marine habitats, because "they live in high density and high biomass and produce defensive chemicals so as not to be invaded by the guy next door. Humans have not evolved to recognize these large unusual compounds. They can be very new to the human body—so, we haven't evolved natural defenses."

This war-torn sea, this red ocean medium of blue-green bacteria, is alien to humans. In textbook terms, "cyanobacteria have evolved elaborate defenses to avoid being eaten by grazers. Many cyanobacteria produce a slime, or mucus coating that makes them taste bad. Their most potent defenses, however, are toxic chemicals strong enough to kill livestock if sufficient quantities are drunk or eaten."[43] In other words, the evolutionary histories of humans and cyanobacteria may be at odds.

In 2005, Tien was coauthor of a study of cyanobacteria collections in Hawai'i, Sweden, and Scotland that suggested that some marine environments contain neurotoxic bacteria that might be linked (through tap water and food chains) to incidence of Alzheimer's.[44] Flowing plankton blooms, nourished by noxious effluent from the land, can channel compounds unfamiliar to humans into our nervous systems, accomplishing a sort of sci-fi alien abduction—complete with memory erasure—through everyday drinking, bathing, and swimming. But cyanobacteria might also be sources of compounds that can ally with human immune systems to fight unfamiliar bugs. The Patterson collection, as assay and salve for human disease, then, holds both poison and promise, the twin potentials of an alien ocean. The alien ocean here is the red ocean of Darwinian contest and hostility to humanity as well as the blue ocean of frontier promise, a site of ecological, green possibility. Death from life and life from death. Symbiopolitics.

Looking at this treasury of blue-green possibility, one realizes that a lot of work—library making, growing cyanobacteria in media, bioactivity screening, changing compounds into units transferable between labs—is required to convert wet wealth into a viable product, something like a marketable cosmetic or nutraceutical. These cyanobacteria are a raw form of what Catherine Waldby calls *biovalue,* "generated wherever the generative and transformative productivity of living entities can be instrumentalised along lines which make them useful for human projects."[45] With the market purposes of MarBEC in mind, the Patterson collection can be called a bank of biocapital—where capital, following Marx, is that accumulated material or labor power employed to produce surplus values like profit or interest.

In *Capital,* Marx describes the circulation of money as capital—in which "more money is finally withdrawn from circulation than was thrown into it at the beginning"—using the formula M-C-M', where M stands for money, C for commodity, ' for the surplus value gained in a profitable exchange of a commodity for money, and M' for the total capital produced by that exchange.[46] For the biotech imagination, we can write an analogous formula to describe the making of biology into capital: B-C-B', where B stands for biomaterial, C for its fashioning into a commodity through laboratory and legal instruments, and B' for the biotech product (or, perhaps, biocapital) produced at the end of this process, with ' the value added through the instrumentalization of the initial biomaterial.

I want to suggest, however, that the sentiment of many biotech boosters has them imagining B' already to be latent in B—believing that biological process itself already constitutes a form of surplus value production. This logic naturalizes biotech. So, it is not only the labor of people like Tien that confers value on the Patterson collection but also a conception of cyanobacteria themselves as little laborers—Falkowksi's "workhorses of the ocean."

Diversa's Mathur, at the Asia-Pacific conference, described marine microbes as "the blue-collar workers of the environment," laboring units that might be taken apart to be put back together again for new tasks. Microbial biodiversity is here configured as accumulated labor power, the products of which can be harnessed to create productive futures. This belief is based, it bears emphasizing, on a metaphor: that organisms are laborers (an equivalence declared even by Marx, who saw the "natural consumption" of eating entailing "production" of the body).[47]

Franklin and Lock understand biocapital to be underwritten not only by production but also by reproduction, arguing, recall, that "generating biocapital is driven by a form of extraction that involves isolating and mobilizing the primary reproductive agency of specific body parts, particularly cells, in a manner not dissimilar to that by which, as Marx described it, soil plays the 'principal' role in agriculture."[48] On this view, biocapital can be derived from oceanic pasturage if the reproduction of the reproductive capacity of marine microorganisms—to make carotenoids, for example—can be channeled into profit-making commodities and accumulation strategies (contrast biocapital with necrocapital: dead matter, like fossil fuel, put to unregenerative, zombie-like work).

But we must be careful not to imagine microbial reproduction as a transparently "natural" process, as though microbes' coming into being straightforwardly designates them as what Marx would have called "means of production already produced," as though their productivity is the essence of their species being.[49] To do so is to see them as natural factories or assembly lines, when in fact they become so only in certain relations. Remember, for example, that for DeLong microbes are stewards rather than blue-collar workers. Seeing microbes as workhorses also suggests that they are workers which, unlike humans, cannot be alienated from their species being, since it is their nature to dumbly labor. As historian Hannah Landecker argues, contemporary biology has become expert at stopping, starting, suspending, and accelerating cellular processes, wedging these dynamics into processes that look like a molecular version of industrial agribusiness.[50] But biotech geese cannot lay golden eggs without daily tending.

Understood as a kind of biodiversity-become-biocapital, the feature cyanobacteria might have most in common with gold is their status as a fetish, an entity thought to have its own life force apart from the relations in which it becomes socially or organically active. Like gold, biodiversity is imagined to be "both symbol and reality of value."[51] In this sense, biodiversity is imagined as a representation of nature as well as the sedimented nature of nature itself. More, since biodiversity is understood *already* to be life, its materialization as a fetish is doubly mystified.

BLUE-GREEN CAPITAL

The comparison of biodiversity with gold founders on a disanalogy. We went off the gold standard a long time ago, didn't we? Don't we all know that money is only the promise of its own redemption? Of course. This is why the value of biotech lies in large measure in promises people make about it. Kaushik Sunder Rajan offers a formulation complementary to Franklin and Lock's, arguing that "in biocapital the 'classical' scientific binaries of truth and falsity are articulated with those of credibility and incredibility."[52] In other words, value for biotech in Hawai'i may be produced by the persuasiveness to investors of suntanned scientists in aloha shirts. If for Franklin and Lock biocapital refers to organisms imagined as reproductive technologies generative of surplus value, for Sunder Rajan biocapital also refers to the market potential of bioproducts in forward-looking commercial futures.[53]

Biotechnologists seek to secure such futures not only through rhetorics of productive nature and laboratory labor like the care of cyanobacteria but also through legal instruments. I got a tutorial from MarBEC's once director of business and development, Kevin Kelly. The relevant history of legal agreements between universities and companies about the commercialization of university property begins in 1980 with the passage in U.S. Congress of the Bayh-Dole Act, which allows universities and their employees to retain rights in patented inventions developed with federal monies and, if desired, to license or sell those inventions to private business. Kelly gave me an example: Let's say a university scientist happens upon a promising compound in a cyanobacterium. Moving a version of this compound "from bench to market" requires fashioning a new, useful, and nonobvious—patentable—variety and then finding a company to license and develop it. Because this is a risky, expensive proposition, Kelly said, part of MarBEC's charter had been to get companies interested in funding projects early on, giving them first access to university patents seeded with company money.

At the center of university-corporate alliances of this kind is a legal instrument called a material transfer agreement (MTA). An MTA specifies that, if the university sends out its material property, the agency to which it is sent will be allowed to do research on the substance in a certain problem area. A sample copy of an MTA between MarBEC and a generic corporate partner states that the university "has a proprietary interest in [X substance or material] from the Hawai'i Culture Collection of the Marine Bioproducts Engineering Center ('MarBEC'), together with any progeny, replications, derivatives or parts thereof (hereinafter RESEARCH MATERIAL) and any related confidential or proprietary information." The MTA specifies that if something is discovered that could lead to a patent, the university must be notified and will be "entitled to a royalty-free non-exclusive worldwide license to use any such new invention for non-commercial use only." U.S. patent laws determine the role and contribution of "Provider" and "Recipient." In this formula, companies fund research from the get-go, engaged in framing research questions from the start.[54]

Once biological materials are covered by an MTA, the Provider can stipulate how "progeny, replications, derivatives or parts thereof" may be used by a Recipient—allowing, for example, an industrial partner to look for pigments in the Patterson collection but not, say, a cure for cancer. The operation of instruments in molecular biology labs thus calibrates to the legal instrument of the MTA.[55] The transfers enabled by the MTA (and recall the chapter 2 discussion of *transfer* as a new biopolitical mode) soften the boundaries between academia and industry, bringing capitalism and its proprietary secrecy into the minutia of laboratory life.[56] Microbes are now broken down into discrete varieties—"progeny, replications, derivatives or parts"—with both biological and legal meanings, with significance as at once created substance and transferable property: as products and products. The standardization demanded turns biomaterials into units that can be exchanged. We must look to lab and legal instruments as interdependent media for realizing the ends of biotechnology, not as neutral conduits facilitating what "biology" or "capital" want to do on their own.

What is required to comprehend biocapitalism are stories of how biology—as discipline, as corporeal substance, as process—is mobilized to make money, of how biology becomes currency, coin, and capital. The *bio-* in biocapital is imprinted with the means and ends of capital itself: the upwardly spiraling symbiosis of production and reproduction. To speak in Aristotle and Agamben's terms, microbial *zoë* in Hawai'i has been infused with the *bios* of capitalism. To discern the specificity of this relation between life forms and forms of life and determine what is particular to marine biotech capitalism, I suggest thinking of capital as manifesting in hues—in the instance at hand, as blue-green capital.

In *Modernity at Sea*, Cesare Casarino compares the circulation of the white whale in *Moby-Dick* to the circulation of money. Both circulations are motivating, mediating forces in social relations of unequal exchange—between Ahab and his contracted but captive crew, and between buyers and sellers, employers and employees. But both forms of circulation also direct attention away from the social relations they enforce; we follow the whale, we follow the money, instead of the animating dealings of people and institutions. This analogy in view, Casarino asks, "What is the color of money?" and immediately answers, "White, of course. The color of money is white."[57] In *Moby-Dick*, the whiteness of the whale signals both absence and excess of color. Casarino suggests that money, analogously, in its function as a universal equivalent, masquerades as an invisible translator of value while also everywhere appearing as the full representation of value as such. This double action defines money as a medium of circulation; it permits the transfer of value from one site to another while also embodying value itself, particularly when it stops for a moment and manifests, for example, as a coin in your hand or a sum in the bank. Such starts and stops are motivated by already existing dynamics of selling and buying that cause money to "flow"—a transformation that hides the unequal social relations that make money "circulate" at all. Capitalism acquires the appearance of neutral exchange through bleaching the movement of money itself of any preexisting history of social inequality. Just so, the imagined circuits of MarBEC capital were supposed to run on neat transfers of e-mailed contracts and FedExed extracts, at once legally transparent and proprietary.

Blue-green capital adds color to this model, accounting for the work of blue-sky speculation and the labor of getting such critters as blue-green algae to produce meaningful substance. Blue-ocean fantasies of life-giving waters are married to the economic fecundity of biodiversity and, in those articulations that see biotechnology preserving nature by sampling only small bits of it, are wed to its ecological, "green" value as well. Blue-green capital keys us into the symbolic importance of the ocean for marine biotechnology. Chris Connery reminds us that "ocean as source and ocean as destiny figure in the ocean's mythological temporality; it is both life-giving mother and final frontier. The conquest of the world ocean being coterminous with the rise of Western capitalism, it is natural that the ocean has long functioned as capital's myth element." Blue-ocean strategies, in which the immensity of the sea stands for unlimited resources, promise what Connery calls an "economic sublime."[58]

But wrapped up in this blue-green capital, and mixed with the yet-to-be-prospected gold of the Patterson collection of cyanobacteria, is another substance: slime. Slime is a sign of that which slips away from containment and calibration and which must be managed to make anything like biocapital circulate. Without stable boundaries—and without ends in the instrumental sense—slime, like the figure of the alien, exceeds and disturbs representation. Though it is true, as Landecker suggests, that "biotechnology changes what it is to be biological"—so that if capital accumulation demands that biotic stuff be rendered "reproductive," then this is what biology *becomes* in that relation—there are also reasons to think that slimy flow may swerve away from full appropriation.[59] Playing with slime, I ask, following Bill Maurer, what biotech tales would sound like if we were to write a "story about an open, porous, seeping, and dripping body of global capitalism . . . ? This would be a different story from the familiar one about the clean lines and fast networks of neoliberal efficiency. Less like a fiber optic network; more like a lava lamp."[60] This is the other side of blue-green capital. Blue-sky dreams reflected in a blue ocean show biotech speculation to be a hall of mirrors (with ricocheting reflections in fact producing escalating investment). And green refers not so much to exploitable reproductivity as to the muck that always threatens to undo capital, a mush of biomedia constantly in need of shoring up, and indeed of being made into media in the first place.[61]

Signs of such muck bubbled up at the Asia-Pacific biotech meeting. After Mathur's enthusiastic presentation on turning microbial species into brand-name products, one audience member worried, "How do you define a species in microbes when they're shifting their genes all around? If you have something that differs by one base from another creature? Is *that* a different species?" Mathur responded, "A similarity of 70 percent or more at the DNA level makes a microbial species, but that's artificial and conventional." No one was satisfied. The anxiety came up again when, a month later, I attended the Assembling the Tree of Life meeting at the American Museum of Natural History, where Ford Doolittle remarked to me that he had overheard someone at the meeting fretting, "If you own a patent on an organism's genome and a taxonomist changes its name, you can lose your patent!" Such questions recall an obsolete, though perhaps newly relevant, meaning of *species:* "a particular kind or sort of coin or money" (OED). Incorporating lateral gene transfer into an account of biology messes up the stability not only of taxonomy but also of alienable species, of entities that can be depended upon as stable currency, exchangeable coinage, and faithful reproducers of their own labor power: capital. Slime-exuding, gene-trafficking microbes erase their own species being, gumming up the gears of blue-green capital—making such instruments as MTAs constantly slip over themselves trying to anticipate all the becomings they mean to channel. Blue-green capital, haunted by slime, never encounters Marx's "means of production already produced" but must always wizard such means out of flux. Marine microbial ecologies do not have latent within them the logic of capitalism; that logic must be folded in.

STATION ALOHA

When I took my leave from the Patterson collection, Tien reflected, "Looking at these samples, it's hard to remember you were once tired, sunburned, or seasick, collecting them." How *are* such samples collected nowadays— especially now that Greg Patterson is no longer taking working vacations? I read in MarBEC's literature about a project called "Antibiotics from Marine Extremophiles and Cyanobacteria" for which researchers hoped to piggyback their sampling efforts on an environmental monitoring project the University of Hawai'i had sponsored for more than a decade, the Hawaii Ocean Time-series (HOT):

> Our proposed research will build on recent discoveries of novel
> microorganisms from Station ALOHA (22°45'N, 158°W), the open
> ocean site of the HOT program. This ongoing (since 1988) research
> program will provide opportunities for sample collection and experi-
> mentation, and for novel microbe isolations. We believe that the
> genomic and metabolic diversity at Station ALOHA provides an
> unprecedented opportunity for research and a great potential for new
> bioproducts. . . . We will target planktonic *Archaea* and other microbial
> groups known to be in high abundance but still not represented in our
> culture collections.[62]

I sent an e-mail to the leader of the HOT project and got myself onto the next boat to Station ALOHA.

The NSF-funded HOT project is dedicated to long-term monitoring of physical, chemical, geological, and biological processes in the open ocean. It means to understand oceanic cycling of carbon through the atmosphere.[63] Microbes are critical catalysts of this process—components of the "biolog- ical pump"—and their absolute and relative numbers provide clues about climate change as well as the productivity of fish populations. Surface waters soak up carbon dioxide, feeding photosynthesizing microbes and plankton which, in turn, are consumed by fish, which, dying, sink to the bottom to be decomposed by more microbes, littering the seafloor with organic debris that makes its way beneath tectonic plates to reemerge through volcanoes as recycled carbon. Every month since 1988, a HOT research cruise has traveled to an ocean location known as Station ALOHA, a circle of 6-mile radius centered 100 kilometers north of O'ahu; "ALOHA" expands as "A Long Term Oligotrophic Habitat Assessment" and *oligotrophic* means "nutrient poor," a condition that characterizes 40 percent of Earth's ocean. The acronym ALOHA (a Hawaiian-language word) also signals—my less patient colleagues will say touristically muscles into— the Hawaiian location of this site.[64] At Station ALOHA, scientists take measurements and water samples germane to this site, meant to represent the North Pacific subtropical gyre, a continent-sized swirl of weather and water.

I headed to ALOHA from Honolulu for four days on the *Revelle,* a research vessel that usually ships out of Scripps. When we set out, surfers in the science party looked wistfully at swells coming in south of O'ahu. Because of the routine nature of HOT cruises, the scientific team of nineteen consisted of postdocs, graduate students, research technicians, and undergraduates doing elementary collecting work. Unlike scientists aboard MBARI's *Lobos,* these personnel would not turn samples into data, much less interpretations—let alone bioproducts. The point of the cruise was anyway not to sample for microbes useful to biotechnology but rather to track another way microbes are hyperlinked to human well-being—in modulating the planet's climate. As Dave Karl, principal investigator on this project, told me, looking at marine microbial genomes could aid us in "understanding the genetic control of the physiology of the sea." Karl guessed that something like one third to one half of microbial genomes might constitute what he called "ecology genes." On this view, the ocean is a genetically modified entity.

Karl argues that the time series suggests that El Niño events are leading to increased temperature stratification of the gyre, contributing to an increase in biomass of a common photosynthetic bacterium known as *Prochlorococcus.*[65] The dominance of primary production by such a tiny creature may squeeze out larger phytoplankton, shifting the size distribution of organisms downward, ultimately reducing the population of larger creatures, like fishes. One solution proposed for mitigating such a thinning out of the sea has been to "fertilize" the ocean with iron, providing a substrate for large-celled phytoplankton (like diatoms), which would mean that nutrient-poor regions could be made to "bloom" (in the bargain, sucking carbon dioxide out of the atmosphere). Unexpected support has come from rock star Neil Young, who owns a house in Hawai'i and has offered his yacht as a vehicle for fertilization experiments. Some University of Hawai'i scientists have suggested a "concert for carbon" to fund this project. Recall Crosby, Stills, Nash, and Young's version of Joni Mitchell's "Woodstock":

> We are stardust
> We are billion-year-old carbon
> And we've got to get ourselves back to the garden.

If, as Young once sang, "rust never sleeps," iron fertilization promises to put heavy metal to work getting Gaia's garden into balance.[66] This project falls within what Gísli Pálsson calls the "regime of the aquarium," the treatment of the sea as a manageable, measurable space.[67] Many biologists, including MIT's Penny Chisholm, are skeptical of this ecological tinkering, arguing that with corporations potentially trading fertilization funding for carbon credits we could see an unpredictable commodity logic reformatting ecological dynamics. On this view, market forms of life do not necessarily guarantee the well-being of life forms. Other researchers argue that fertilization may defend against climate change in a world where policy makers in the United States are unlikely to curb fossil fuel emissions.[68]

If many samples gathered by HOT cruises might be transformed into data bearing on global warming, others, as the MarBEC website suggested, could end up as raw material for biotech—though students on this cruise might not know whether, when, or if they are bioprospecting, collecting biological resources useful for biotech. One sampling episode I witnessed aimed to detect phytopigment in the water but would also pick up microbes with interesting carotenoids. These are substances MarBEC named as sources for bioproducts, from antioxidants to cosmetics. Whether individual microbes from any particular sampling moment on this trip would be leveraged into biotech applications later is probably unknowable—though I admit to daydreaming about "following the microbes," tagging along as they moved from bottle to freezer. Such a task would likely see me still in Hawai'i, living in a lab in a sleeping bag, killing time rereading Latour's *Science in Action.*[69] As it happened, I *did* see samples collected for Ed DeLong, destined for the sort of DNA libraries I had seen earlier at MBARI (see figure 8).

As work was completed at ALOHA, I chatted with students and learned that several were on this boat against the desires of their parents, who worried that marine science was a vacation, not a vocation. One told me her dad pestered her about the significance for fisheries management of her zooplankton studies. A Hawai'i-born Korean student told me her doctor parents pointed to peers who had become accountants, asking why she was not doing something similarly "relevant." A student of Japanese descent said he became a computer science major so he could claim a "useful" skill while working in marine science. Speaking to a history of migration to Hawai'i from the East, and to aspirations for model minority career advancement rather than Euro-American rejuvenation, he had told his tale for an Asian-American Career Center website.

The different seas eddying through these narratives prompted me to ask an obvious question of the science party: "Where *are* we?" "What do you mean?" a grad student in a Grateful Dead T-shirt objected. "It's obvious. We're at 22°45' N, 158° W, Station ALOHA." Would he know that without already knowing our coordinates? No, he said, though he could probably figure it out from the stars. I asked the student of zooplankton. She took the question seriously, saying that if she walked onto deck, she could guess we were in a tropical ocean, supposing from the blueness of water that these were oligotrophic seas. It struck me that I had been doing what anthropologists call "multi-sited fieldwork" simply by staying in one place, 22°45' N, 158° W, for there were other ways of thinking about where we had been.[70] From one point of view, we had been in the Exclusive Economic Zone of the United States, which made any microbes collected here part of a pool of American natural resources. The travel of samples from Station ALOHA into labs and—if MarBEC had been successful—to market would depend on maintaining this mapping of nation onto nature. I would soon find, however, that other cartographies were possible. To follow samples toward the speculative markets of biotech, I had to navigate the choppy seas of Hawaiian state law in which bioprospecting contracts turned out to be both constituted and contested.[71]

BIOTECH INSTRUMENTS VS. SUBMERGED HAWAIIAN HISTORIES

On June 4, 2002, just after the Asia-Pacific conference, the Diversa Corporation had announced the signing of a biodiversity access and MTA with MarBEC, an announcement that generated controversy about the rights of Native Hawaiians to the bioresources of the archipelago and about whether marine microbes could be turned into legal tender at all. That controversy was still very much alive when I got back from Station ALOHA. Diversa had declared that the contract gave it "the right to discover genes from existing material collections and from environmental samples collected by MarBEC researchers in and around Hawaii with the intent of commercializing resulting products."[72] On June 5, 2002, *Pacific Business News* reported that "MarBEC is especially interested in simple species that can easily be modified."[73] In a follow-up article, "Weird Science: Company Contracts with UH for Access to Strange DNA," the same newsletter reported that, "though details are face down, the split [of profits between Diversa and MarBEC] depends on what kind of compound it is. It might be an enzyme, a protein or a small molecule. Diversa might find a commercial buyer for a new vaccine made from Hawaii samples, or a reagent." The article quoted Diversa's Mathur: "We'll probably do a bioprospecting expedition, with maybe two people from Diversa, some people from MarBEC, a couple of support staff."[74]

Bioprospecting entails not only searching for biotic substances that might provide leads for bioproducts but also securing access and rights to such materials through contracts drawn up between corporations or research groups and those agencies that "own" or "represent" such biodiversity, often nation-states or indigenous NGOs.[75] Bioprospecting has become controversial because organizations scouting for intriguing organisms often hail from northern industrialized nations, whereas the biota in which they are interested are frequently sited in so-called developing nations in the tropics and global South. Though bioprospectors have promised money, profits, and education to those designated as local stewards of biodiversity—indigenous healers holding ethnobotanical knowledge are iconic—such pledges have had less than utopian outcomes. Some opponents have seen in bioprospecting the legacy of colonial relations, terming the activity "biopiracy."[76] Vexed questions of who can be recognized as a remunerable representative have bedeviled the practice wherever it has traveled.

The MarBEC-Diversa agreement drew the attention of Native Hawaiian organizations that believed they were wrongly written out of benefits that might flow from the arrangement. Native concern turned on the history of land and sea ownership in Hawai'i and, in particular, on the legal disposition of resources from territories called "ceded lands." Before the overthrow in 1893 of the Kingdom of Hawai'i by a consortium of American businessmen, Hawaiian lands pertained to the Hawaiian crown and government or were held as private property by people, including foreigners, who had been granted royal patents.[77] After the 1893 coup, crown and government lands were placed under the ownership of the new regime, called the Republic of Hawai'i. With annexation by the United States in 1898, the Republic ceded these lands to the federal government. In 1959, when Hawai'i became a U.S. state, these ceded lands were transferred to the State of Hawai'i to be held "as a public trust for the support of the public schools and other public educational institutions, for the betterment of the conditions of native Hawaiians"[78] and "as a public trust for native Hawaiians and the general public."[79] Some of these lands were underwater—"submerged lands," extending to one marine league seaward (3 miles out from shore). The Office of Hawaiian Affairs (OHA)—a quasi-governmental organization formed by an amendment to Hawai'i's constitution in 1978 to manage "any lands received for the sole benefit of native people [from, e.g., the federal government], including lands awarded as reparations"[80]—now oversees ceded lands and seeks to ensure they are used in accordance with the constitution. OHA receives 20 percent of any revenue—from rent, duty-free shopping—generated by activities on ceded lands.

An early entry into the Diversa debate appeared in the September 2002 edition of the OHA newspaper, in which a trustee wrote: "A University of Hawai'i Department or its marBEC division have contracted an arrangement with a Dow Chemical affiliate or the Diversa Corporation to gather plant and animal material samples to evaluate their potentials. . . . The intellectual property right for natives should . . . be a consideration. . . . submerged lands are also in the ceded land inventory and OHA should be entitled to at least the indicated 20 percent income there from."[81]

Bill SB643 was introduced on January 17, 2003, into the Senate of the Hawaiian legislature proposing an act to establish "a moratorium on bioprospecting [on trust lands] and a temporary bioprospecting advisory commission to address issues related to bioprospecting, including equitable benefit sharing." According to one of the native attorneys responsible for the bill, Le'a Malia Kanehe, "The bill relates to bioprospecting on all public lands and submerged lands. Bioprospecting includes collecting samples drawn from natural resources and developing the small molecules and enzymes for pharmaceutical, agricultural, chemical, and industrial markets. When bioprospecting goes unregulated and resources are taken without consideration for the rights of the State, the public or the Native peoples it is called 'biopiracy.'"[82]

Writing of the University of Hawai'i agreement that gave "Diversa exclusive rights to discover genes from existing material collections and from environmental samples collected by MarBEC researchers in and around Hawai'i," Kanehe continued, "This action constitutes a sale of public trust assets from both the Ceded Lands Trust and other public lands. One department in UH has effectively sold the biodiversity of the entire marine ecosystem of Hawai'i."[83] The bill would put a three-year moratorium on all sampling.[84]

In a statement of support, Hawaiian lawyer Mililani Trask—a founder of Ka Lahui Hawai'i, an organization advocating a sovereign Hawaiian nation to be created on ceded lands (and often at odds with OHA)—added that "the state has kept the complete contract secret and is refusing to release data on who will make a profit from the sale. . . . The biological resources of our State's public trust are being sold illegally."[85] I was able to speak with Trask on the phone and in our conversation she emphasized that Hawaiian waters were *not* a public domain waiting to become property; rather, they were part of a public trust upon which Native Hawaiians had claims.[86]

An amendment to the bill proposed by its opponents tried to claim the opposite: waters were not part of the trust. Kanahe, in response, wrote, "Ocean resources appurtenant to those submerged lands and the water column above such lands are as much a part of the Ceded Lands Trust as any other type of natural resource. . . . Furthermore, the impetus for this resolution and the Commission came in the context of the sale of ocean biological resources by MarBEC. It would therefore, be a grave mistake to exclude these resources, which have already been specifically targeted for bioprospecting."[87]

Kenahe strategically claimed biodiversity for the people of Hawai'i: "Hawai'i's most valuable resource is our biodiversity. Of more than 22,000 known species that inhabit our islands, 8,850 are found nowhere else in the world. Diversa has now captured our marine biogenetic resources for its sole benefit."[88] In Kathleen Ann Goonan's sci-fi book *The Bones of Time*, Native Hawaiian sovereignty activists discover how to leap through time by accessing the fractal character of the universe. In one of the few time travel books I have read with a happy ending, they rewrite the past to resurrect the kingdom. Kenahe's statement has a bit of that time-traveling flavor, porting the scientific concept of biodiversity into a project grounded in Hawaiian history. A statement in support of the bill by civil rights counselor Jill Leilani Nunokawa, however, pinpoints the dangers of adopting words with that seductive *bio-* prefix: "Different language describes our islands' ancient ahupua'as and mokus. They are now called biospheres. The life that once flourished within the ahupua'as is now referred to as biodiversity. The taking and marketing of these precious resources for research or commercial development is now called bioprospecting."[89]

Nunokawa's statement highlights the different languages in play and is notable for calling attention to the *bio-* that marks so many scientific pronouncements about nature these days. Hayden explicates the work done by this prefix: "In the optimistic view of its proponents, the 'bio' in bioprospecting sanitizes, 'launders' the image of the hauntings associated with that *other* kind of prospecting [mining]."[90] Nunokawa's testimony, which concludes that bioprospecting might rather be called biopiracy (a fitting metaphor, given the maritime context), asks who is authorized to remove sunken treasure from the sea. In both bioprospecting and biopiracy, *bio-* stands for the good, the ethical, life forms with moral meanings for forms of life.

Biotech advocates opposing the bill, with its moratorium and call for a commission, also pronounced on matters of cultural difference. BioPhoriX, self-described as "a biotechnology start-up company whose mission is to develop life-saving and life-enhancing drugs from terrestrial and marine sources" (talk about promissory), offered this: "At the suggestion of the U.H. ethnobotanists, with whom we are now initiating a partnership, we have already agreed *not* to sample plants generally recognized as native to Hawai'i until the issues of fair access and equitable benefit have been settled."[91]

The idea here was that sampling nonnative plants would avoid offending native organizations. A university chemist submitted in testimony that "most of the biodiversity that is currently found in Hawaii is part of the common global biodiversity that is not owned by any particular person or nation."[92] According to this logic, ceded and submerged lands might be fair game as long as one looks only for nonnative species. Naming native plants off-limits in order to designate everything else as part of a common global biodiversity allows bioprospectors to pose themselves as enacting an ethical form of life by prospecting away from politics. However, such arguments not only assume that scientific distinctions between native and nonnative have cultural meaning for Native Hawaiians but also conflate native people with native biota (itself a contested category)—a mix-up I treat in detail in chapter 4.[93] Behind its protestations of sensitivity to native issues, however, BioPhoriX in its opposition to the moratorium bill makes clear its bottom line: "With passage of this bill it is unlikely that we will be able to attract the large investments needed from venture capitalists."[94] Targeting only nonnative biota represents an attempt to keep the currency of blue-green capital flowing by bypassing problematic politics and people.

OHA, meanwhile, argued not on classificatory terms but on legal and historical grounds. Notable in native testimonies in favor of a moratorium on bioprospecting was a commitment to properly interpreting the state constitution. This approach made sense if one had in view the possibility that OHA, a state agency, might administer benefits deriving from bioprospecting. A more radical approach would have been not to recognize the state at all—or even the 1893 overthrow of Queen Lili'uokalani or the 1898 U.S. annexation[95]—arguing that full control (not just a 20 percent cut) of resources in Hawaiian waters should belong to descendents of citizens of the internationally recognized kingdom legally consolidated by Kamehameha III in the 1840s. Such a position would be consistent with the platform of today's Council of Regency of the Kingdom of Hawai'i, a sovereignty group that holds that the kingdom was in fact never legally dissolved.[96]

The moratorium bill was defeated, as was a similar one in the House (HB572). University scientists were worried that research "deriving new compounds from newly discovered bacteria and archaea in the open ocean" would, under the terms of the bill, become illegal.[97] But they also made a case that biological sampling was involved in activities as benign as monitoring drinking water and doing science on invasive species.[98] The moratorium, they said, would put an end to activities that had nothing to do with profit, like gathering clues to global warming on HOT cruises.[99] Students on *Revelle* would certainly have been shocked to find themselves accused of biopiracy (even if later scientific work—by people piggybacking on the HOT program—might enlist their samples into such unanticipated futures).

A later version of the bill drafted with the participation of native lawyers introduced another issue: "A sharing mechanism between native Hawaiians and the people of Hawaii should be established, on a statewide policy basis, for the ownership rights of any organisms, minerals, genetic codes, or other resources discovered through bioprospecting on Hawaii's natural resources." The university now argued that ownership was a separate question from the recognition that Native Hawaiians might be entitled to benefits. The university wrote: "If ownership rights are to be established, then a sharing mechanism for ownership rights must be established. This should also include full participation in funding research that provides the discoveries that can reap a tangible or intangible benefit."[100] In less neutral words, as one scientist unsympathetic to the native position put it to me, "If they want to own it, why don't they fund it?"—by which he meant the research that would capitalize this property. On this view, ownership is about having means to purchase rights. This position overlooks the processes through which Native Hawaiians' means have historically eroded as land slipped into foreign ownership—one reason Hawaiian contests over bioprospecting were organized around the question of ceded lands. It also misses the more contemporary fact that, as Sheila Jasanoff has argued, "in promoting rapid transfers of knowledge from academy to industry, the Bayh-Dole regime privatized discussion of the larger social purposes of knowledge creation."[101] In the opinion of my interlocutor, ownership was a matter of the transfer of money. In this model, property is nothing if not alienable. For many boosters of marine biotech in Hawai'i, such transfers of ownership and rights are unproblematically good because they can "bring money into the state." Such practices, these scientists say, could enrich Native Hawaiians, who might now find ethnobotanical knowledge transmutable into capital (unless the biotech people are chasing microbes).

This market sentiment and faith in alienable property echoes the calls of those nineteenth-century foreigners in Hawai'i who pressed the monarchy to privatize land holdings, which was disastrous for the kingdom. Linda Parker writes, "Aliens supported their demands for unrestricted rights of buying and selling land by arguing that such a policy would promote Hawaiian prosperity, [and] encourage native acquisition of the habits of industry by foreign example."[102] Rewriting this sentence as "*Marine biotechnologists* supported their demands for unrestricted rights of buying and selling *organisms and their elements* by arguing that such a policy would promote Hawaiian prosperity, [and] encourage native acquisition of the habits of industry by *scientific* example" makes it clear that today's epistemology of alienable property is not so distinct from earlier practices. Not just the alien ocean, then, but also the alienable ocean. Fungible banks of cyanobacteria are, like plantations, patches of biotic matter legally and economically organized to produce capital—as long as the legal instruments can be made to work, something the 2003 legislative debate put in question.[103]

ALIENABLE OCEANS

At work in the bioprospecting debate were different notions of property, ownership, and invention. For marine biotechnologists and some of their colonial forebears, maps of land and maps of genes—grounding representations for royal patents and bioproduct patents—secure the potential for ownership, for the alienability of things from themselves.[104] But, as Hayden points out, "royalties, in the amount deemed acceptable to participating companies . . . are not up to the task of mediating the complex histories and futures of inequality into which prospecting interjects."[105] If money is not up to the task, it is notable that native activists felt legal instruments might be. As Sally Merry and Don Brenneis have argued, some Native Hawaiians accept the legal structure (if not full legality) "of the colonial state in Hawai'i . . . because the Hawaiian Kingdom adopted these legal forms and practices before the colonial takeover by the United States, not after."[106] Law adapted from the Anglo-American legal system was a foundation of the nineteenth-century Hawaiian nation-state—"a route to sovereignty in an imperialist world"—and some Hawaiians are hopeful that the law can provide a path toward increased recognition for Native Hawaiians, including the reclaiming of land and the awarding of reparations. A few hope use of the law at the international level can open a path toward the resurrection of a sovereign Hawai'i.

The dispute about Diversa-MarBEC reveals different sentiments about what constitutes the nature of the ocean. For Diversa and MarBEC, the ocean exists as a site and source of microscopic, microbial life, molecularly understood. For these scientists, the boundaries of the sea are not those of politics but those between the visible and subvisible world, and those of ecosystems such as the North Pacific subtropical gyre. Such a view comes with a political dividend, however: because they are so cosmopolitan, microbes, like fishes, can be found in many places, and the politics of locality can be outmaneuvered.[107] An additional attraction of marine microbes under this regime—call it the *nonterritorial sovereignty of science*—is that one can bypass the sort of ethnobotanical knowledge that might demand recognition if one were dealing with, say, a seaweed known by Hawaiians to cure some disease. The native side of the debate saw things differently: nature was embedded in the history and politics of territory. On this view, scientists must abide by the laws of the state within which they forge agreements. Simply claiming that the law cannot keep pace with science will not do. Even if the University of Hawai'i is a state institution, these activists urged, it does not therefore have the right to act *as* the state.

The language of the bioprospecting bill as of 2005 eliminated any *historical* accounting of state obligations to Native Hawaiians, substituting instead questions of "culture" and "indigenous knowledge." In this process, the legal and political history of Native Hawaiians was erased and they became generic "traditional, indigenous knowledge holders," who, as one House resolution had it, should be given "meaningful participation"—not, note, political participation—in matters pertaining to conservation, sustainability, and bioprospecting benefit sharing insofar as these impinged on indigenous knowledge. The separation of culture from politics accomplished in these framings spirits Native Hawaiians out of history, on the pretence of engaging them as cultural equals in what is in fact a setting of political economic inequality.[108] The emergent biotech economy bears resemblances to earlier plantation politics, with mainland haoles and a few Japanese-descended people in charge, while Native Hawaiians find themselves on the outside, using the legal system to protest.[109]

What happened to the Diversa-MarBEC agreement? Diversa neglected it, especially once MarBEC began to spiral into ruin. In one sense, this comes as no surprise. In his talk at the Asia-Pacific Marine Biotechnology conference, Mathur had been at pains to speak carefully about the political dimensions of how Diversa obtained rights to genes. He detailed the benefit-sharing plans Diversa had with concerns in foreign nation-states. As he put it, "Gone are the days when you can just send your scientists on vacation to collect samples. Now, you must have rights from governments to look for and commercialize biodiversity." He said Diversa kept in view the need for sustainable use and equitable benefit sharing and that some revenue should always go to conservation. As he put it, "Exploration of biodiversity should be contrasted with biopiracy; we should protect, not pillage."

Even so, Diversa enjoys the option of leaving difficult situations, not unlike multinational corporations that skip around the world to take advantage of shifting labor laws, wages, and exchange rates. Not only can Diversa go elsewhere, it can work with samples it already possesses, using computer techniques to explore the space of possible genetic and metabolic pathways. It can bioprospect in databases, in the biomedia of cyberspace, in the uploaded ocean, leaving local places behind:

> Diversa's DirectEvolution technologies enable modification of genes and proteins, providing particular value for modifying enzymes and antibodies. DirectEvolution['s] suite of methods includes Gene Site Saturation Mutagenesis™ (GSSM™) and Tunable GeneReassembly™ technologies. Diversa's GSSM technology is a patented method of creating a family of related genes that all differ from a parent gene by at least a single amino acid change at a defined position. By performing GSSM technology on a gene encoding a protein, all possible single, double, and/or triple amino acid codon substitutions within a protein are created, removing the need for prior knowledge about the protein structure and allowing all possibilities to be tested in an unbiased manner.[110]

One biologist I met in Monterey was appalled by Diversa's DirectEvolution method, telling me, "It's like patenting nature." This complaint makes sense only if one views nature as a space of recombinatoric possibilities—which might be accurate enough if we recognize that the practices of biology in the age of biotech seek not so much to describe the world as to break it down to build it up anew. We might better say that Diversa wants to patent "future nature"—reserving to itself the right to make promises to stockholders and at the same time to hedge its bets about its pledges. Diversa attached the following disclaimer to its announcement of the MarBEC deal: "Statements in this press release that are not strictly historical are 'forward-looking' and involve a high degree of risk and uncertainty. These include statements related to the discovery of genes, the discovery of unique molecules from environmental sources, the commercialization of products, the future success of Diversa's discovery program, . . . all of which are prospective."[111] We are back to the promissory nature of blue-green capital.

In *Promising Genomics*, Fortun asks, "In a territory of forward-looking information, how does one decide between fact, speculation, puffery, hype, straight dope, and bald-faced lie?"[112] A good question—but one that Native Hawaiian opponents of the Diversa-MarBEC agreement may not have needed to answer in order to argue that something was amiss in the way the blue-sky promises of the bioprospecting contract flew right over Hawaiian laws about those old-fashioned territories, the land and sea. We could say that OHA's effort to thwart bioprospecting sought to reveal the currency of biotech as counterfeit—unauthorized by the state, not legal tender. Maurer writes that "a counterfeit is only known when its circulation, its flow, is halted." But such inauthenticity is a retroactive effect: "If it circulates, even if it is 'false,' it is nonetheless 'true' in the now of the transaction."[113] In promissory capitalism, after all, money need not be made off marine microbes but can sometimes just as well or better be made off promises about the sunken treasure that will be extracted from them in a possible, artificially selected, blue-ocean future. In the event, Diversa was able to zoom off into other forward-looking futures and places. It did not need Hawai'i.

MarBEC, eager to bring money to the state of Hawai'i, could not release itself from imagery of locality the same way. A parallel strategy for marine biotech in Hawai'i, then, might be to find innovative means to localize scientific activity. One still-in-formation idea when I left was to promote "native bioprospecting." The past twenty-five years have seen a resurgence of interest in the maritime techniques used by Polynesians to settle the Pacific. The material expression of this interest has been the construction of facsimiles of ancient canoes by such organizations as the Polynesian Voyaging Society. Outrigger sailboats such as *Hokule'a* have become famous not only for demonstrating possible paths of early settlement from Tahiti to Hawai'i but also as vehicles for forging new cultural ties among peoples of the Pacific. Anthropologist Ben Finney calls such recuperative cultural work "identity voyaging."[114] University scientists and native schoolteachers are now suggesting that voyaging canoes be used as platforms for teaching children about microbiology, weaving together native and scientific knowledges about marine biology, something the school is beginning to support through its coordination of the Center for Microbial Oceanography Research and Education (C-MORE), which, in line with NSF calls for diversity in the sciences, is aggressively seeking to increase participation of underrepresented groups. Such cultural revivalism might also include retrieving traditional knowledge about biomaterials—like boat-sealing agents. A high school teacher I met at the Asia-Pacific meeting first told me about such practices, about what could be called "identity bioprospecting." He speculated that Hawaiian chants might contain information about properties of plants in medicine and boat building.

On the one hand, such searching could add a biological component to Hawaiian revivials—and the seemingly apolitical language of the latest bioprospecting bills could begin to look different, with indigenous knowledge now a potent political instrument of contest. On the other hand, valorizing culture while sidelining the legal and political history of Native Hawaiians could grab attention away from difficult questions of who should use land and sea when and how. Such a project, a fusion of the alien and the native sea, mixes gray areas with blue skies and oceans. But for now this anthropologist will have to defer—no forward-looking statements here. We can only know how identity bioprospecting would look if it happened—and at the moment there does not seem to be any institutional push in this direction. C-MORE's efforts to bring marine science together with native education are emphatically not commercial, focusing rather on working with K–12 teachers to develop curricula that fuse histories of Hawaiian practices around land and sea with lessons about water-quality testing, chemical cycles, and food chains.[115] C-MORE's stirring of nation into ocean—on the model of multiculturalism—mobilizes different sentiments and instruments than MarBEC's market-based mix of alienable oceans and sovereign scientists. The relation of both of these models to new waves of thinking about a native ocean remains to be seen.

4 Alien Species, Native Politics

Mixing Up Nature and Culture
in Ocean O'ahu

Descending the escalator into the Honolulu airport baggage claim in June 2003, I found myself gliding toward a Hawai'i Department of Agriculture poster reading "HELP PROTECT HAWAII. Undeclared fruits, vegetables, plants, and animals can cause damage to Hawaii's fragile environment." Later that evening, at a store stocked with dog-eared books near the University of Hawai'i campus, I happened upon a 2001 volume titled *Hawai'i's Invasive Species,* in which I read that the archipelago, the most isolated in the world, has been the location most "invaded" by organisms introduced by humans.[1] Hawai'i is a poster island chain for invasion biology, a life sciences specialty that studies species out of place, organisms often called *alien species* or, sometimes, *introduced species* to highlight the role of human agency in moving living things between ecosystems.[2]

Nonhuman organisms have long been transported across widely separated localities, but invasion biologists argue that transfer rates have shot up steeply in recent decades.[3] Airplanes like the 757 I flew in on are relatively minor vectors. Of greater concern to biologists of island zones are less monitored paths of transfer associated with the ocean; boat hulls and ballast water have hosted organisms—molluscs, seaweeds, crabs—across distant ports, with a variety of deleterious effects (see figure 19). The University of Maryland Sea Grant College has produced a documentary about such species, titled *Alien Ocean.*[4]

According to the Global Ballast Water Management Programme, headquartered in London, "shipping moves over 80% of the world's commodities and transfers approximately 3 to 5 billion tonnes of ballast water internationally each year."[5] Taking on ballast water compensates for the decreased weight of ships after cargo holds have been emptied. With recent improvements in ship design, seawater now replaces such solid ballast as

FIGURE 19. West Coast Ballast Outreach flyer.
©1999 Ed Lindlof. Used with permission of The
Regents of the University of California. All rights
reserved.

sand and metal. With the efficiency of water ballasting, however, has come
an unforeseen environmental result: microorganisms picked up in one part
of the world can be transported and released into distant ports. Microbes,
small invertebrates, and larvae of miscellaneous marine creatures are
tiny enough to fit through the grating of ballast intake pumps. The
International Maritime Organization, created in 1948 to set global shipping standards, in 2004 named *Vibrio cholerae* and *E. coli* as indicator bacteria for measuring the health risks of transported ballast water.[6] Heavily
trafficked sites like Hawai'i's Honolulu Harbor are attracting increased
attention. The transfer of microbes across maritime territories edges into

the literal a metaphor offered by Gro Harlem Brundtland, former director of the World Health Organization, in a lecture he gave at the World Economic Forum in January 2001: "In the modern world, bacteria and viruses travel almost as fast as money. With globalization, a single microbial sea washes all of humankind."[7]

Exactly which creatures are moved from one place to another depends on the scale, speed, and geography of global trade. The movement of so-called alien species maps prevailing transoceanic economic, military, and tourist connections. For example, of the 4.6 metric tons of ballast water discharged in California in 2000, 50 percent originated in the waters off China, Japan, and the Koreas, with Mexico coming in second at 30 percent.[8] In Honolulu, from 1997 to 1998, the only period for which there are data, ballast water came from California (35 percent, mostly with container ships); Japan, Korea, and China (34.2 percent, primarily from cargo ships); and Indonesia and Alaska (9.1 percent and 5.8 percent, respectively, mostly with petroleum shipments).[9] Regulations in many maritime states require ships to swap out ballast water while at sea, outside the 200-mile limit of Exclusive Economic Zones.

In this chapter, I analyze scientific and popular discussions of so-called alien aquatic organisms in Hawai'i, a subject I was prompted to look into after attending the Marine Bioinvasions Conference at Scripps in March 2003. Microscopic plankton are a prominent subject of worry for those concerned with alien aquatic organisms. But although we could zero in on invasive marine microbes as embodiments of something like microbial globalization, biologists in Hawai'i spoke to me more often of another genre of floating ocean organism: seaweed.[10] Though microbes may be good guides into the permutation of the alien ocean suggested by invasive aquatic species, looking at creatures farther up the size scale of the planktonic chain—like brown, green, and red algae—allows access to a visible, tactile realm in which scientists are joined by lay people in arguing about what should count as alien at all. People outside of biology labs can care very much about how and with what instruments—taxonomies, gene sequences, maps, tastebuds—species are defined as native or alien.

Anthropologist Arjun Appadurai, in *Modernity at Large*, proposes that the modern world is constituted by global "flows" and identifies five conceptual territories across which such flows move: ethnoscapes, technoscapes, financescapes, mediascapes, and ideoscapes.[11] The language of "flows," however, widely adopted in studies of globalization, often consigns to the background the vehicles that allow such movement to take place—vehicles such as the lab and legal instruments in chapter 3 that enable biotech substances to flow as currency, for example. As an alternative to the metaphor of flow, Anna Tsing, in her account of the overlogging of Kalimantan rain forests in Indonesia, employs the figure of "friction" not only to make palpable the political causes of the 1997 forest fires that devastated the region but also to explore the combustible tensions consequent upon international extractions of timber for global markets.[12]

It would be easy to imagine that routes of global ocean commerce and transport represent a literal instantiation of flow—after all, we are talking about boats channeled though shipping lanes often tracked along ocean currents.[13] But political and regulatory frictions constrain these movements; boats cannot dock wherever they like. Further, worry about aquatic invasive species suggests another sort of agitation: turbulence.[14] For many scientists and policy makers concerned with alien aquatic organisms, the suction and pumping of ballast water symbolize a disturbed mixture of elements from the orders of nature and culture. With its jumble of marine organisms, industrial shipping practices, and paths of international commerce, the seaborne introduction of alien species describes a flow that puts human and nonhuman agencies, forms of life and life forms, into a confusing spin cycle. Alien species are alternately given agency (and responsibility) when they are described as "invaders" and deprived of agency when they are named as "introduced"—a seesaw dynamic that makes it unclear whether invasion biology is a natural or a social science, or both.[15]

As the phrase "global ballast management" implies, invasive species are considered a matter of global import, a cause for concern for parties participating in the forging of networks that sew together such objects as global commerce or global ecology. Writing of such global, but locally inflected, phenomena as the traffic in human organs and stem cells, anthropologists Stephen Collier and Aihwa Ong in *Global Assemblages* argue that unlike, say, kinship practice, whose validity makes sense only within a given cultural worldview, "global phenomena . . . have a distinctive capacity for decontextualization and recontextualization, abstractability and movement, across diverse social and cultural situations and spheres of life. . . . Global forms are limited or delimited by specific technical infrastructures, administrative apparatuses, or value regimes, not by the vagaries of a social or cultural field."[16]

Although Collier and Ong are correct that things like traveling livers and cells look different depending on the contexts within which they move, I am less sure about the separation of technique and administration from the social and cultural. "Global phenomena" like alien species are at once technical and social objects, and the ways they are defined with respect to "cultural" questions matter a good deal—not only, in this case, for how alien microbes or algae are interpreted but also for how they are delimited as such.

Weeks after my arrival in Hawai'i, the state Department of Land and Natural Resources released a draft of an aquatic invasive species management plan. The plan detailed dangers of alien species, extolled virtues of native species, and offered the following contrast: "Aquatic invasive species (AIS) include species in marine, freshwater, brackish water, and estuarine environments, whose introductions cause or are likely to cause economic or environmental harm, and/or harm to human health. AIS are a serious problem in Hawai'i, posing a significant threat to Hawai'i's native plants and animals, as well as their associated ecosystems."[17] In *Alien Invasion: America's Battle with Non-native Animals and Plants*, journalist Robert Devine, writing of nonhuman flora and fauna, suggests, "Hawaii's natives are vulnerable because of the rapid change brought on by modern civilization"—including introduced, alien species.[18]

How do biologists define *alien*? The way into this question, for this chapter, is through asking its complement: how do biologists define *native*? Far from being a straightforward matter of biological classification, this is a taxing taxonomic question, especially in Hawai'i, where the word *native* resonates with descriptors used by and for the indigenous people of Hawai'i, known as Native Hawaiians. This chapter, drawing on interviews I conducted with scientists at the Hawaii Biological Survey of Bernice Pauahi Bishop Museum, at the University of Hawai'i, and at the Waikīkī Aquarium, maps the ways biologists sort these things out, especially in the aquatic realm. Classificatory issues become particularly vexed when sited in the sea that weaves in and out of this part of what Europeans have called Polynesia. *Alien* and *native*, of course, are relational categories, defined in terms of one another. To anticipate my argument: native species are often allied with nature, whereas alien species are associated with culture—but the boundaries of nature and culture are themselves subject to frequent shifts in the politically charged geography of Hawai'i and, more, in the epistemologically turbulent zone of Hawaiian waters.

In the first part of this chapter, I report on conversations about aquatic alien species with scientists who identify as invasion biologists. I turn then to scientists, educators, and activists keen to enmesh their classifying activities in Native Hawaiian practice. I show how the definition of *alien species* is continually under negotiation. It is not that the phenomenon is decontextualized and recontextualized, as Collier and Ong would have it, but rather that context itself is continually reinvented to calibrate to what will count as alien and native. Following Marilyn Strathern, I scrutinize *context* as an analytic concept emergent from scientific, cultural practices of classification.[19] In other words, this chapter does *not* offer an exposé of a "cultural" difference between scientists and nonscientists and the ways they parse the "nature" of the alien ocean. Rather, I show how definitions of *nature* and *culture* themselves, of figure and ground, are put into flux by the very idea of alien species—and all the more when these creatures are denizens of the ocean.

HOW SCIENTISTS THINK: ABOUT "NATIVES," FOR EXAMPLE

Let me present summary definitions used by invasion biologists in Hawai'i to think about native and alien species, on land or at sea. I collect the following from *Hawai'i's Invasive Species:*

> *Native species*—found naturally in an area, not introduced by humans; includes both indigenous and endemic organisms
>
> *Endemic species*—naturally restricted to a particular place and found nowhere else
>
> *Indigenous species*—naturally occurring in a given area as well as elsewhere
>
> *Alien species*—nonnative, that is, a species introduced to a place accidentally or intentionally by humans
>
> *Accidental introduction*—a species introduced by humans by chance, without intent, or through carelessness, and often with unfortunate results
>
> *Intentional introduction*—a deliberate introduction of a species (either authorized or unauthorized) by humans, involving the purposeful movement of a species outside its natural distribution
>
> *Invasive alien species* (= invasive species)—alien species whose introduction and rapid, aggressive spread do or are likely to cause commercial, agricultural, or environmental harm or harm to human health.[20]

This categorization is organized around the presence or absence of human agency and, once we arrive at the category of *invasive,* around what particular groups—small farmers, developers, or agribusiness, for example—designate as harmful. This is a metataxonomic category, transcending the classificatory grid of biological nomenclature. The Hawaii Biological Survey maintains a "Checklist of the Marine Invertebrates of the Hawaiian Islands," which can be searched online using Linnaean categories of

TABLE 1. Classification of Native and Alien Species by Natural and Cultural Valence

Nature	*Culture*
native	alien/introduced
endemic	accidental
indigenous	intentional
	invasive

phylum, class, order, family, genus, and species but which also allows perusal by "biogeographic status," including "native," "introduced," or "cryptogenic" (which designates species whose status as native or alien/introduced is uncertain).[21] Biogeographic status is not a Linnaean category; it is more historical than evolutionary. *Nature* and *culture* are the organizing rubrics in this metataxonomy (see table 1).

On the surface, this classification offers a simple mapping of natural versus cultural agency, but note that the category *invasive* requires a further step: social judgment of harm. It might seem this category could be pegged as social rather than a matter of objective taxonomy—and, indeed, one author in *Invasive Species in a Changing World* argues that "the 'noxious invasive' of one group is the 'desirable addition' of other groups."[22] But many biologists argue that there are in fact measurable, recognizable characteristics of invasive species. These characteristics might be employed to move a creature from *cryptogenic* to *introduced*. According to *Hawai'i's Invasive Species*, such organisms are

adaptable to and capable of thriving in different habitats;

tolerant of a range of conditions (light, temperature, moisture);

able to eat and survive on a diversity of food resources;

fast growing, thereby able to displace other plants or animals;

disturbance tolerant, able to proliferate in places disturbed by humans or natural events;

easily dispersible to new localities.

Reproductive features are also important. Invasive species

are able to produce many seeds, larvae, or juveniles and begin doing so early in life;

(plants) are able to reproduce vegetatively as well as by seed;

(animals) have clonal or hermaphroditic reproduction, which can make them especially invasive;

have long breeding seasons, or even breed year round;

have seeds, eggs, or larvae easily dispersible, for instance by animals, wind, or accidentally by humans;

(plants) have seeds with no special germination requirements (e.g., a period of heat or cold exposure, soaking in water, drying out).[23]

The formal characteristics enumerated mean to define invasives with respect to biological concepts—of robustness, fecundity—but are also criteria which, as biologist Banu Subramaniam points out, resonate with anti-immigration language—take as an explicit example New Zealand's National Centre for Aquatic Biodiversity and Biosecurity, which worries about "devastating foreign invaders."[24] Subramaniam argues that

> the parallels in the rhetoric surrounding foreign plants and those of foreign peoples are striking. . . . The first parallel is that aliens are "other." . . . Second is the idea that aliens/exotic plants are everywhere, taking over everything. . . . The third parallel is the suggestion that they are growing in strength and number. . . . The fourth parallel is that aliens are difficult to destroy and will persist because they can withstand extreme situations. . . . The fifth parallel is that aliens are "aggressive predators and pests and are prolific in nature, reproducing rapidly." . . . Finally, like human immigrants, the greatest focus is on their economic costs because it is believed that they consume resources and return nothing.[25]

Anthropologists have made similar observations. Tsing finds in panicked media discussion of "Africanized bees" entering the United States a displacement of fears about violent and hyperfecund immigrants.[26] Jean and John Comaroff suggest that nation-states obsessed with "aliens and alien-nature . . . share a common feature: all are former labour importers and centres of capital—and as such, nexes of wealth within a vastly unequal world economy—into which job-seekers and fortune hunters are popularly imagined to be pouring, usually across ill-regulated borders, in order to take scarce work and resources away from locals."[27]

But in Hawai'i, rhetorics of invasive alien species partake of a different politics. This is because organisms introduced during the original human settlement of Hawai'i, beginning as early as 400 C.E.—"pigs, taro, yams, and at least thirty other species of plants"[28]—are sometimes tightly associated, even identified, with the people who introduced them, frequently designated by their descendents and others as "native." This suggests that not all human agency is to be treated identically.[29] We must look at particulars. The alien ocean is not a unified field; its outlines depend on whom you ask.

Definitional matters are complex in Hawai'i; many Polynesian introductions are considered native, a fact that came into view during my conversations with scientists working on the Hawaii Biological Survey at Bishop Museum. The museum has as its mission "to record, preserve and tell the stories of Hawai'i and the Pacific, inspiring our guests to embrace and experience our natural and cultural world."[30] I spoke with five scientists at Bishop. None identified as native and all were careful not to speak on behalf of indigenous people; all would be designated as haole in the Hawaiian terminology of ethnic belonging. Their categorizations of native and alien species were instructive for meditating on how scientists think—about "natives," for example.

I am inspired to this phrasing by the title of a book by anthropologist Marshall Sahlins, *How "Natives" Think: About Captain Cook, for Example.* In this disquisition on how to think about cultural categories historically, Sahlins maintains that eighteenth-century Hawaiian cosmological categories were capacious enough to consider Captain Cook an instantiation of the chiefly akua, Lono.[31] Sahlins argues against Gananath Obeyesekere, who holds in *The Apotheosis of Captain Cook* that accounts portraying Cook as godlike for Hawaiians are cases of European myth-making.[32] Hawaiians, Obeyesekere maintains, employed a universal human practical reason to see Cook for what he was: a human being, not an entity partaking of both natural and supernatural orders. Sahlins responds that Obeyesekere's argument evaporates cultural difference, turning all people into bourgeois rationalists. At stake in the Sahlins-Obeyesekere debate are questions of the translatability of categories and the meaning of practice in varied historical and political settings.[33] Sahlins's argument is that cultural rationalities unfold within symbolic systems that shape what counts as empirical and interpretive, natural and supernatural. We can say the same for scientific definitions of *alien* versus *native.*

Questions of classification, of alignments between the empirical and the interpretive, have been at the heart of anthropology. Classifications—of kin, land, and food, to take a few examples—have been understood to afford neat points of access to cultural difference. But it has been the classification of things in that realm Western culture has named "nature" to which anthropologists have appealed most sharply to press points about the social character of thinking about the world. Thus, in his 1967 "Why Is the Cassowary Not a Bird? A Problem of Zoological Taxonomy among the Karam of the New Guinea Highlands," Ralph Bulmer explored how the Karam can get away with not thinking of cassowaries as the avians European taxonomy takes them to be.[34] Bulmer argues that in the local imagination cassowaries are sisters and female cross-cousins to Karam men and so cannot fall in with the birds. Of course, to pose his question at all, Bulmer must hold to the grid of Western zoology, taking biology as an idiom into which all cultural systems can eventually translate. Harriet Ritvo, in *The Platypus and the Mermaid,* has advanced a more culturally located view of the creation of European biological categories, outlining the variety of principles used in classification before Darwin settled on genealogy.[35] Many such principles—vegetative form, diet, physiology—survive in present-day taxonomic practice.

So, how do scientists think, about "natives"—and "aliens"—for example? Adapting Sahlins's argument, I say that biologists make their way through the world with a sense of the empirical nested in frameworks of interpretation. Such a contention suggests there is no one way scientists think; rather, we must attend to contexts called into relevance in various accounts.

Over lunch in Bishop Museum, marine biologist Ron Englund, living in Hawai'i for about eleven years when I met him, provided me with a taxonomy of the *native* and the *introduced,* with reference to Hawai'i.[36] In his classification, *indigenous* fluctuates, sometimes excluding any human introductions, sometimes including only Polynesian ones.

Native: "Things that got here under their own power, without humans"

Endemic: "Creatures can be endemic to the Hawaiian Archipelago, to just one island, or even one volcano."

Indigenous: "Native to an area but also widespread elsewhere, like other islands or continents."

Introductions, Polynesian: "Anything brought here during the original settlement of these islands. What these are called depends on whom you talk to. A Native Hawaiian—and I don't want to speak for Native Hawaiians—might also call these things indigenous. Why? Because they don't cause any problems and are economically beneficial. Some examples are taro and sweet potatoes. These are in a different categorization because they're culturally important. They are not invasive because they are considered highly desirable."

Introductions, nonindigenous, also *nonnative,* sometimes *alien:* "Anything brought by man within the past two hundred years, since Cook arrived [in 1778]. Cook is a cut-off point because after him the quality and quantity of introductions increased. And if you're native, you have to trace your heritage back before he arrived."

Invasives: "Introductions that are harmful, outcompeting local organisms."

Another scientist, also recognizing that this is complicated definitional territory, offered that "most people [in Hawai'i] who are native are also part Caucasian or Filipino," suggesting that the claim of nativeness might be strategic, not warranted by the biology of the matter (though categories like Caucasian or Filipino are hardly biological either).[37] When I reported this analogy to Englund, he found it problematically racialist, though noting that entanglements with native politics were anyway unavoidable.

Such associations were not necessarily always essentialist or divisive; often they helped strategically direct public attention toward the problem of introduced species. The back cover of *Hawai'i's Invasive Species* carries an approving blurb by Hawaiian navigator Nainoa Thompson, known for his work with the Polynesian Voyaging Society and his piloting of the traditional double-hulled canoe *Hokule'a:*

> We are all so privileged to be able to live in Hawai'i, with its rare, beautiful, and fragile natural environment. Each of us has an obligation to *malama* this place—ensure that we practice stewardship at its highest level. . . . *Hawai'i's Invasive Species* is a wonderful contribution to that effort. With their straightforward explanation of the fragility of our environment, the harm brought by alien and invasive species and the action each of us can take to preserve Hawai'i's biodiversity, the authors have done much to *malama* Hawai'i.

As I discuss in chapter 3, the link between native species and Native Hawaiian politics also features in legislative debate about marine bioprospecting. University of Hawai'i botany professor Will McClatchey, in connection with the proposal for a state Senate bill calling for a moratorium on bioprospecting, argued that Hawaiian species should be off-limits to bioprospectors: "The major categories of species that should be of concern are ALL indigenous and endemic plants and animals that were introduced by the ancient Polynesians or developed over the history of the Hawaiian people. These plants and animals are the natural inheritance of native Hawaiians and should not be freely used by others without consent and recognition of the long-standing relationship."[38]

TABLE 2. McClatchey's Classification of Native and
Alien Species

Nature	Culture
native	alien
endemic	accidental
indigenous	intentional
introduced by ancient Polynesians	invasive
developed over history of Hawaiian people	

Noteworthy in this view is that "indigenous" and "endemic" plants and animals can be at once "introduced" and a "natural inheritance of native Hawaiians," a definition that does not draw the same lines of agency as categorizations above but does, in fact, open up the possibility that what is considered native or indigenous can change over time. This view starts to complicate—even, perhaps, unravel—the view offered in the testimony of the biotechnology startup BioPhoriX, which posited a one-to-one relation between native people and native organisms.

I would map McClatchey's parsing as in table 2. Here, Native Hawaiians are still allied with nature (and the implicit symbolic alignment here has *native* to *nature* as *alien* is to *culture*). Indeed, they are positioned as *part of* nature. What is interesting, though, is that there is now room for their "natural inheritance" to be "developed over . . . history." So, nature might be complex, admitting of historical entwining with culture. In *In Amazonia*, Hugh Raffles argues that Amazonian "nature" has been reinvented many times, from early days of European colonialism to contemporary environmentalism. But whereas the nature of which Raffles writes is continually reinvigorated for participants as pristine, McClatchey's articulation allows a historical imagining of the transformation of a nature always under revision.[39]

Not all scientific statements align on how to think about native nonhumans. Bishop Museum scientists with whom I spoke are inclined to say that Polynesian introductions are . . . introductions. For instance, Steve Coles, a marine invertebrate zoologist: "Obviously, for an island, everything is introduced. The ancient Hawaiians altered the environment, too, and not always for the better as far as native plants and animals were concerned. But with the arrival of European ships and commerce, the scale increased dramatically and the consequences have been much more invasive and severe."[40]

Another invertebrate zoologist told me that, when it comes to things like taro, "it depends on whether you talk to a person who says they're Hawaiian or a twenty-first-century biologist, who says they're introduced. If you go back far enough everything is an introduction. Native is something you could think would be here pre-Polynesian days. But you have to work on definitions. It depends." The first sentence is ambiguous; it is not clear whether "Hawaiian" refers to organisms under definitional scrutiny or to a hypothetical Hawaiian person. The saturation of this sentence by two possible readings is a sign of the complexity of speaking in the same breath about biology and politics in Hawai'i. Another scientist at Bishop reflected, "This is a very culturally sensitive issue. We prefer to call them 'canoe species.' They're still introductions. There are about thirty canoe species, and some of them are unintentional introductions, like the rat." This scientist suggested that rats were invasives, placing the agency of canoeing humans in the same realm as later seafarers.

Ancient Hawaiians, then, for the three scientists whom I have just quoted—distinct from McClatchey—are not part of nature. The context of Hawai'i requires attention to questions of scale as well as sensitivity quite literally to "the natives' point of view"—but would not seem to require a thoroughgoing redefinition of the textbook meaning of *introduction*. Here, the guiding epistemology holds that, as Brian Rotman glosses it, "the context needs to be controlled as background in theoretical and laboratory accounts, but plays no important and certainly no constitutive role in the thinking process."[41]

Richard Archer,[42] a molecular biologist, expressed concern with attempts to recognize Polynesian introductions as native to Hawai'i. He told me, "Native plants are those species that found their way to Hawai'i without human intervention, or evolved here from those founding populations. Cultural revival movements often want to retrieve practices promoting the return of Hawai'i to a natural environment, but not always are these a part of the prehuman Hawai'i. Thus their concept is something pre-1778 rather than before about 1,500 years ago." A story presented in *Alien Invasion* offers the kind of alignment between nature and culture that Archer considered problematic. Devine recounts a conversation with a Hawaiian healer, Kapi'koho Naone III:

> "With so many alien species, some of them real destructive, traditional practitioners are afraid," said Naone. Tradition and the need for high-quality plants demand that Hawaiian healers can gather their specimens in the wild. . . . Naone also worries about the effects of alien species on Hawaiian culture in general. "Hawaiian traditions have been

gaining ground in recent years," said Naone. "Our plants and animals are very important to those traditions. Our environment and our culture are all intertwined." Nainoa Thompson, the master navigator . . . shares Naone's concern. He writes, "Each time we lose another Hawaiian plant or bird or insect or forest, we lose a living part of our ancient culture. Stopping alien pests is about choosing our future and saving our past."[43]

Archer found such reasoning too pat. More, thinking of Hawai'i's islands as one geographic entity to which plants might be native had led, he said, to accelerated shufflings of plants from one island to another. In 1997, the Department of Land and Natural Resources declared that a Department of Agriculture inspection would be required to transport plants between the Islands.[44] But recent years have seen organisms freely crisscrossing the chain—often under taxonomic names favored by Native Hawaiians. "Ironically," Archer told me, "the encouragement of people to grow native plants has created a mechanism for bringing in alien species." He continued:

> Until the later part of the twentieth century, State policy discouraged people from growing or transporting native plants. This policy on the one hand promoted the widespread use of alien species within the islands but on the other kept the local island gene pools relatively pure. Relaxing of these policies over the past decade has aided in the survival of rare species but has worked to break down the unique island gene pools on the other. There is also a growing trend towards restoration of the natural environment, or "attempts to return to the old Hawai'i."

He saw this as a more general issue in indigenous epistemology:

> Most aboriginal cultures believe that they are in tune with the land. These peoples have generally remained in one place for a long time; they believe that their lives are in balance with their environment. This balance is a matter of human survival. However, while it may appear true on a generational scale, they are in fact moving *with* the very gradual changes they make within the landscape. Introduced species that become invasive species tend to be landscape-altering, which goes against the idea of a culture living in harmony with the environment. So, for example, if you now suggest that an introduced plant is invasive, it may be considered offensive and insulting to those cultures.

Recounting a meeting with taro farmers, Archer remembered, "They were saying 'We have to restore Hawai'i to what it was.' There is now a period of 'let's try and recreate what we lost.' So, there is a lot of speculation and invention going on. A lot of wishful thinking is involved in recreating this past." People were thinking of introduction ahistorically.

So, how do these scientists think, about "natives"? For McClatchey, plants and animals can be native if introduced by native peoples; there is a relevant difference in kind here. For people like Archer, native plants and animals belong to the realm of nature, a space outside human agency. On this view, humans, by virtue of being humans, cannot themselves properly qualify as native, especially on an island. Nativeness for humans becomes a matter of political strategy or a romanticized invention of tradition.[45]

In Archer's view, native peoples adapt to their landscape using practical reason, but this does not render their agency different in kind from other humans'. Archer had keen ideas about how to stabilize "nature" amid native renamings. Speaking of Hawaiian names for plants, he said, "Perceptions change of what cultivars or varieties are, and we need a way to pin them down. Cultivar names are not set down in a Linnaean scientific way. Likewise, common names change from place to place. In the future, we'll want to know that the Hawaiians had *these* species, and we will be able to determine them with *these* DNA fingerprints." Nature, transcribed into the language of genetics, would for Archer provide the basis for discerning which creatures were native. To borrow a phrase from philosopher Hans Reichenbach, we might say Archer appealed to the genetic assay as an authoritative "context of justification."[46] Archer found any other context irrelevant and focused on the text of DNA as the code that could adjudicate how nature sorted itself out.

We might take the opposite view: invasive species undo our concepts of the natural itself—just as lateral gene transfer may disturb concepts of natural classification flowing from the organization of species trees around genes. A contributor to *Invasive Species in a Changing World* writes: "The general global picture is, then, one of tremendous mixing of species with unpredictable long-term results. While many introduced species have special cultivation requirements that restrict their spread, many other species are finding appropriate conditions in their new homes, while many more may invade their new habitats and constantly extend their distribution, thereby representing a potential threat to local species. All of this calls into question the concept of 'naturalness.'"[47]

The scientists of whom I have written so far think about natives in ways that call upon *nature* and *culture* even as they demonstrate the instability of this distinction. In this process of either affirming or denying culture as a conditioning frame for understanding nature, these biologists produce the very idea of context that allows them to parse the world in this way, and, importantly, to see some organisms as native and others as alien. Boundaries between nature and culture are further blurred when ecologists argue that anthropogenic global warming promotes range extensions for marine organisms—something Al Gore called attention to in his 2006 film, *An Inconvenient Truth.*[48] Here, the alien ocean is a side effect of humanity's overreliance on fossil fuel.

Another means of approaching the question of how scientists think about natives would be to ask how scientists think about Captain Cook. Cook marks a difference in either degree or kind in the character of biological introductions to Hawai'i. Those who think Cook's arrival only accelerated the rate of a process already in motion—a difference in degree—will not class canoe species with the natives. Those who think Cook's arrival ushers in a different regime of introduction—a difference in kind—will be more likely to class Polynesian introductions with natives. Much turns on ideas about the history of nature and culture. Devine, in *Alien Invasion*, writes: "By nature's standards, the pace of introduction has become extremely rapid: hundreds and thousands of times faster than is natural. Consider Hawaii, whose isolation made it exceptionally difficult to reach by natural means. Researchers calculate that prior to the arrival of humans, a new species arrived on this chain of islands about once every 70,000 years. Today, despite efforts to keep non-natives out, a new exotic becomes established every 18 days. That's about a million and a half times faster than the natural rate."[49] Posed in these terms, the question becomes one of a threshold rate at which nature is left behind. Is it after European sailing ships? Steam? Airplanes? The opening of Western trade with China in the 1970s?[50] GATT? The question is about scale and point of view—about context.[51]

A story about navigator Nainoa Thompson illustrates how such contexts condition what count as alien oceans:

> In mid-April 1995, the traditional Polynesian sailing canoe, *Hokule'a*, and its five sister canoes had set sail for Hawaii from French Polynesia. The crew members were retracing the migration route that some of their Polynesian ancestors had taken many centuries before in similar vessels. But just a day short of completing their 6,000-mile voyage, the canoes stopped. . . . Crew members had been bitten by midges. After radioing to Hawaii for advice and consulting with his crew, the *Hokule'a*'s master navigator, Nainoa Thompson, and the captains of the other canoes made the decision to come to about 200 miles from the Big Island of Hawaii. They halted at such a distance to ensure that the wind couldn't possibly blow any stray midges to land; then crew members gathered every bit of organic matter on the boats that could harbor midge larvae and heaved it overboard. Meanwhile a Coast Guard plane flew out from Hawaii and airdropped canisters containing cans of aerosol insecticides. Crew members sprayed their big, double-hulled

canoes inside and out. . . . Outside the port, the voyagers stopped again. A fumigation team met the boats and tented them. Only then did *Hokule'a* enter the harbor. . . . The object of all these precautions was a gnatlike insect almost too tiny to see: a midge called a punkie. These wee beasties had stowed away on the sailing canoes in the Marquesas, where the punkies have tormented people for decades.[52]

In this episode, we see a fluid stratigraphy of time, nature, and culture. *Hokule'a* reenacts a historic voyage using contemporary technologies to recreate the floating ecosystems of ancient canoes. Rather than see this as generative of contradictions between nativeness and the exigencies of the contemporary, we could think about how tradition travels through time; "reinvention" is not the only term to describe what is happening.[53] As educator Manulani Meyer writes, "Hawaiian epistemology is a long-term idea that is both ancient and modern, central and marginalized. . . . It shifts, it is metamorphosed, it is changed by time and influence."[54] It is this flexible view of time that suffuses the work of the next set of scientists of whom I write, for whom context is not always an a priori condition but might rather be seen as an outcome of classifying practice and politics.

ALIEN ALGAE

Many haole scientists I interviewed were cautious when speaking about native affairs they saw as delicate and controversial, even as they ventured informed opinions about how the politics of local nature and culture entered into metataxonomic issues. I became curious to think about these issues in a domain where Native Hawaiian practice might be more directly entangled with science. I found that zone in discussions of alien algae.

Rebecca, an undergraduate working at the Waikīkī Aquarium, monitors the growth of invasive algae off Waikīkī beach.[55] Having met Rebecca on the HOT cruise described in chapter 3, I took her up on her offer to lead me on an underwater expedition looking for alien algae. After we donned snorkels and waded into Waikīkī's waters, she pulled out of the ocean a sample invasive species, *Gracilaria salicornia*, a knobby-tipped macroalga recognizable by its orange-brown color and clump-forming morphology. This was a species native to the Philippines, introduced to Waikīkī in the 1970s by a university botanist experimenting with aquaculture of this species to harvest carrageenan. The plan was to promote *G. salicornia* in the South Pacific, where it might be grown as a profitable crop. But although the *Gracilaria* experiment was started, it was never finished; the

FIGURE 20.   Logo for Waikīkī alien algae
cleanup event.

algae were abandoned in the water. Over the years, the seaweed grew and
formed tumbleweeds that traveled to various places around Oʻahu.

Rebecca pointed out a green filamentous alga that looked like a tangle of
fettuccini, *Ulva reticulata;* this species was growing abundantly, attached
to *G. salicornia.* As we glided over *Gracilaria*-covered coral, the water
began to look murky. Surfacing, Rebecca pulled up more alien species,
including *Acanthophora spicifera,* a spiny seaweed that arrived on the hull
of a barge from Guam in 1950. Algal identification, she said, is mostly done
phenotypically, in the water, although other algae are difficult to identify
while wet. Some algae can be identified only on land, under a microscope.
As we proceeded farther from shore, we saw occasional living coral, though
more impressive was a sense that we were swimming through a vast coral
graveyard, a landscape of sculptural forms flowing into one another like
porous bones, written over by the agency of algae.

Back on shore, Rebecca told me about public events aimed at cleaning up
alien algae (see figure 20). Scuba volunteers begin by uprooting algae.
Snorkellers yank it up and place it on rafts. On the beach, volunteers pull it
out in bucket brigades. The process can involve a hundred people. Flyers
advertising these events pitch them as community gatherings. In these
cleanups, researchers find native species alongside alien; one in particular—
Gracilaria coronopifolia—is used in poke, a popular marinated raw fish
and seaweed dish. Rebecca told me, "When we do the alien algae cleanups
in front of the Waikīkī Aquarium, this is the primary species that we
retrieve and return to the water so it can continue to grow." Rebecca's
words hint at how algae are linked to local practice.

It is in practice that Linnaean and local taxonomies interact. I read about
Hawaiian taxonomic systems in a book by the world expert on Hawaiian

algae, Isabella Abbott, a Chinese Hawaiian phycologist and ethnobotanist. Stating that "[all edible] algae had Hawaiian names before Western science started to describe and name them,"[56] Abbott reports that, about a century ago, "Minnie Reed, a science teacher at the Kamehameha Schools, made a significant contribution to the knowledge of Hawaiian *limu* by emphasizing taxa used for food by Hawaiians. . . . For all these taxa, she listed Hawaiian and scientific names. . . . The listing of Hawaiian names was particularly important, for at that time the students in local public schools were sometimes punished for speaking their native language."[57]

Abbott offers a historical accounting of classifications of limu: "The word *limu* has come to mean 'edible algae,' although the original meaning, according to Hawaiian dictionaries, applies to edible and nonedible organisms growing in wet or damp places with, generally speaking, relatively simple structures. Aside from the edible algae, these include animals such as the toxic soft coral *Palythoa*, known as *limu-make-o-Hāna* (the deadly *limu* of Hāna), all jellylike lichens and fungi, and certain liverworts."[58]

In spite of this transformation in meaning, Abbott points out that native names for limu have been stable compared to scientific ones: "It is ironic that, through oral traditions, the Hawaiian names have been perpetuated and usually *accurately applied* to the individual species, whereas three-fourths of the scientific names have been changed in the 90 years since Reed published them."[59] This irony was not lost on me after Archer had told me that DNA fingerprints would finally "pin down" identities of Hawaiian plants.

When I visited Waikīkī Aquarium, I heard Abbott's observation repeated by resident scientist Cindy Hunter; scientific names are constantly changing as scientists veer from morphological to reproductive characteristics to genetics and back. Hunter said one must always think about alien species as case specific—and even within the same case definitions of good or bad might differ markedly.[60] As one of the organizers of the alien algae cleanups, Hunter had an astute sense of the contingency of how aliens are evaluated; again, much depended on context. One of the most publicized fights against alien algae took place in an artificial lake on O'ahu and was aimed at protecting introduced sport fish. Some introduced species have appealing applications for Native Hawaiians; Hunter reported that one invasive species dredged up during cleanups was favored by taro farmers for use as a fertilizer.

Hunter invited me to join an Aquarium workshop aimed at introducing high school teachers to a culturally informed science curriculum about algae. Addressing the gathering, Hunter explained, "The point is to get the word out about aliens, but you need to know the natives first." An education professor I sat next to, Pauline Chinn, told me we were going to hear about a "Trojan horse for getting Hawaiian knowledges into the classroom." The Hawaiian renaissance of the past couple decades—which has drawn attention to the political economic and colonial plight of Native Hawaiians (known increasingly by the name Kanaka Maoli in some corners of the sovereignty movement) and which awakened interest in indigenous knowledge systems—would now come into science education.[61] Chinn told me that Hawaiian knowledge about limu had historically been women's lore, passed from mother to daughter. Later, I found this account:

> Many plants and animals were forbidden to women to touch (taro, *Colocasia esculenta*, the body form of the god Kāne) or eat (taro, coconut, shark, turtle, whale, dolphin, and other foods). It has been suggested that, except for big fish and large mammals, anything in the sea was *noa*, or free from *kapu*, so women became the specialists of *limu* (algae) and small invertebrates, which they came to name, recognize and harvest. When native Hawaiians were asked for native names of *limu*, men told the interviewers to ask women, because they were the ones who recognized the species and could give the names for them.[62]

Retrieving this knowledge would not be only of interest to native politics, Chinn said, but could also be a feminist issue.

Hunter spoke about joining "indigenous knowledge with Western science so that we can *malama* [care for] our natural resources." About thirty high school teachers were in attendance, most in their thirties or forties and most women. The room held a mix of haoles, some people of Asian ancestry who grew up in the archipelago,[63] and a few Native Hawaiians. We were given plastic bags and sent to the ocean in front of the Aquarium, where I had snorkeled with Rebecca, to gather limu. Participants kicked off their sandals and waded. Some searched by color, others by shape. A few spoke together in Hawaiian.

Back in the classroom, a science instructor (a self-described "limu lady") led us through a curriculum called "Investigating Limu." For the first unit, "Looking at Limu," we were divided into four groups. Arrayed around the room were twelve bowls, each containing a different limu. We were to match descriptive cards with each sample. Each card listed a Hawaiian name, a scientific name, the habitat of the limu, and a clue about what it should look like. Someone exclaimed, noting that one card lacked a Hawaiian identifier, "Does it have a Hawaiian name?" "No, it's alien!" Although I did not see anyone put the limu in their mouths, we were told, "Some clues are in taste."[64]

Participants linked Hawaiian limu names to cultural uses—in eating, in medicine, in family activities. People familiar with these smiled when they identified their favorite limu and shook their heads when they came upon alien algae, often repeating the story of the aquaculture experiment gone wrong. One Native Hawaiian teacher, identifying *Kappaphycus*, recognized it as an introduced alga that had caused distress to Kāne'ohe Bay, a site of several significant traditional Hawaiian fishponds, and expressed incredulity that haole biologists had been arrogant enough to experiment with this plant. Participants slid between taxonomies, thinking through their experiences, positive, negative, using both Linnaean and Hawaiian names. In no case did people identify plants themselves as agents of invasion; attention was always on those whose ignorance or hubris had been the impetus for travel.

When, sometime later, I had lunch with Isabella Abbott, I got a new angle on these matters. Abbott frowned when I asked about alien species. Even though in her long career (she was 83 when I met her) she had written an article titled "There Are Aliens among the Algae, too—or Limu malihini," she was impatient with language of alienness; "there is so much work to do on indigenous species."[65] She alone had named six new species in the past year. There are more data about Hawaiian marine flora to examine before jumping to conclusions about origins. Sampling is incomplete. Some places are hard to get to for humans, though easy for algae. "Since 1997," Abbott told me, "I've received collections from the Northwest Hawaiian Islands." Someone from the National Marine Fisheries was picking up lobster traps and thought Abbott would be interested in classifying seaweed on them. She said, "I have found extended distributions of these seaweeds; I have found that they are more widespread—like from Hawai'i to Australia—and they may be connected not by currents [specific to hemispheres] but by eddies [which can cross hemispheres]. I'd rather find *that* than find a new species. It makes this big world smaller. And it means that we phycologists have to get along. So, alien species? I'd prefer to be quiet until I have more information."

Twenty-five years earlier, at meetings of the Hawaiian Botanical Society, Abbott had objected to the word *alien*. "It reminded me of all these aliens from outer space!" The fellow who introduced the word said, "That's why I like it!" He wanted people to be concerned. But Abbott felt the term assumed phycologists already knew everything about the distribution of plants and were now risking transposing their own values (and ignorance) onto the nature of plants themselves.

So, how do *these* people, these scientists and teachers interested in algae, think about natives (and aliens)? They think of limu as bound up with native ways of life and with the Hawaiian renaissance. And these ways are dynamic, not a static tradition to be merely retrieved or reinvented. Returning to McClatchey, who argues that a "natural inheritance" can be "developed over . . . history," we could think of relations with "native" algae, in the words of Meyer, as "metamorphosed, . . . changed by time and influence."[66] Such a view, in which context is produced rather than posited, is in tension with what Archer argued to me about algal systematics: "The memories of what organisms really are is fragile. Native Hawaiians used the same name for vastly different species because they based names on texture, on taste. We want to give these things names and forensics can help. DNA is a real good tool to sort these things out." With molecular data, he said, scientists sequence genes and "find out where things originally came from." But this origin would be one grounded in instrumentally partitioning nature and culture, assigning origins only to a nature that does not include humans. In the limu lesson, humans are crucial players in creating the taxonomies that make limu as such appear. Nature and culture, life forms and forms of life, exist in tumultuous relation.

THE ALIEN VS. THE AMPHIDROMOUS

Bishop Museum biologist Ron Englund directed me to someone who might speak for a constituency interested in the fate and uses of Hawaiian marine creatures—Kaipo Faris, a Native Hawaiian instrumental in reviving taro farming on the windward side of Oʻahu. The town where he lives, Waiahole, has been at the center of the "Waiahole Ditch fight," a dispute in motion since the 1990s about the flow of water from the Koʻolau mountain range through Waiahole stream. Waiahole stream historically traveled down the windward side of Oʻahu, but, with the rise of sugar plantations in 1916, in what was heralded in the American press as a feat of engineering but regarded by taro farmers as a theft of their waters, the stream was diverted by a system of ditches to the leeward side of the island to irrigate commercial crops. This diversion deprived windward farmers of water they had used for taro fields. Before diversion, 30 million gallons flowed down the mountain a day; eighty years later, this had slowed to 3 million. In the 1990s, when plantations began to close (recall the Hawaiian Sugar Planters Association building from chapter 3, with a new name and new tenants), a movement arose to channel water back to earlier paths.

People in the Waiahole water movement did historical and legal work to find out where taro used to grow and how streams once flowed, while the corporate interests of the Robinson Estate, Campbell Estate, and American

Factors fought these activists, trying to keep water moving leeward toward tourism, development, and golf courses.[67] In 2000, the Hawai'i Supreme Court decided in favor of the call to revert Waiahole water back to its historical path. Earth Justice summarizes the decision: "The Court vacated permits the Commission had issued to leeward interests and directed that the Commission reevaluate the level of flow that must remain in windward streams. . . . Healthy stream flow is needed for 'instream uses'—to restore native stream life, such as 'o'opu and hīhīwai; protect traditional and customary Native Hawaiian gathering rights; support the productivity of the Kāne'ohe Bay estuary; and preserve traditional small family farming, including taro farming."[68]

Faris became involved in the Waiahole project in the 1990s. Becoming interested in taro farming led him to look at streams that might irrigate the fields and, in that connection, at creatures living in streams. When I visited him at his home, he introduced me to a key creature, the limpet, which comes in three varieties: hīhīwai, hapawai, and pīpīwai. These are molluscs that dwell in different portions of a stream; *hapawai* means "half water," indicating that it lives halfway upstream. These organisms, he told me, as we walked to the stream behind his house through a garden of native plants, are amphidromous—creatures born in fresh water that in their larval stage go out to sea, becoming part of the plankton, and then return to fresh water for the remainder of their lives. In their early days, they look like little jellyfish. When they circuit back to fresh water they metamorphose, forming a shell, and crawl upstream. These creatures living in Hawai'i's streams and oceans have been endangered because their habitat has been reduced (they like swiftly flowing waters, not trickles, in freshwater streams). Faris found several hīhīwai on the island of Moloka'i and brought them back to O'ahu to place in Waiahole stream, as an organismic instrument to measure the health of the water—to provide evidence that streams were no longer healthy in the wake of massive water diversion.[69] As we stood by the stream, he said, "I'm not a scientist, but I grew up with my grandparents and they taught me that the best indicator of a healthy stream is the native life that is in it."

Corporate lawsuits tried to block Faris's restorative placement of the hīhīwai into the stream, arguing that he was aiding an "alien introduction." Faris maintained that these organisms were endemic to the Hawaiian Archipelago, and that this should count rather as a "reintroduction." After research into how he might prove this claim, he sent samples to Louisiana State University for genetic studies, undertaken in January 1999 by biology professor John Michael Fitzsimmons and graduate student Melanie

Bebler.[70] As Faris informed me, these tests indicated that there were no significant genetic differences between Moloka'i and O'ahu limpets.

Thus, Faris called upon genetics to underwrite a project connected to imperatives of Native Hawaiians; unlike the molecular biologist arguing that Hawaiian names should be replaced with scientific ones based on DNA, we find here an activist using both classificatory systems in concert. Faris's was a use of genetics to wage a claim about the indigeneity of marine creatures, a language the courts, fond of DNA evidence, could understand.

Implicit in Faris's work is an argument that Island waters are part of Hawai'i and can be shown to be such through delineation of a biogenetic kin network for limpets. We can detect here an archipelagic imagination mobilized in the service of a project of indigenous resistance—and a project of amplifying what counts as endemic.[71] The siting of nativeness in a network of fresh and salt water offers an articulation of nativeness that shifts attention from land to the fluid domain of the sea.[72] Terms of debate about nativeness change when transported into water. Archer's complaints about the pliability of "native tradition" do not acknowledge that something can be both native and mobile, as we see in this instance of a metamorphosing mollusc.[73] In the Pacific, activist links between native people and water may be politically easier to secure than landed claims; native territorializations of the sea may work in part because the ocean has been historically regarded as aterritorial by the European powers that colonized this part of the world, as Epeli Hau'ofa, of the University of the South Pacific in Fiji, has argued in his call to reconceptualize what Europeans have called "Oceania" as "our sea of islands."[74] It is no coincidence that Native Hawaiian nationalism has the voyaging canoe as one of its symbols.

FIGURE/GROUND/LIQUID

How do scientists think—about "aliens," for example? Much relies on their view of nature, culture, and agency. For many scientists, humans are either distinct from nature by virtue of culture or are all equally part of nature. Others have a more fluid account of how nature and culture are entangled with organic and social history. The evolutionary biologists I interviewed for the first portion of this chapter viewed the question of native and alien as a metataxonomic question—an issue about categorical frames transcending existing classifications. Scientists holding to this view enumerate properties characterizing invasive species, hoping to work backwards from cryptogenic creatures to stable evolutionary moorings. At times, though, they say the issue cannot be so neatly adjudicated; what is or is not invasive—and, indeed, alien—depends on context. Scientists like Abbott and those leading the limu lesson (along with people like Faris), meanwhile, frame the issue as one of what I call parataxonomies—setting scientific and Hawaiian taxonomies alongside, athwart, one another, sometimes allowing for translation of one form into the other, but not always. We might think of the parallel classifications offered in the limu lesson as animated by parataxis: "the placing of propositions or clauses one after another, without indicating by connecting words the relation (of coordination or subordination) between them" (OED). If we think of a taxonomy as a proposition— "something put forward as a scheme" (OED)—parataxonomies become paratactical principles of classification that can coexist without necessarily being squared with one another.

It should be clear that the metataxonomic conundrums faced by invasion biologists are also, in their way, parataxonomic (and, etymologically, since *meta-* means "with" or "after" rather than "above," this is not surprising).[75] On this view, *context* might be reimagined not so much as "the parts which immediately precede or follow any particular passage or 'text' and determine its meaning" but rather, in tune with an obsolete but useful meaning, as a "weaving together" (OED), a weaving that happens contingently, not deterministically—as when "alien algae," for example, are written into native farming, or genetics is stitched symbiopolitically into amphidromous belonging. I offer parataxonomic thinking as a conceptual instrument to show how nature/culture, native/alien, science/politics always shape one another.

And so, the relation between *alien* and *native* is one that depends on how parameters are tuned to create contexts that ground the difference. It summons up questions of the perceptual frames we employ to decide what will count as foreground and background. Inspired by shape-shifting alien aquatic species, like limpets, I suggest that we are presented here with a question of figure and ground in which the ground is liquefying: figure/ground/liquid. Percolating up into the figure of the alien species, this liquid disturbs taxonomy. Like the lateral gene transfer examined in chapter 2, the flowing territories of amphidromously inhabited waterscapes call attention to the fluidity of our foundational categories. Whether such received classificatory systems as Linnaean nomenclature or trees of life are simply broken down to be built up again—in other languages, at other scales, across novel territories—or radically rerouted depends on who is doing the classifying and with what means. To swim in the turbulent waters of alien oceans is to put at risk any unitary account of the nature of the sea itself. It is to acknowledge the location of scientific signification in the shifting streams, pressures, and politics of practice at all scales, from the grandest to the teensiest, from the mobile native to the unevenly global. If the alien ocean is a space of organisms-out-of-place, *place* is a matter of how *space* is imagined. There exist a range of alien oceans.

5 Abducting the Atlantic

How the Ocean Got Its Genome

I am listening to J. Craig Venter lecture to an MIT audience in March 2004 about his scheme to sequence the microbial DNA of the world's oceans, "categorizing the earth's gene pool." Venter, the man whose company, Celera Genomics, was credited in 2000 with speeding the sequencing of the human genome, has started small, embarking on a modest research craft—his own 95-foot yacht, the *Sorcerer II*—to gather the genes of microbes in just one environment. The scale of his first sampling effort, however, speaks to the scope of Venter's ambition; his lecture today reports on "Sequencing the Sargasso Sea."[1]

At each of four sites in the Sargasso, off Bermuda, Venter and his team collected 200 liters of seawater from which they extracted microbial biomass. DNA libraries prepared from this material were sequenced at the J. Craig Venter Science Foundation in Rockville, Maryland, an organization Venter founded with monies from his success as head of Celera. Venter explains that the sequencing of this DNA has yielded from Sargasso waters 1.045 billion base pairs, 1.2 million previously unknown genes, and at least 1,800 "genomic species" or "phylotypes." Many sequences belong to the most abundant photosynthetic marine bacterium in the world, *Prochlorococcus*, discovered in 1988 by MIT biologist Penny Chisholm, sitting now in Venter's audience.[2]

Venter has deposited his bounty of data online at GenBank, a public database of DNA sequences hosted by the National Institutes of Health, so that marine microbiologists might double-click for matches with their favorite organisms. Venter is eager to recruit his listeners into a project of immense, eventually global, reach—into making sense of what all this code can say about elementary life forms in the sea. He is plainly taken with the scale of his undertaking, learning about a vast region of ocean from a few data points. To underscore his universe-in-a-grain-of-sand vision, Venter offers a quotation from Maronite poet Khalil Gibran: "In one drop of water are found all the secrets of all the oceans." Some months later, a popular article in *Wired* magazine summarizes Venter's grand plan: "To sequence the genome of Mother Earth."[3] Venter later explains his vision this way: "By sequencing multiple sites we might be able to compile an actual sequence database of the ocean's genome."[4]

In this chapter, I tell the tale of How the Ocean Got Its Genome, how swaths of sea came to be imagined as meaningfully summed up by the genes of resident microbial communities. Just as in Rudyard Kipling's *Just So Stories*, in which readers learn, for example, How the Leopard Got His Spots, this story tracks leaps of logic and sometimes-wishful reasoning. Formatting much of Venter's tale is what geographer Erik Swyngedouw calls a *scalar narrative,* an account that links entities and events across scales in order to make claims about causality, consequence, and conditioning contexts.[5] Microbes rise to the status of ecological representatives when their disembodied genes are treated as proxies for the environmental action of microbial communities. When microbial sequences are placed into abstract spaces like GenBank, the sea acquires an informatic body double: a genome.

A distinctive logical operation animates this scaling up from genes to sea. When marine microbial genetic sequences are uploaded into cyberspace—transformed into data that can be transferred and recombined ad infinitum in digital format—the ocean is enlisted, I claim, into a process semiotician Charles Peirce called *abduction.* Whereas induction is reasoning by inference from particulars toward general conclusions, and deduction the drawing of conclusions from known principles or theories, abduction is reasoning from premises that *may* materialize in the future. Venter's taking of digital snapshots of marine microbial genomes rests upon the promise of yet-to-be-invented techniques and theories for making sense of such information. Abduction, as Peirce writes, constitutes "a method of forming a general prediction without any positive assurance that it will succeed either in the special case or usually, its justification being that it is the only possible hope of regulating our future conduct rationally."[6] Abduction, in other words, joins hope to reason, present texts to future contexts, contemporary life forms to scientific forms of life yet to come.

There is more than one way to abduct one's data. Writing of human groups, Benedict Anderson argues that "communities are to be distinguished, not by their falsity/genuineness, but by the style in which they are imagined."[7] So, too, with microbial communities, which can be distinguished by the style in which they are abducted. Venter's uploading of the sea leaps from gene to globe, peeling marine microbiological information away from located ecosystems, hoping that others will do the work of placing sequence data in context.

There are other modes through which marine microbiologists gaze up to the blue-green globe to locate the ultimate referent of their work. To place Venter's vision in perspective, I report on a voyage to the Sargasso I joined with biologists trained by Penny Chisholm, scientists keen to embed their data in such material circumstances as ocean currents, seasonal weather, and time of day. As they sample, sort, and interpret drops of water, these scientists combine genomic approaches with curiosity about environmental presents and futures. They use genomics but do not claim to be sequencing the ocean's genome.

To see Venter as well as his critics in another light, this chapter concludes with an account of a visit I made to a microbial observatory on Sapelo Island, one of the Sea Islands off the coast of Georgia, a site where people descended from Africans enslaved on nineteenth-century American plantations coexist uneasily alongside marine biologists. Here, uploading salt marsh microbial genomes may reinforce and reorganize on-the-ground social geographies, including the racial divisions that mark the topography of the Sea Islands, divisions that fashion Sea Island residents as "local" while visiting scientists and their data move within the circuits of "global" science. Both *local* and *global*, I show, are effects—not preconditions—of how such genome science is narrated. And, indeed, the uploaded, abducted ocean can be downloaded into empirical and theoretical activities that absorb scientists in the intricacies of putatively local affairs. This chapter plays with the idea that these are alien abductions, forward-looking modes of reasoning detoured by elements that come out of the blue. Here, the more common meaning of *abduction*—an unexpected capture against one's will—haunts the logical meaning.

During his talk, Venter tells us that he has already embarked on the next stage of his plan, to sail around the world conducting an "Ocean Microbial Genome Survey." He has modeled his circumnavigation on the journey of the HMS *Challenger* and on Darwin's voyage on the HMS *Beagle*, and, as befits this second comparison, his presentation features slides of *Sorcerer II* in the Galápagos Islands. Venter even throws in snapshots of finches, favorite birds of Darwin, commenting that the mutations behind beak sizes will become legible through sequencing. Venter has himself undergone something of a mutation since his days as Celera

FIGURE 21. Venter on *Wired*. Photograph by Ian
White, design by *Wired*, ©2004 Condé Nast
Publications. Reproduced with permission.

president. In the 1990s, he stirred the ire of researchers in the U.S.-gov-
ernment effort to map the human genome. By taking on the entire
genome himself, he upset the collaborative ethos endorsed (if not always
perfectly enacted) across labs mapping different chromosomes. More
galling to most scientists, though, he sought to secure Celera patents for
segments of DNA. His ocean project means to be different; all data go to
GenBank, and no sequences will be patented. Even if he is still ambitious,
Venter's code of conduct has changed. His dress code, too. He appears
before us not in the business gear that once marked his public persona but
in the casual wear of a busy scientist. He has also grown a tidy beard,
which, combined with his baldness, gives him the aspect of a trimmed
Darwin or a no-nonsense sea captain. The look makes flashy press; *Wired*
puts Venter on its cover later in 2004 (figure 21). The magazine's
"Fantastic Voyage" framing recalls the B-movie of the same name, in
which scientists shrink down to enter a human body. *Sorcerer II* scales the
ocean's body, sequencing its DNA as it sails along.

Venter winds up his talk with a line calculated to generate envy: "I leave in two days for French Polynesia, and as I've said in other places"—he quotes himself from a *New York Times* profile—"it's tough duty."[8] Venter's initiation of a venture growing from his personal curiosity is reminiscent of researches undertaken by Prince Albert I of Monaco, who in 1891 also used his own boat to set out on a career of oceanographic exploration. Albert, whom historian Jacqueline Carpine-Lancre has called an "oceanographic sovereign," similarly hired collaborators to help him realize his vision (including a pair who built "a box for microbes").[9] Venter as oceanographic CEO exits the MIT stage with a flourish: "We hope to leave exciting new knowledge in our wake."

Venter's words whirl into conversation among participants at this conference sponsored by MIT's Earth System Initiative. Venter is not trained as a marine scientist, and various coffee-drinking clusters complain that his approach leaves out essential contexts for understanding microbial sequence data. He has not controlled for seasonality. He has not always taken measurements of sea temperature and salinity. He has ignored how ocean currents complicate the representativeness of samples, whose locality is not the same as their coordinates at any one time. The discussion is reminiscent of 1990s disputes over how to gather genetic data for the ill-fated Human Genome Diversity Project (HGDP). In that debate, one side advocated a random sampling of people worldwide while the other hoped to zero in on specific populations. Venter might be seen as following something like the first approach, with opponents desiring historically anchored sampling. A political comparison suggests itself, too: scientists of the HGDP got in hot water for their naïveté about the politics of taking blood from indigenous people, injudiciously designated as "isolates of historic interest."[10] Venter will get into similar trouble trolling without license for marine microbes in the national waters of Ecuador and French Polynesia.

Most bothersome for conferees is their suspicion that because *Sorcerer II* is not a full-fledged research vessel Venter's samples may be contaminated by effluent from his boat, such as sewage containing *E. coli*. Less delicately, a few graduate students joke, Venter's genomes may be full of shit. Or, as a response to Venter's research publication in *Science* phrases it a few weeks later, "marine microbes associated with organic particles, dead bodies, zooplankton feces, etc., can create hotspots of bacterial growth that bias estimations of diversity."[11] But Venter's talk spawns not only complaint but also grudging gratitude for his bringing media and funding attention to marine microbial biology. Even if marine microbiological research has been under way for a while, Venter has created an inescapable frame for future study. What will it mean to follow in Venter's wake?

I am fortunate to find out. Through connections I have developed with Chisholm since joining MIT, I sign on three months after Venter's talk to an oceanographic cruise headed for the Sargasso Sea. Chisholm needs samples for work on *Prochlorococcus,* and one of her former postdocs, now a professor at the University of Georgia, has organized an expedition to Bermuda. I agree to do some sampling for Chisholm's lab—of seawater containing marine viruses that infect *Prochlorococcus.* The trip on R/V *Endeavor* will also take me to a site famous in modern folklore for its association with alien abduction: the Bermuda Triangle, a zone where ships and planes are reputed to have disappeared. These travels triangulate me back to another outpost of the University of Georgia, the Sapelo Island Microbial Observatory. Together, these trips help me make sense of the cultural and scientific geometries of the microbial, genomic, bioinformatic, and uploaded Atlantic.

BACTERIA ON BOARD

As *Endeavor* cruise no. 393 steams out of its home port in Narragansett, Rhode Island, on a three-week journey to Bermuda and back, I sit on deck reading the NSF grant funding the trip's project, "*In Situ* Pico-Cyanobacterial Growth Rates in the Sargasso Sea Based on Cell-Specific rRNA Measurements."[12] The grant promises to investigate how quickly some tiny light-eating microbes called *Prochlorococcus* and *Synechococcus* reproduce in the Sargasso. Such phototophic critters—"primary producers" at the base of the food chain—use solar energy to make carbon copies of themselves, providing a steady stream of nutrition up a narrowing pyramid of bigger creatures, from ciliates to cetaceans. Tracking the growth of bacterial populations by looking at their rRNA might yield an estimate of the speed at which light gets turned into life. These microbes provide a contrast to deep-sea hyperthermophiles; they are imagined as beaming with vitality rather than dwelling in mystery and shadow, even as I find that here, too, the apparition of the alien makes an occasional appearance—as the viral, the just out of view, the sensorially perplexing. Here, the alien ocean is a space of invisible infection, a subaqueous, subvisible realm floating just beneath the scale at which human senses operate.

Berthed on board the 185-foot vessel are a crew of fourteen and a science party of ten: two marine biology professors, a lab technician, two graduate students, four undergrads, and me. The marine researchers are based in Georgia, though the professors—Brian Binder at the University of Georgia in Athens and Liz Mann from the Skidaway Institute of Oceanography in Savannah—trained at MIT and consider themselves northerners. The college students, Georgians all, are here to learn about science at sea and earn class credit. The graduate students are collecting dissertation data.

Prochlorococcus and *Synechococcus* are cyanobacteria, blue-green algae (named for the color of their photosynthesizing pigments, though some employ red or pink), and are ubiquitous in the open ocean's photic zone (the upper 200 meters, where light gets through), particularly in the latitudinal band between 40° N and 40° S. Why travel to the Sargasso if such microbes are so cosmopolitan? *Prochlorococcus* does well in open-ocean, low-nutrient waters; the cell's small size and high surface-to-volume ratio allow it to take in nutrients efficiently, especially in zones where light is not blocked by other organisms, as in richer coastal environs. In earlier days of biological oceanography, scientists believed the Sargasso representative of open-ocean ecologies generally. Many marine science departments arrayed projects around this sea, easy to access from the East Coast. As a result there is a lot of context into which to place research findings— some of which, I have been instructed, Venter would have done well to consult before sampling. Owing to a monthly time-series study begun in 1954 by the Bermuda Biological Station for Research at a location south of Bermuda called Hydrostation S (supplemented in 1988 with another site, BATS, for Bermuda Atlantic Time-Series, a companion to the Pacific-based Hawaii Ocean Time-series, HOT, discussed in chapter 3), there now exists fifty years of such context.[13] On my trip, I will find scientists concerned not only with gathering samples and data but also with calibrating "context"— a stable semantic frame for making meaning out of Sargasso microbes at local and global scales. Collecting, filtering, scanning, preserving, and distributing water samples represent techniques that ready the sea to be uploaded, abducted—to databases, to future theories—broken down to be built up again.

A SCALED-DOWN HISTORY OF LIFE IN THE SARGASSO SEA

The Sargasso Sea is an elliptical swirl of subtropical ocean bounded on its north and west by the Gulf Stream, on its east by the Canary Current, and on its south by the North Equatorial Drift. It is the only sea in the world defined not by enclosing lands but by encircling surface currents. The size of Australia, the Sargasso is noted for calm waters and the *Sargassum* alga that gives the sea its name, a golden seaweed kept afloat by berry-sized air bladders. Portuguese sailors named this plant for its floats, which resembled a grape they called *sargaco*. The sea overlaps the "Horse Latitudes," a windless zone named for the practice Spanish sailors were rumored to have, when becalmed, of slaughtering their horse cargoes for food.[14] But if maritime legend has been fascinated by the Sargasso as a sea of stuck ships—a reputation reinforced by tales of the Bermuda Triangle—marine scientists find this area compelling because the clockwise currents that define the sea isolate it from the encompassing Atlantic. The Sargasso Sea is a massive lens of warm water, 3,000 feet thick, floating atop a colder and deeper ocean that touches bottom 3 miles down. In the Sargasso, warm water spirals toward the center, cutting off nutrient supply from cold upwellings below, corralling creatures closely within the sea's boundaries. In mornings and afternoons, its waters glow a gorgeous azure, empty of the phytoplankton that make coastal waters greener.

These features make the Sargasso a relatively transparent site for research into how algae condition the growth of larger organisms. In 1889, German oceanographer Victor Hensen visited the Sargasso on his Atlantic Plankton Expedition to learn about the distribution of zooplankton and phytoplankton, which he considered "primitive food for marine animals" and from which he hoped to deduce the productivity of world fisheries. Hensen in 1887 had "established the term *plankton* for the floating animals and plants of the sea"[15] and called plankton the "blood of the sea." Some of his disciples took the bodily metaphor further. In 1892, Franz Schütt had written of the sea's "metabolism": "Life in the sea appears to be a superorganism [*grosser Gesammtorganismus*] in which each individual organ has its own unique function, and which furthermore activates, hinders, furthers, regulates all the other organs and influences their functioning, thereby also exerting influence upon the totality of living phenomena, upon the metabolism of the whole [super]organism."[16]

German oceanography fell into decline after World War I, and by the 1920s Britain was the center of biological oceanography and the guiding metaphor for plankton now drew from farming. E. J. Allen envisioned the sea as "a blue pasture for the raising of marketable fishes."[17] The rural comparison would surface later in an American idiom. Marine biologist John Teal wrote in his 1970s diary, "The sea was like a prairie with six-foot hills of slightly uneven blue ground stretching on and on under a bright sun. This prairie quality is one of the most pleasant of Sargasso aspects. I can almost see spring flowers blooming on the swells."[18]

Over the next decades, biological oceanography moved into an inductive, mathematical frame, particularly in the United States, where limnologist Evelyn Hutchinson championed quantitative models. Hutchinson was skeptical of seeing biotic communities as organisms. He was curious about how Darwinian processes like competitive exclusion might work in watery worlds. He was puzzled by what he in 1961 called the "paradox of the plankton," recently glossed as the question of "how so many different kinds of phytoplankton can coexist on only a few potentially limiting resources, when competition theory predicts just one or a few competitive winners."[19] Marine viruses, I would discover, might provide one stimulus for such multiplicity.

Recent years have seen growing interest in ever-smaller planktonic life forms. Chisholm, who received her Ph.D. in 1974, began her career examining the silicate bodies of diatoms under microscopes. Her attention then moved to *Synechococcus* and the tinier *Prochlorococcus*, now understood, with abundances of up to 70,000 cells per drop of seawater, to be perhaps the most plentiful organism on the planet. One sign of the importance of the Sargasso for ever-miniaturizing phytoplankton studies is the prominence in marine microbiology of biotic types first identified in this sea, though found worldwide, such as SAR-11, the first group of uncultured marine microbes discerned through ribosomal RNA sequencing and a bacterium named for its geographic provenance rather than phylogenetic address.[20]

Organisms like *Procholorococcus* came into view only in the 1980s when devices for detecting cells in solution were brought on board ships. During my first days on *Endeavor*, chief scientist Brian Binder told me a tale emblematic of the movement toward tinier subjects for biological oceanography. Brian's first sampling trip in the 1970s saw him snorkeling above the depths of the Sargasso, wielding a syringe into which he hoped to suck colonies of a bacterium called *Trichodesmium*. These assemblages were visible to the unaided eye—barely—as balls of floating fuzz. Brian laughed when he remembered that moment before the plankton he studied winked out of sight into worlds scaling down to the microscopic and molecular.

SAMPLING THE SARGASSO

How is Sargasso water sampled for its microbial ballast and its genomic contents? On the *Endeavor* cruise, the primary sampling device is a circular assembly called a CTD-rosette, cast into water over the side of the boat

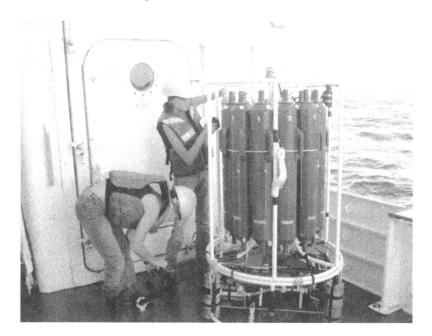

FIGURE 22.   Students on deck gathering water samples from the CTD-rosette. The slack cable rising from the top is used to raise and lower the assembly into the water. Photo by the author.

at various times during our trip. The "CTD" is a package of sensors measuring the conductivity, temperature, and depth of segments of the water column. The "rosette" is a collection of twelve polyvinyl cylinders called Niskin bottles—each a meter tall—arrayed in a ring around the CTD core and held in place by a circular metal armature (figure 22). These cylinders, each with 12-liter capacity, are open at their base but can be sealed shut by remote control—"fired," as scientists say—as the CTD-rosette descends, allowing water to be captured from different depths. As with MBARI's *Ventana*, communication with the device is through a tether outfitted with fiber optics that can unwind thousands of meters. Much of the cruise consists of casts of the CTD-rosette, bringing up water (and microbes) from specified depths.[21] *Endeavor's* CTD is also outfitted to measure chlorophyll, which suggests depths at which photosynthetic plankton tuned to specific light frequencies may reside. Data gathered by the CTD produce ecological context for pronouncements about the sampled microbial sea.

The undergraduates have been enlisted to cast and retrieve this device, a process that involves managing wires when the bulky package is moved about on deck and demands that somebody inside the ship watch video trained on the grappling students. The students try to make the routine work fun, dancing around for the camera while wearing required lifejackets and hardhats. Brian tells students to label their hats, generating jokes about silly nicknames. One student has been reading Norse mythology, inspiring the pillage of Norse names. The Nordic connection strikes me as appropriate for the three college women, all blonde. I have just been handed pseudonyms: Frida, Agnes, Emma. Everyone agrees that my graduate student cabinmate, into heavy metal, should have a "brain bucket" bearing the badass name Bröndólfr, "swordwolf." I'll call him Brandon.

One of the ship's crew, Ben, who has begun regaling us with sea shanties, laments that there are no songs for modern sailors. These days, the sound of work at sea is the noise of mechanical hums (though maritime nostalgia finds a potent icon on this boat: a wooden relic from Captain Cook's *Endeavour* is framed in this *Endeavor*'s lounge, next to a signed photo of the crew of the Space Shuttle *Endeavour*). First mate Fred adds that scientists these days lack the skills they once had. No one knows how to tie nautical knots. The students and I look at each other sheepishly. We write a stupid shanty about the CTD.[22] A day of Sargasso sampling is punctuated by the punchiness and boredom that come from doing a task over and over. During breaks, students sunbathe, read novels, nap. A couple of days on the cruise see the undergraduates casting the CTD hourly— twenty-four hours a day—cataloguing samples from each deployment, generating hundreds of vials of seawater.

This seawater, with its resident microbes, can have a few destinations. One is through a flow cytometer, an instrument that counts cells to deliver a picture of the sizes and sorts of microbes in a sample. Howard Shapiro, an inventor of cytometers, explains the aim in one of the whimsical songs he is known for writing (and, occasionally, at conferences, for singing):

> . . . cells are classified by cell and nuclear shape and size
> And texture, and affinity for different types of dyes,
> And almost all of these parameters can quickly be
> Precisely measured by techniques of flow cytometry.[23]

I watch as Brian hooks test tubes of seawater up to the cytometer, the size of a photocopier. It sucks water into its innards and streams it through a laser beam. Because cells scatter light distinctively depending on size, the cytometer is able to partition size classes, detecting a difference between *Prochlorococcus* and *Synechococcus*.

The abundances of these Lilliputian phytoplankters are often inversely related in geographic, seasonal, and vertical distribution; as one goes up, the other goes down. Within a global frame, their dynamics may be indicators of shifts in climate patterns.[24] *Prochlorococcus* and *Synechococcus* also contain different photosynthetic pigments, which fluoresce uniquely under laser light. Flow cytometers are machines for producing inscriptions to be interpreted according to semantic conventions that bind light to signs of life. Software packages such as CellQuest Pro arrange cytometric data into graphs representing cells as scatters of dots, plots that can then be translated into depth profiles for different sorts of cells.[25] All this data crunching is labor intensive, and so scientists freeze many samples to process on land. The use of flow cytometers for marine microbiology is still being fine-tuned; the technique was developed in medical diagnostics for looking at larger entities, such as blood cells. What Brian terms "homegrown modifications" have been necessary. Being at sea adds the task of keeping the cytometer's laser from wiggling. Still, Liz tells me, flow cytometry has revolutionized biological oceanography. Without it, scientists may never have been able to see seawater swimming with so many cells.

Between CTD casts, undergraduates preserve samples for shore work. When I join the students at the bench, they protest that I will not learn much from them because they do not really know what they are doing. "But you're *learning*," I say, putting on my best professorial tone. "What are you learning?"

Frida starts, "How to do stuff, but we still don't really understand why we're doing these things. We have to figure it all out when we get to land. But, you know, that's kind of why I like it—it's like a mystery."

"But," says Emma, "it makes me suspicious that people can just keep running their data till they get what they want!"

"But if you don't get what you want, even that's interesting," offers Frida.

They are still puzzling through not just techniques but how to think about the tangle of deduction, induction, and abduction that shapes interpretations of the samples they gather.

The media ecology of CTD-rosettes, flow cytometers, and humans delivers a portrait of the sea scientists on this trip believe offers a context that Venter's genetic snapshot lacks. Such context frames the "text" of flow cytometry and later, of sequences. It also anticipates the kinds of explanations that will be persuasive to the research community. Brian has a related observation about the molecular techniques brought to bear on microbial marine life, techniques he thinks may be increasingly used by the generation of students he is training. He tells me,

> The new molecular techniques that have become available have shaped the questions people ask, which is only natural. But, while molecular techniques can give you lots of answers—*good* answers—these are *not* always the answers to the questions that are most interesting. But don't get me wrong; people *are* answering ecological questions with these techniques. We're trying to on this cruise. It's funny, because in a way, we've returned to the old days of natural history. People are not just using these tools for experimental questions but are also using them simply to *examine* the environment, saying, "Look, there are all these genes floating out there."

In other words, hypothesis testing is often put on hold with new techniques, though our work on this cruise has been organized partially around Brian's hypothesis about the relation between growth rates of *Prochlorococcus* and *Synechococcus*, now addressed—this is important—with reference to "all these genes floating out there." Brian tells me that environmental genomics is taking off in marine biology because it is easier to get genes out of the sea than out of soil: "They're already in solution." If the sea is becoming a genomic body, it is also already a test tube. In a way, microbes in nature are already "in culture."

SCALING UP OCEANIC LIFE

If "life" in the sea these days is scaling down to the microbial, it is scaling up, too. Shortly before the *Endeavor* cruise, Chisholm gave a talk at MIT titled "*Prochlorococcus:* How to Dominate the Oceans with 2000 Genes," emphasizing links between the genomic and global and giving her take on the recent sequencing of *Prochlorococcus:* "I consider this the minimal life form—having the smallest number of genes that can make life from light and only inorganic compounds. It is the essence of life."[26] After explaining the place of *Prochlorococcus* in Earth's "biological pump"—it is responsible for 25–58 percent of chlorophyll production in the North Atlantic, for which the Sargasso serves as a microcosm—Chisholm concluded by offering that *Prochlorococcus* should guide biologists to "think of life as something with properties similar at all scales, a system of self-stabilizing networks. Life is a hierarchy of living systems." Over coffee with me, she pointed out that phytoplankton fixes—continually incorporates—nearly half the world's carbon dioxide. What she called this "forest" of phytoplankton could in time-lapse images of the planet be seen to "breathe," like rain forests of the land. The image of Earth breathing prompted her to say, "I believe the earth is a living entity." The definition of *life* here has it as a hierarchy of genomic and global systems, a web of genetic bare life networked to a Gaian globe. In this scalar narrative, genes are abducted—a logical operation animated, remember, by the "*hope* of regulating our future conduct *rationally*"—into a buoyant relation with our life-sustaining planet.[27]

We can detect in this vision of the microbial sea two approaches to complexity that the philosopher Chunglin Kwa calls "romantic" and "baroque."[28] Romantic visions gaze upward to a unified, integrated, overarching, holistic system, like Gaia. Baroque approaches to complexity tunnel downward to the intricate, convoluted, endlessly and multifariously recombining—to, for instance, networks of marine microbial consortia.

Chisholm's image of phytoplankton as a forest has much in common with philosopher Alfred North Whitehead's holistic, romantic vision of trees: "In Nature the normal way trees flourish is by their association in a forest. Each tree may lose something of its individual perfection of growth, but they mutually assist each other in preserving the conditions for survival."[29] Chisholm's claim for life as self-similar across scales is anchored in a key finding in phytoplankton ecology. One of the best-known scaling relations in the sea is the Redfield ratio, named for Alfred Clarence Redfield, who discovered in 1934 that the relative proportions of nitrogen, phosphorus, and carbon in the open ocean were identical to their ratios in the bodies of marine plankton.[30] This finding posed marine microbes as microcosmic indicators of macrocosmic processes, prompting Redfield to pronounce, "Life in the sea cannot be understood without understanding the sea itself."[31] The implication of Redfield's work was that oceans would taste, smell, and look different were it not for the life they contained. This idea, powerfully articulated in Lovelock's Gaia hypothesis, rests on the concept of the *biosphere*, first formulated in 1875 by Austrian geologist Eduard Suess and turned into the cornerstone of ecology by Russian geochemist Vladimir Vernadsky. Vernadsky's definition of the biosphere—the full complement of life on Earth animated by the transformation of solar radiation into organic molecules—was introduced in the 1940s to the English-speaking world by Hutchinson.[32] DeLong, in a talk he gave before joining the MIT faculty in 2004, brought holistic connections into the DNA age. He opened his lecture saying, "The genome lives in the context of the environment. *Genome* is not a noun; it's a verb. We need to look at the *dynamics* of genes," and he concluded with a gesture toward Gaia as the final frame within which to think of such dynamics.[33]

Alongside the romantic vision of the sea, and often entwined with it, are baroque descriptions, interpretations that burrow down into nonlinear connections. Microbial oceanographer David Karl, a founder of Hawai'i's Center for Microbial Oceanography, gave me a definition of the ocean that mixed temporalities, spatialities, scales: "The ocean is a microbial soup, a vestigial reflection of what was on the earth billions of years before multi-cellular organisms evolved. In that sense, it's a relic—even though it is also thoroughly modern. These organisms are still with us. The ocean is a bio-geochemical reaction; it assimilates carbon, in a Gaian sense. It can accept pollutants and degrade them. It's a flowing medium. This reaction sets up everything from climate to global fisheries. Fisheries are a small part of the carbon cycle that is available for harvest."

Karl sees the sea as a baroque jumble of chronotopes. The vestigial coex-ists with the contemporary. And the context within which we should understand the sea depends upon the tuning of parameters—the ocean is a relic / the ocean is modern; the ocean "can accept pollutants" / the ocean offers nutrition to fish-eating forms of life. Venter's favored context is far less baroque and more romantic, abducting genome to biome in one fell swoop: "sequencing the Sargasso Sea."

A romantic-baroque vision of oceans is illustrated by the International Census of Marine Microbes (ICoMM), a project meant to catalogue single-cell sea creatures and to "place that knowledge into appropriate ecologi-cal and evolutionary contexts."[34] ICoMM, still just getting started in 2008, aims to understand lineages—phylotypes—and functional groups—ecotypes—as well as to comprehend the distribution of "microbial assemblages" across world seascapes. But because of the complexity of sort-ing marine microbes in time and space, ICoMM's envisioned database, MICROBIS, will permit a variety of molecular, taxonomic, geospatial, and environmental angles into the data. ICoMM does not take the gene to be the coin into which all must be translated. MICROBIS exemplifies Geoff Bowker's claim that "the trend in databases has been to hold off on decisions about classifications until later and later in the information process."[35] Although the framers of MICROBIS, in their call for "reproducible and strictly congruent sampling and analytical strategies," do entertain a dream of a common scientific language ("Is this a job for the UN????" they ask in exasperation in one report), they are also mindful that the database will not exist without people constantly revising it, like the collectively authored online encyclopedia Wikipedia. No wonder one of the databases to be linked to this bit of cyberspatial biomedia is called *biopedia*.

LIGHT AND LIFE

It is afternoon and Frida has quit sunbathing, having sizzled herself pink. Liz fishes some *Sargassum* out of the sea. It is the first we have seen. The weed is a world unto itself, full of crabs, snails, shrimp. The etymological links between *planet* and *plankton*—both kinds of wanderers—make sense. I return to reading a paper about *Prochlorococcus* and arrive at a sentence suggesting a connection between this bacterium and the rotation of our planet: "The *Prochlorococcus* cell cycle is highly synchronized in the field, with DNA synthesis taking place in the afternoon and division taking place after dusk."[36] It dawns on me that *Prochlorococcus* cells are probably embarking upon their DNA synthesis right as I read the article. I ask Brandon whether the prochlorococci are making more genetic material just now and, looking at his watch, he says, "Yes, they started about an hour ago." I am impressed. At dusk, I gaze at the sea, trying to imagine massive cell division. The timing of our science is calibrated to the cycles of *Prochlorococcus*. Episodes of sunbathing—whether the undergraduates realize it or not—are not randomly sequenced. Newly tuned to the tempo of our days, I can predict when the senior scientists and graduate students will undertake particular tasks. Liz comes on deck. I ask what attracts her to *Prochlorococcus*. She says, "It's interesting to see how things on a cellular level affect global levels. It's neat to be able to think on all these scales at the same time. What appeals to me about oceanography is the environ-mental context. *Prochlorococcus* is at the base of the microbial food web and has an important impact on biogeochemical cycles."

When she thinks about them, does she visualize them as nodes in a net-work?

"No, I don't think of them as individuals. I think of them as groups. When you look through the flow cytometer, you can see how they are sorted out."

"But," I ask, "you are looking more at traces of them than at the things themselves, right? Do you think about all the *Prochlorococcus* when you look out at the surface of the ocean—like right now?"

"No," she laughs. "That really kind of freaks me out—and I'm only sort of joking. I don't want to be swimming and then think 'Oh, My God, there's ten to the sixth bacteria per mil of water—and I just drank some.'"

Liz does suggest that looking out at the Sargasso at particular times of day might help her guess how the ecology works. For example, phyto-plankters are divided up by the level and color of light they prefer. The first *Prochlorococcus* strain isolated was called "Little Greens" because its chlorophylls were similar to green algae, absorbing red and violet light, making their structures appear green.[37] Since then, *Prochlorococcus* strains with greater spectral range have been characterized. Indeed, "the natural vertical gradients of light and nutrients may provide more stringent conditions for speciation within the *Prochlorococcus* genus than does geographic distance. . . . Among *Prochlorococcus* strains, those belonging to the 'high-light clade,' which inhabit the nutrient-depleted surface layer, appear to be the most recent to diverge."[38]

Some times of day, then, offer frames of reference for making claims about population boundaries and diversification. In fact, it is now commonly thought that *Prochlorococcus* is a misnomer, since the name was based on a pigmentation shared with other prokaryotes that seems to have originated independently in different lineages. Such lineages are not quite genera or species, and scientists like Liz now speak more of high- and low-light "ecotypes" of *Prochlorococcus*.

Light, then, offers a register within which to sort out phytoplankton. Light and color are often the only signs of these lives invisible to the unaided eye. This was early enunciated by Schütt, who in 1904 "showed how color of the water, a property that might have been beloved of Goethe or Schelling, was the product of its physical structure and the physiology of the organisms that lived in it, including . . . bacteria."[39] But learning to see and interpret the color of water as a symptom of life takes apprenticeship (though sometimes color is not available as a starting point; one of the graduate students on *Endeavor* told me he was blue-green colorblind and uses other, still visual, cues—graphs, diagrams—to access the world of the cyanobacterium).

Light is one conduit marine biologists use to get at the world of the sunbathing microbe, a pathway into its semiosis—its strategy for making sense of its world, a world that microbiologists glossed for me as alien to our own, in which surrounding water seems to microbes as syrupy as molasses, in which light is experienced as a kind of tickle.[40]

A detour into biosemiotics, a field of inquiry dedicated to understanding how living things perceive and interpret their environments, whether these are mediated through language, chemical gradients, or intensities of light, is illuminating here. This area of study was inaugurated by early twentieth-century zoologist Jakob von Uexküll, who viewed organisms as possessed of a range of signifying and sign-searching practices.[41] Uexküll held that each organism summons forth its own *Umwelt*, a perceptual and experiential envelope, and he described his attempts to represent the sensorial apprehensions of sea urchins and jellyfish as "excursions in unknowable worlds."[42]

Spectrally tuned ecotypes of *Prochlorococcus* are keyed to distinct frequencies of light. When scientists describe such tuning, they typically do so in terms of numbers, quantities. But they also translate such measures, via human semiosis (using lasers, spectrometers, eyes, and literal and figurative language), into the register of color: red and orange fluorescing pigments, for example, or the little green microbes and blue pastures and prairies of the open ocean. The biosemiotics that has humans squeezing significance out of light-eating microbes mixes what Charles Peirce would have called "indexes" of life—physical traces, like chlorophyll fluorescence—with icons of life—representations, as of cells as scatters of dots—with symbols of life—conventional signs, such as light, standing by historical association for vitality (with particular colors, like the blue-green hues of the pastoral sublime, adding a healthy glow).

Anthropologist Franz Boas's 1938 reflection on his 1881 physics-geography dissertation, "Contributions to the Understanding of the Color of Water," aids in considering such varied sign seeking: "In preparing my doctor's thesis I had to use photometric methods to compare intensities of light. This led me to consider the quantitative values of sensations. In the course of my investigation I learned to recognize that there are domains of our experience in which the concepts of quantity, of measures that can be added or subtracted like those with which I was accustomed to operate, are not applicable."[43] Boas concluded not only that angles of vision affect perceptions of color but also that discernments of color are conditioned by habits of mind. Later anthropologists elaborated their descriptions of such habits to include symbolic associations with colors—red can mean danger, but also good fortune. Whether one is oceanographer or ethnographer, understanding such associations is approached through fieldwork, gathering perspectives that can be diffracted through one another.

Apprehending the qualities of light on the sea fuses scientific and socially conditioned responses, romantic and baroque attitudes. When I run into Liz on the fantail of the ship again, at night, her sense of the ocean has modulated. Now, the ocean is eerier. She tells me that at times like this, when we cannot see but can still sense the expansive sea around us, it becomes clear to her that "the ocean is foreign to us. It's like outer space. If you fall overboard, you're pretty much dead. It's the closest you can get to outer space without leaving Earth. It's scary. Being in the open ocean with 4,000 meters of water below you is not the same as having a picnic in a green meadow." John Teal, in *The Sargasso Sea*, reports a similar sensation. He's just seen a shark, at night: "The shark is so alien and the sea so foreign that I really get the shivers when I think of diving at night amongst them. It must be a lack of solid surface underfoot, and the inability to sense things clearly through the water. The Sargasso at night has a weird, other-world quality that is too much for me. The senses really break down in the face of all this strangeness."[44]

Looking at the scatter of stars now above the *Endeavor*, I remember flow cytometer engineer Howard Shapiro's claim, delivered at a Chisholm lab meeting, that marine microbiology might think of its task as "cellular astronomy."[45] The pastoral sea of the daytime has turned into the alien ocean of the night. It is dark. And we have arrived in the Bermuda Triangle.

VIROID NIGHT LIFE

It is midnight, and I am about to do the work I promised the Chisholm lab, extracting marine viruses from this northern sample of the Sargasso Sea. In a way, this project follows in the wake of Venter, who wrote in his *Science* article, "We concentrated on the genetic material captured on filters sized to isolate primarily microbial inhabitants of the environment, leaving detailed analysis of dissolved DNA and viral particles on one end of the size spectrum and eukaryotic inhabitants on the other, for subsequent studies."[46] Our research on the small end of the spectrum is one such study, though also a consequence of work already under way in Chisholm's lab.[47]

Brandon, my lab mate, has put on techno music—the Chemical Brothers' *Brothers Gonna Work It Out*, British white guys sampling American black guys (reminding me of the deterritorializing effects of sampling in any medium). In *Moby-Dick*, Melville has whalers bonding with one another as they stick their bare hands in whale spermaceti. These days, we wear latex gloves and handle plastic containers of marine life we cannot see. Instead of sea shanties we have techno (or soon, Brandon's favorite, Black Sabbath). We label the vials that will hold our samples. I review the lab protocol I was given at MIT. I am to filter Sargasso seawater through a 0.2-micron mesh Acrodisc Syringe Filter. This mesh allows only viral-sized items into the 1.2-milliliter cryovials arrayed before me. Once I fill the vials, I am to place them in liquid nitrogen for cryogenic transport back to Cambridge. Brandon and I saunter outside with bottles to get water from the CTD-rosette, which, to our amusement, has surfaced bearing a squished Styrofoam head, the kind used to model wigs. Sending these down is a favorite pastime; the compressing effect of pressure on the porous heads generates a weird thrill of bringing something back from another world.

Back in the lab, I have first to filter samples from four different depths. Four times, I pull 5 milliliters of water up from my bottles into a plastic syringe, affix a filter, and squeeze water samples from 15, 45, 75, and 90 meters

FIGURE 23. Microscope image of microbes from Bermuda Atlantic Time-Series study site. Bermuda Biological Station for Research, *Currents* (Summer 2002). Photo by Rachel Parsons and Craig Carlson. Reproduced with permission of Craig Carlson.

down into temporary holding tubes. I think about the viruses slipping through the filter and am told that the resistance I feel when pressing the top of each syringe is due to the abundance of cyanobacteria blocked by the filter's mesh. The water, of course, looks clear, nothing like the spacey cellular astronomy pin-up on the cover of the newsletter of the Bermuda Biological Station (figure 23).

The next step is dosing filtered water—as well as parallel, unfiltered samples, for cellular studies—with preservants. One preservant I have is gluteraldehyde, which arrests cellular processes and kills cells. It does not bind to DNA, allowing genetic material to stay intact for sequencing. I also have dimethyl sulfoxide, a solvent that preserves membranes in unfiltered samples, saving cells for cryofreezing for possible cultures. If I spill either preservant on my skin, Brandon tells me, it will attack my surface cells. But no big deal. These parts of my body are dying all the time, he reassures me.

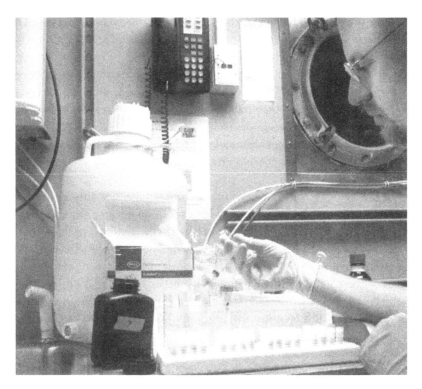

FIGURE 24. The author at sea, squeezing Sargasso viruses into cryovials. Photo by Susan Dennard.

The gloves I wear are to protect Sargasso viruses against contamination by *my* skin cells or, indeed, any bacteria I may have unwittingly collected on my fingertips. I do not want to get genes from my body mixed up with genes from the ocean's body. I do not—I hope—have to worry about being infected by seaborne Sargasso viruses.[48]

I have arrayed on a lab bench thirty-two cryovials—a vial for each depth, filtered and unfiltered, treated with gluteraldehyde or dimethyl sulfoxide (and made in duplicate for each combination). I remove all the vials' caps and place them upside down, making sure they do not spiral off the table as the ship rolls. Using a P20 Pipetman, I add 6 microliters of gluteraldehyde to the relevant caps, and then, with a P200 Pipetman, add 75 microliters of dimethyl sulfoxide to the appropriate cyrovials, staying with one Neptune Brand disposable barrier tip so long as all my sampling sources stay the same, and then switching, so as not to forward contaminate (figure 24). I am excited by how organized I am.

Once the seawater is distributed, I let the gluteraldehyde-treated samples incubate in a box. "Incubation" is an odd word to use for this preserving practice since gluteraldehyde is actually killing organic matter—jarring, too, to think about in connection with viruses, rhetorically positioned between life and death.[49] Viruses, entities imagined as other to the body and its health, as foreign material that poisons the familiar space of the self, are alien to vitality yet enmeshed with it.[50] Viruses operate by employing the replicative genetic apparatus of the hosts they infect to make more copies of their own genetic material, a propagation they are unable to accomplish on their own.

In the baroque history of evolution, viruses have not only or merely parasitized organisms in which they have taken up tenancy but also laterally contributed—think tangled tree of life—to the genomes of those creatures, as viral material has been transduced into host DNA. As Deleuze and Guattari have it, "We form a rhizome with our viruses, or rather our viruses cause us to form a rhizome with other animals."[51] Philosopher Keith Ansell Pearson calls such reticulated relations *viroid life*, the continual production of infectious difference. In "Viruses in the Sea," marine microbiologist Grieg Steward puts it this way: "In addition to acting as executioners, viruses can also cause non-lethal infections which influence the physiology or evolution of their hosts."[52] Viruses can be vectors of differentiation in the microbial sea: the phages that infect *Prochlorococcus* are agents of gene transfer between high- and low-light ecotypes of *Prochlorococcus* and also between *Prochlorococcus* and *Synechococcus*.[53] For this reason they may hold the key to Hutchinson's paradox of the plankton: viral incursions into prevalent strains constantly reshuffle Darwinian chains of dominance.[54]

Science studies scholar Theresa MacPhail forwards the concept of the *viral gene* to describe the recursive relations between the genomes of hosts and viruses, asking, "Could viruses in the genome change the way we conceive of genetic agency? In other words, do genes partially constructed by leftover viruses ultimately script life? . . . Where would that leave the boundary between coding and non-coding, gene and virus, living and non-living?"[55] If the alien genes encountered in chapter 2 scramble lineages, worrying the tie between genealogy and classification, viral genes usher the liminal, putatively nonliving, into the genetic center of "life itself."

In the ocean, with ten billion viruses per liter, gene transfer, suggests Jed Furhman, "would have the effect of mixing genes among a broad variety of species, with wide-ranging effects on adaptation and evolution. Although transfers of these sorts may be extremely rare, the typical bacterial abundances of 10^9 per litre in the euphotic zone and the huge volume of the sea (3.6×10^7 km^3 in the top 100 m), coupled with generation times on the order of a day, implies that an event with a probability of only 10^{-20} per generation would be occurring about a million times per day."[56] Fuhrman cautions that lateral transfer "should also be considered in the evaluation of the potential spread of genetically engineered microbial genes, or of antibiotic resistance introduced by intensive fish farming."[57] This is the viral gene scaled up to the viral sea, with the alterity of viruses providing a channel for the amplification of the alterity of humanity, resulting in an infectious natural-cultural instantiation of the alien ocean. Life forms, fed through a heedless biotechnological form of life, become other than they were.[58] If the sea is life, it is also death, as well as a symbiopolitical swirl that blurs their boundaries.

None of this is apparent to me during my sampling. I make a record of my activities on a log sheet, crucial context for scientists down the line. I am mindful that what Mike Fortun calls "the care of the data" is important for scientists in Chisholm's lab.[59] I sometimes get confused about what time it is and have to recopy information. This is because the ship's clocks provide Greenwich Mean Time as well as Bermuda time, two hours earlier. And until yesterday, *Endeavor* had held to the Eastern Standard Time Zone from which we had departed, putting waters around us one hour into the future. No wonder people get confused about time and space at sea. Charles Berlitz, grandson of the founder of Berlitz language schools, postulated in *The Bermuda Triangle* the existence of a time anomaly in this area, as if time could fold into itself, fusing and confusing past, present, and future— just like alien genes that crosswire marine lineages and warp evolutionary chronologies.

After sampling, I head to the bridge, where, in swaying darkness, I talk with the first mate about the Bermuda Triangle. "It's a place of unexplained phenomena," he tells me. "The latest theory is that methane bubbles coming up from the ocean floor are causing ships to sink. The boats suddenly lose buoyancy and—whoomp!—down they go." I later find this story soberly considered by earth scientist Richard Corfield: "The ocean— and indeed the air above it—can suddenly lose the ability to support objects such as ships and aircraft because the density of water and air are suddenly reduced. This idea proposes that the density change is caused by the submarine convulsions of a strange and little-understood material buried underneath the seabed: methane hydrate."[60]

If MBARI's message from the mud told us of life support, this missive reads of trap doors, an ocean that abducts unwary seafarers. Writing of extraterrestrial visitors to the Bermuda Triangle, Berlitz suggests, "Perhaps . . . alien intelligences are . . . taking specimens they will preserve as an example of earth life as it was before the planet destroyed itself."[61] With worries that the ocean is dying beneath our exhaust pipes, the scientists making salvaging snapshots may be the alien intelligences here. Indeed, our work might be seen as one style of alien abduction, a venture in which humans, as strangers to the sea, employ a mixture of logic and last-ditch hope to make sense of something unfamiliar. With alien and viral genes in the mix, the aliens in this brand of alien abduction might be both agents and objects of such investigation.

VENTING ABOUT VENTER

After my microbiological all-nighter, I find Liz at the CTD-rosette, collecting water. She will be doing "DNA work" on marine microbes once she gets back on shore. "Once you get a bunch of DNA," she tells me, "you can do anything you want with it." Trying to get a temperature read corresponding to this CTD cast, she is reminded of Venter's Sargasso work and picks up in the middle of a sentence she started days ago: " . . . and that's something Ventura doesn't have: *context.* Wait, am I saying his name right?"

"Venter," I laugh. Ventura—Jesse Ventura—is the professional wrestler who surprised the United States by becoming governor of Minnesota in the late 1990s. Liz's slip might be appropriate, since Venter is another case of someone muscling his way into a field outside his primary expertise. I tell Liz I have just received over shipboard e-mail a message from my wife, Heather (in the middle of her own anthropological research, with artisan cheesemakers, who have their own microbial preoccupations), telling me that Venter gave the commencement address at Boston University two days previous. The university president introduced Venter:

> Recently you have been sailing around the ocean trolling for new microbes, which in turn provide new genomes to sequence. You have already identified thousands of new species of microorganisms and millions of new genes. Even though your discoveries are invisible to the naked eye, you are in the tradition of the heroic age in which Charles Darwin sailed through the South Seas. He was exploring a macrocosm and you, on the sea and in the lab, are exploring microcosms. Your discoveries are as likely to transform the world as his did.[62]

Liz continues now, with spirited sarcasm, "Venter, Ventura. Perhaps I should just call him *Darwin!* But Darwin came up with a unifying theory.

Venter only has data. It is very interesting data, but until he comes up with a theory, people should stop this comparison in its tracks." I ask her about the context Venter's study lacks. She tells me that proper context comes from doing flow cytometry to know what abundances look like, from using CTD data to know where the chlorophyll maximum is, from taking ocean temperatures to look at water column stratification, from knowing time of day and season.[63] Liz points out that the sampling sites pictured in Venter's *Science* paper were visited at different times of the year, which the map elides. No seasonality is apparent in the graphic.

Liz continues, becoming precise:

> He *does* have some temperature and chlorophyll fluorescence data from the three stations that were sampled in February, when the water column was fairly well mixed. Unfortunately, there is no background data to go with the May samples. This is really too bad because the water column was probably more stratified in the spring and the distribution of at least the cyanobacteria may have been different. A lot of effort went into sequencing, but some simple measurements, like temperature and nutrient concentrations, that would have provided some context, were not taken.

Brandon walks onto deck and chimes in:

> He's only collected a shitload of water and sequenced it. He's just skimmed the surface. He's kinda like a guy who has found a bunch of computer parts. But who knows if he'll be able to put it all together? It could be really cool, but we'll have to wait to find out. He's smart, but in a way he seems like he just wants to get notoriety in this new field— even if it means doing work that other people—oceanographers—have already been doing and pissing them off. Sometimes, he seems like one of those scientists at the end of their career who go just completely off the deep end with some insane project.

A paper published in *Science* responding to Venter evaluates his project this way: "It is simply impossible to understand patterns of community structure from random sampling of the world's oceans."[64] The science party on *Endeavor* is trying to do things right. When I return to shore, I hear that Venter's deposit of Sargasso data took up fully half of the GenBank database. Because the sequences were not partitioned from the rest of the archive, the rumor is that people working in other fields—like mouse genetics—soon found the database telling them they might find genetic matches in the Sargasso Sea!

The next year, DeLong concluded that "the Sargasso Sea dataset represents a very useful resource when placed in the appropriate context and

interpreted properly."[65] Like Jean Rhys's *Wide Sargasso Sea*, which provides an alternative, colonial back-story biography for the madwoman in
the attic of Charlotte Brontë's *Jane Eyre*, an expanded context for the
Sargasso dataset may reveal explanatory histories, conditions that reframe
the stories scientists wish to tell. The question of what counts as appropriate context, of course, calibrates to what one wishes to know, whether one
is in search of, for example, the holistic complexity of the Sargasso or evidence of the baroque paths of ship effluent. Such commitments, as Bill
Maurer has suggested in another context, may lead one to see any other
context "revealed as confidence-tricks—con-texts—not the purchase of
new perspectives," misplaced expressions of confidence in other frameworks, and therefore not appropriate contexts at all.[66]

HOW THE OCEAN GOT ITS GENOME

We now have tools for solving the puzzle of How the Ocean Got Its
Genome. To begin, it was granted a *body*, an image with a scientific warrant
as far back as the German biological oceanographers. Nowadays, however,
the ocean's body is animated not by Hensen's "blood of the sea" but by
"genome sequences from the sea."[67] Because the sea is a *body of water*, it
is also now imagined literally to embody that scientific abstraction known
as the "gene pool." More, in the days of gene databasing, when gene pools
are not so much abstract as they are virtual—a cyberspatialized aggrega-
tion of nonlinear data that can be materially manipulated in various
ways—a vision of the ocean as a space of flow comes to be mirrored in the
hypertexty database. When bits of the sea's pool of genes are mapped and
uploaded into databases, a body of water becomes the blueprint for a body
of knowledge. Sequence maps become the territory.

When researchers like Venter describe the objects of environmental
genomics as environmental genomes (or even an ocean genome), they slip
from a description of their method into a claim that their sample represents
a bounded entity in the real world. In this process several things drop out,
elements that *must* vanish for the genome to be conjured as such, for the
ocean to be identified with its microbial genes. Fishes. Seaweed. But also seasonality, ocean currents—all the things microbiologists interested in context might care about. As with Kipling's "just so" story of How the Whale
Got His Throat, this tale substitutes fanciful, fantastical logic—indeed, a
kind of fable—for more detailed narrative and numerical accountings.[68]

Equally important in conjuring something like an "Ocean Microbial
Genome" is the tale of Venter's Darwin-styled circumnavigation, which,

by dint of looping the globe, summons the world itself as the object of study—even if the trip is local at all points. Venter takes the world as an information superhighway.

For scientists not so eager to follow in Venter's wake, the Sargasso might rather be seen as the pertinent culture medium within which genomic data from uncultured marine microbes make sense. Such microbiologists are like anthropologists asking for better fieldwork, for a thicker description of a local culture—a more precise understanding that may make Venter's environmental genomes dissolve, like the imaginary islands that once peppered old maritime maps.[69] But *culture* is hardly a transparent term in anthropology, and the analogy needs qualification. The standard concept of culture, as a collection of ideas, customs, and artifacts, is, like the Sargasso genome, a theoretical construct, a heuristic that does not preexist particular descriptions. Traditional models of culture saw it as a thing-in-the-world that could be abstracted from practice, a form that might be mapped in databases like the Human Relations Area Files (HRAF), a tool launched at Yale in 1949, aimed, as its website has it, at permitting anthropologists to "search for specific information cross-culturally" using "a vocabulary of over 700 cultural categories" (e.g., kinship, sorcery).[70] HRAF allows something like bioinformatics' crosswise comparisons, but for culture (understood, like Venter's ocean, as an integrated system whose logic scientists will figure out later).

Venter's claim to be sequencing the Sargasso Sea depends upon assumptions that the samples he took represent a unitary, meaningful entity[71]—assumptions not dissimilar from those of early anthropologists describing, say, "the Trobriand Islands" as a bounded culture, a description based on partial observations and predicated on leaving out conditioning and cross-cutting contexts like colonialism (as well as forgetting that space itself is not a neutral grid but emerges from interactions).[72] These days, anthropologists configure culture as dynamic in time and space, as a matter of practice rather than code. As Brian Street has it, resonating with DeLong on genomes, "Culture is a verb" rather than a noun.[73]

Fieldwork, too, has been subject to query when "the field" often emerges from the anthropologist's work rather than precedes it, when the researcher is subject to alien abductions, detours, unanticipated effects, emergent in part from her own tentative abductions about her future work.[74] Even if the Sargasso is still a field for Brian and Liz, it may be more fitting to say they seek an account of the ocean not so much as a culture medium but as culture in practice, a set of conditions whose description is always in the making.[75] Today's marine microbiologists are less interested in microbial colonies (whose home is the lab) than in what they now call microbial assemblages (whose ecology is the field), a shift that resonates with anthropology's move from working in colonial circumstances to following heterogeneous, often ephemeral, social and institutional configurations, forms that social theory types have lately come to name *assemblages*.[76] Indeed, one anthropological strategy of *Alien Ocean* is inspired by DeLong's style of environmental genomics: to map networks of association in ever-recalibrating contexts rather than in fully characterized "cultures."[77]

PLAYING GOD IN THE GALÁPAGOS

I tell Liz that Venter has also run afoul of people other than those trained in marine biology. As he set out on his journey, he discovered that permission would be needed to sample in the national waters through which *Sorcerer II* passed. Venter recounted his difficulties in his MIT lecture:

> These studies aren't as easy as it might seem at first. We have a team of three full-time people that just have to work with the U.S. State Department and each of these countries to be able to take 200 liters of seawater from their waters. We have to have import and export permits and people get very excited about this. We did an MOU [memorandum of understanding] with Mexico and Chile and we're doing others along the way. Each country wants to try and patent these sequences. We're insisting that they all go in the public domain. So, it's not as simple as taking 200 liters of seawater. In fact, we're dealing right now with a group that's protesting us taking biological samples in Ecuador. The other thing—it was a big surprise to me—there's very little international waters left in the world. Here, I thought I was just out sailing free in the ocean and somebody's claimed it all.[78]

Venter's map of the ocean initially excluded such bodies as the United Nations, whose Law of the Sea Convention has been largely responsible for delineating national nautical territories. Most marine scientists know this seascape well, doing clearances routinely, but Venter started his project undereducated, like many Americans, in history and geopolitics.

Venter's Captain Nemo fantasy about the free seas ("Ah, Monsieur, one must live—live within the ocean! Only there can one be independent! Only there do I have no masters! There I am free!")[79] and naïveté about the politics of sampling drew attention from an international civil society organization based in Canada called the Erosion, Technology, and Concentration (ETC) Group, formerly the Rural Advancement Foundation

FIGURE 25. Playing God in the Galápagos. Cartoon by Rey Pagé for Erosion, Technology, and Concentration Group (2004).

International (RAFI). The March/April 2004 issue of ETC's newsletter was dedicated entirely to one article, "Playing God in the Galapagos: J. Craig Venter, Master and Commander of Genomics, on Global Expedition to Collect Microbial Diversity for Engineering Life." It featured a cartoon of Venter on the prow of a ship, dressed as a Victorian naturalist, overseeing the work of walking mops—a quotation of the fairy tale about the sorcerer's apprentice—as they load biological samples onto *Sorcerer Too*, outfitted with satellite dish and pirate flag (figure 25).

The article condemns Venter's Galápagos sampling, quoting a spokeswoman for an environmental advocacy organization in Ecuador: "Venter's institute has flagrantly violated our Constitution and several national laws, including the Andean Pact Decision 391 on access to genetic resources. . . . When negotiations on access to genetic resources take place behind closed doors, in the absence of public debate or information, and in the context of opening the doors for monopoly patents—we call it biopiracy."[80] Acknowledging that "Venter promises that intellectual property on raw microbes and their gene sequences will not be sought," the newsletter cautions, "there is nothing to prevent monopoly patent claims on commercially useful results derived from collected diversity." Venter's scaling from genome to ocean can work only by skipping over human agencies and institutions.

His scalar narrative places in shadow not only organisms as agents in their own history but also all the social relations—of predation, cooperation, governance, regulation, scholarship, medical research, monetary exchange, pollution, proprietary research, international travel—that range between and enable the link between microbial genomes and that thing the space race, the cold war, and environmentalism taught us to call the "globe."[81]

Genes are potent symbols of life, to be sure—but placing them afloat in a boundless body obscures questions of how such substances actually travel into human nervous systems, coastal drainage systems, and international contests over, say, gene prospecting. Like "Northern environmentalists [who] extend their 'global reach' when they describe a problem of deforestation or pollution as planetary, rather than as regional or local,"[82] Venter's romantic "Ocean Microbial Genome Survey" skips over the baroque processes that intervene between genome and ocean. If Venter's earthly gene pool is not exactly what Deleuze and Guattari would call a "body without organs" (a space of unrestricted flow and desire—or maybe it *is* that for Venter), it is certainly a body without government and nongovernmental organizations.[83] He sites his vision of life forms within a form of life that takes the colonial exploring expedition as its model for scientific practice.[84]

Spotlighting another of Venter's projects, to engineer microbes that might scrub up greenhouse gas, ETC asks, "Will microbes collected in the Galapagos form the genetic template for Venter's new, artificial life forms?" In addition to his eponymous science institute, Venter heads the Institute for Biological Energy Alternatives (IBEA), "dedicated to employing the tools of genomics to develop cost-effective biological fuels and other biological approaches to greenhouse mitigation and exploring solutions for carbon sequestration using microbes, microbial pathways, and plants."[85] IBEA has garnered funding from the U.S. Department of Energy's "Genomes to Life" program, interested in using microbial materials to create biofuels—which may indeed open up questions of the proprietary status of microbial life Venter gathers. Unlike the artificial life conjured by computer scientists at New Mexico's Santa Fe Institute for the Sciences of Complexity in the 1990s, this synthetic biology does not exist only in the notional zone of cyberspace or in the eyes of individual, cloistered researchers who consider computer viruses as alive; it is made up of elements drawn from an amalgam of real-world seascapes, cyberscapes, and lawscapes.[86] Interestingly, in the IBEA mobilization of a utilitarian rather than ecological context, the unity of any "ocean genome" liquefies, no longer the sole grail of Venter's project but simply a reservoir for recombinant bioengineering. If

Venter's ocean genome is romantic, his envisioned synthetic genomes pre-pare baroque linkages between prospecting, engineering, and the politics of climate change.

Still, ETC's alarm reads as hyperbolically as Venter's self-promotion. The audacity of Venter's magnanimous, suntanned Darwin persona is matched by ETC's view of him as a pirate Frankenstein. The connections among sampling, sequencing, archiving, publishing, and patenting are more tenuous in practice than ETC's commentary suggests. Nonetheless, ETC does have a point: the ocean described by genomics is increasingly abstract, not always understood with respect to geographic, geopolitical knowledge but through traces uploaded into databases—the eventual uses of which are never crystal clear but which may indeed be leveraged into new kinds of property, both substantial and promissory, later. This is the ocean abducted. Outsiders are left to follow in the wake of the scaled-up agencies of powerful traveling scientists who move quickly from one place to another.

Even Venter cannot keep up with himself, quoting his own words from the *New York Times*. The *Wired* profile does little to scale back his larger-than-life image. In an article titled "Craig Venter's Epic Voyage of Discovery" we read, "He wanted to play God, so he cracked the human genome. Now he wants to play Darwin and redefine the origin of species—then reinvent life as we know it."[87] This article makes him come off as still more self-aggrandizing. Running into complications around sampling in French Polynesia, he mocks international politics, as though being a scien-tist allowed him to glide above human conventions: "It's French water, so I guess they're French microbes."[88] When the French state department sug-gests that *Sorcerer* might entertain a visit from a French officer, Venter offers, "How do you say 'fuck you' in French?"[89] It is no wonder Venter's project gets beaks bent out of shape. He's always getting away. For all my own attempts at contact, I never did catch up with this great white male.

TACKING INTO THE COLOR LINE

What is the status of *Endeavor*'s Sargasso samples? To work in Bermuda's Exclusive Economic Zone, Brian Binder had received permission from Britain's Foreign and Commonwealth Office, since Bermuda is a U.K. dependency. The foreign office wanted a record of data but did not care about samples, since the expedition aimed only at generating material for scientific publication.[90] Some Sargasso samples, I learned, would make their way into classrooms at the University of Georgia, where students would be assigned to sequence their microbial components, even depositing data in GenBank.

When I meet up with the three college women as we approach Bermuda, one tells me she took just such a class, recalling that "it was way cool to iso-late a cell's DNA." Frida, wearing a "Protect the Rainforest" T-shirt, says she is interested in marine biology because "so much of the ocean is unex-plored. And economically speaking, it's going to be really important—like in aquaculture." From rural Georgia, Emma's ideal job would be at an aquarium. She's embarrassed by how giddy she is out here, reporting that she has dreamed of becoming a marine biologist since she was nine, when she loved *National Geographic* programs. Agnes, smoking, interrupts to joke about her makeshift sunbathing outfit—old gym shorts and a tank top—which, she comments, is "so *ghetto*, it's not even funny." Soon every-thing is ghetto, from scuffed hard hats to ragged life preservers. These white kids tease each other with the word *ghetto*, making light of how dif-ficult it is to stay presentable at sea, but the word, to my ears, is accented by its appropriation from African American slang.[91] I ask: What's the racial mix of students at the University of Georgia? Frida shakes her head and says, "UGA is pretty whitebread. It's really weird. All it is is these upper middle-class white kids whose parents send them there." Emma agrees and says her high school was 40–60 black-white, so it was a shock to arrive in Athens, where only 5 percent of the student body is black, in a state that is about a third black.[92]

Marine biology in the United States, while fairly international, is pretty white. According to the American Society for Limnology and Oceano-graphy (ASLO), whereas underrepresented minorities account for about 24 percent of the U.S. workforce, American Indians, Native Alaskans, African Americans, and Hispanic Americans represent less than 5 percent of earth, atmospheric, and marine science Ph.D.s awarded to U.S. citizens and permanent visa holders.[93] My thoughts tack from the color of seawa-ter to the pronouncements of Boas's contemporary, sociologist W. E. B. Du Bois, who wrote in his 1903 *The Souls of Black Folk* that "the problem of the Twentieth Century is the problem of the color line."[94] Racism in the United States has indeed often turned on questions and perceptions of color—with linear and nonlinear effects on urban, suburban, rural, and coastal geographies. Drawing on off-the-shelf sociologies of black-white economic and residence patterns, ASLO speculates that

> most aquatic scientists probably develop an affinity for water long before
> their academic interests mature. Exposure to lakes, creeks, or the oceans
> in childhood helps ignite initial interest. These sorts of opportunities are

often not as available to minority youth. . . . Additionally, the legacy of segregated facilities means that parents and grandparents who were kept away from swimming pools, beaches, and state parks, are less inclined to introduce their own children to water-related activities. While minority youth are probably aware of various aquatic systems from the media, this does not replace direct experiences with these systems.[95]

Although this account points to key questions of experience and access—and elsewhere takes up educational funding, recruitment, and retention as well as perceptions that careers in oceanography are less economically secure than other scientific paths—it does not disaggregate the various ethnic and racial histories entangled in American ocean science nor explore ideologies of nature through which "lakes, creeks, and oceans" call forth feelings of curiosity in the first place.[96]

Marine microbial samples also come to Athens from an island outpost of the University of Georgia in the Sea Islands, an archipelago fringing the coasts of South Carolina, Georgia, and Florida. The Sapelo Island Microbial Observatory (SIMO) is located on a Manhattan-size island south of Savannah, a landmass also home to a population of about seventy residents who identify as Geechee and who, like the more widely known Gullah, descend from Africans brought to the United States under slavery. Brandon tells me that he has led undergraduate sampling trips to Sapelo, a delicate operation since the kids stay in buildings near the old island mansion of R. J. Reynolds Jr. (son of the tobacco magnate) and treat the voyage as a camping trip: "They get rowdy and run around and I have to calm them down." In the two trips he has taken to Sapelo—forty kids total—none of the students has been black, not surprising, perhaps, given the university's scandalous racial ratios.[97]

As it happened, I had myself taken a trip to Sapelo a few months previous, having learned about this site by clicking around on the Web and landing on hypertext describing this out-of-the-way research site dedicated to exactly the genomic, bioinformatic microbial marine biology I had been following. Brandon's tale helped me frame my own trip—a final data point in my tale of uploading and abducting the Atlantic.

NATURE AND CULTURE ON SAPELO

The Sapelo Island Microbial Observatory was formed in 2000 under an NSF grant and operates through the University of Georgia Marine Institute, whose campus has been sited since the 1950s on Sapelo's southern tip. After a few e-mails with the observatory director, who works in

Athens, I arranged a hurried visit, but, owing to my own speedy schedule as well as hers, we missed each other. I did manage to speak with the incoming director of the institute and a just-retiring biologist, though meetings were brief, having to fit into scientists' commuting schedules, which took them off Sapelo by ferry in midafternoon.

I stayed on Sapelo in the community of Hog Hammock at a self-catering guesthouse operated by Cornelia Bailey, of the last generation to be born and schooled on this island. Bailey is at the center of a movement to revitalize Hog Hammock with heritage tourism aimed at people who want to research their Geechee or Gullah roots or learn Sea Island history—a curiosity primed in part by Julie Dash's 1991 film *Daughters of the Dust*. Novelist Alice Walker has been one noted guest. A figure of some renown herself, Bailey is the author of a folkloric memoir, *God, Dr. Buzzard, and the Bolito Man: A Saltwater Geechee Talks about Life on Sapelo Island, Georgia*.[98]

Bailey's book reports that, in the nineteenth century, sugar farmer Thomas Spalding purchased West Africans in Charleston, North Carolina—people taken from what is now the Gambia, Sierra Leone, and Liberia—and set them to work as slaves on Sapelo land drained for plantation cotton, sugarcane, and corn. During Reconstruction, 370 descendants of these people remained on the island, purchasing land from Spalding's heirs and becoming independent fisher-farmers. When in the early twentieth century white businessmen acquired island territory to build country estates, a wage economy drew the Geechee into renewed patterns of racial segregation and subordination. R. J. Reynolds Jr. was among those who set up house. The University of Georgia arrived in the mid-fifties, transforming Reynolds's compound into its campus.

After learning that I studied scientists, Bailey told me that when the marine institute people first showed up in 1953 she and her childhood friends called them "the mud people." Bailey reflects on the scientists' arrival in her book:

> There was a change coming. . . . Richard Reynolds had invited the scientists over to live on Sapelo and study the marsh. We were used to white people being here, but it had always before been Reynolds, his family and friends and the white people who worked for him. . . . The scientists were less threatening than the men Reynolds had working for him, and that was good. They were not here to meddle into your business and tell you how to live and we liked that. We were respectful of them too, to a degree, because they were well educated. But still, it was not like they were Christopher Columbus and we were the native people running down to the water to welcome them. . . . We tried to

ignore them, but they were riding around in Jeeps and digging holes in the earth to get their soil samples or standing out in the marsh getting all muddy. You couldn't help but see them. "Why those crazy people digging holes all over the place?" the old people said.[99]

Bailey elaborated, telling me, "We're a show and tell people. And after a while, it's time for a show and tell of what you've been doing. After fifty years, show us what you've been doing." From her book: "The scientists who were here full-time worked on projects for an average of five to seven years but they didn't ever explain what they were doing or try to interest young people over here in marine biology. Some of the scientists were kinda standoffish and Aunt Mary and others would go, 'Oh, those people don't want to bother with people in the community here.'"[100]

Bailey remembered with approval a scientist who worked on sea pansies, discovering that they might contain compounds active against cancer. This reminded her of work done on horseshoe crabs, up in Woods Hole, harvesting their blue blood as a screen for diseases.[101] She argued that, if scientists were doing research in sites other people called home, they should speak to community concerns. Scientists at the marine institute often kept apart from Hog Hammock community, although several Hog Hammock residents worked for the institute in maintenance. With all the scientists Euro-American or European, the racial economy was clear. Still, scientists were preferable to other newcomers: "But a few of the scientists handled it just right, and didn't treat us like we were here just to service them. Those were the ones some of us got to know while taking care of the lab to keep it clean, working with them in the marsh, building things they needed or going with some of them on the ocean."[102]

Ecologist Steve Newell was a man Bailey recommended as a scientist able to speak to both science and the history of Sapelo. Unlike most of his colleagues, he had lived on the island for twenty-five years. When I met with him, he quickly told me that he was "not a DNA person" but a field biologist who "ground-truths all that stuff." Newell, about to retire, specialized in the fungal ecology of salt marshes.[103] In Sapelo salt marshes, fungi and microbes decompose a cordgrass called *Spartina alterniflora*. This decayed grass, often covered with microbes that move in mucous-lubricated "slime nets," constitutes the basis of the marsh food web. These days, as the fungi are better characterized, molecular techniques are employed to find previously unknown microbes in the salt marsh system—and, indeed, these organisms are the subjects of the Sapelo Island Microbial Observatory. Newell thought something was lost in this move from a sustained presence in the field to looking at data in labs often physically far from the environment in question. He cared about the well-being of these specific marshes and had a sense that, even if he had a different history than people in Hog Hammock, they shared this care: "The local people *love* this island, especially the older people."

Bailey agreed that concern for the marshland constituted a link between scientists and Hog Hammock residents. The marsh is dying and scientists have called this mortality the result of red tides—blooms of toxin-producing dinoflagellates, phytoplankton that give a red tint to the water. Bailey said that her father called it "the freshes," when the rain comes and washes away the salt water. The scientists, especially ones who did not live here, "never asked *us*, who'd lived on the island for generations. We see, we understand—even without scientific language or degrees from Boston. Test tubes are fine—don't get me wrong—but you need to blend the old ways with the new."[104]

Part of the issue is the structure of academia, in which students come to places such as Sapelo for a while and then leave. Bailey told me about a scientist who worked to keep oyster beds flourishing on Sapelo. But, because he was here for just a few years—and only during summers—the project was abandoned. To be sure, scientists are not often rewarded for engaging with ostensibly local—or social—matters. But with the institute's move away from aquaculture and toward interest in microbial matters—a path eased by funding for genomic approaches—there is less in the way of local connection (though such connections are not necessarily positive; it might be good for Hog Hammock that scientists are vanishing). When I met the new director of the University of Georgia Marine Institute, Bill Miller, he agreed: "The real need for scientists to be here has lessened because the science has changed. If you're going to do microbial ecology, you don't have to set up an experimental grid in the marsh, bring stuff to the lab to look at under a microscope, then run back outside for more observations. You don't have to have that traditional local presence in the field. With genomics, people can 'grab and go' and just don't have to *be* here anymore."

What remains, then, are fieldtrips to the island, often by students from the university, to collect samples. Brandon told me that he made sure to bring students to Hog Hammock to patronize a restaurant run by Bailey's neighbors, but this was not an audience for heritage tourism. For marine science students, the island was more about nature than culture.

Indeed, plans for "development of heritage tourism at Hog Hammock to promote Geechee cultural traditions and provide an adequate economic base for community members"[105] are not always harmonious with the plans of the National Estuarine Sanctuary (which wants minimal development), the

R. J. Reynolds Wildlife Reserve concern (which sanctions wild turkey hunting on the island), or the marine institute itself. Brandon had put it to me this way: "The island is a laboratory and you don't want people tromping all over the lab." Though the language would be different, Bailey might agree about the crowds, if not their configuration. For the marine institute scientists, Sapelo stands for nature, accessed through the culture of science, which makes the place a lab; for Bailey, Sapelo stands for culture, entangled through practices such as fishing, farming, and hospitality with a particular nature, which makes the place a home.

A variety of views on the nature of the surrounding sea are in motion here. From Bailey's memoir: "We were surrounded by water, yes, but the old people always were worried about their children drowning. They'd tell kids, 'Stay out of the water. Stay out of that water. Don't go in that water.' . . . It was like they distrusted the water because that water had carried our ancestors here from their home in Africa."[106] With this story, Sapelo becomes visible as a point on the black Atlantic—that analytic and historical unit Paul Gilroy has suggested can link histories previously kept apart by such continental or national terms as *African* and *African American.*[107] The distrust of water described by Bailey should remind us of another Atlantic geometry of abduction, one more brutal, less fanciful, than the Bermuda Triangle (and certainly of a different semiotic order than the Peircian variety): the ocean of the Middle Passage—that line in the triangle trade that saw captive Africans crossing the Atlantic toward servitude, if they did not—as perhaps as many as half did—suffocate, mortally sicken, or drown along the way.

As Bailey's designation of herself as a Saltwater Geechee suggests, however, the sea has also been an intimate, positive presence in the black Atlantic, particularly in the Sea Islands. Water is a buffer from mainland politics. Baptism has been symbolically associated with swimming escapes from slavery.[108] Bailey's book recounts childhood times collecting shells, playing in the surf. To be sure, there is nostalgia in these tales, "memory-haunted versions of the [black] rural south" that serve as a magical realist salve for northern readers longing to find a counternarrative to the all-too-present racial lineaments of the urban North.[109] I walked to a beach celebrated in Bailey's book and found it covered with sand dollars.

The reproductive doings of the sand dollar fascinated African American marine biologist Ernest Everett Just, who spent his childhood on the Sea Islands and in the 1930s investigated "the cortical response of the egg to insemination and the role of a substance called fertilizin in straight and cross-fertilization."[110] Dividing his time between Howard University, the

Woods Hole Oceanographic Institution, the Stazione Zoologica di Napoli in Italy (home institution to Uexküll, before World War I), and the Kaiser-Wilhelm Institut in Berlin, Just exemplifies the black Atlantic subject posited by Gilroy. Just loved the ocean, in part because it connected him to Europe and the Mediterranean, places where he escaped the overwork and underpay of Howard. In a poem Just wrote during a sea voyage from Naples to Bremen, he sighed, "The purple sea is home for me—Harbors are my bane. . . . O, wine-dark sea, I hold to thee—Take me utterly."[111] Like marine geologist Kathleen Crane, who during the cold war collaborated with Soviet scientists when the American oceanographic infrastructure blocked women from working on navy ships, Just sought transoceanic relief from prejudices that threatened to derail his research.

Just was interested in the inward propagation of biochemical waves from cell walls and considered this motion an impetus for cellular development motivated not by nucleic acids but by cytoplasm.[112] This biosemiotics hears significance emerging from embodied analog dynamics.[113] These days, the genre of information that concerns many marine biologists is digital, bioinformatic. And at Sapelo much of this information is delaminated from the organism, away, indeed, from Sapelo itself. Thus, students in Athens, Georgia, sequence genes gathered from microbial samples taken far away from the lab.

In *Liquid Modernity*, Zygmunt Bauman writes of delocalization as a tendency of modern power and its forms of life: "The prime technique of power is now escape, slippage, elision, and avoidance, the effective rejection of any territorial confinement with its cumbersome corollaries of order-building, order-maintenance and the responsibility for the consequences of it all as well as of the necessity to bear their costs."[114] Where once the ancestors of people like Bailey were brought into subordination through their deterritorialization from originary lands, now, "in the fluid stage of modernity, the settled majority is ruled by the nomadic and exterritorial elite."[115] Bioinformatics often services this nomadic geography, deterritorializing ocean ecologies and scientists. The move to microbes as mascots for a translocal, even globalized, ocean also often displaces those imaginations of the sea that operate at intermediate scales, relegating them to the realm of "local knowledge."

My argument here is not animated by nostalgia for local places, as though these are sturdy, self-evident things; *locality*, like *the global*, is an effect rather than a preexisting condition.[116] After all, the Sea Islands are as global as they are local, part of the "rhizomorphic, fractal structure of the transcultural, international formation" Gilroy describes as the black Atlantic.[117]

I hope, rather, to deliver a description of new networks within which ocean life is being made to travel—ones distinct from, say, the nineteenth-century circulation of whale products or the arrival from far-flung places of barnacle samples to Darwin's country home.[118] Those were social networks through which organisms traveled; new webworks increasingly limn the ocean itself as a network, as a sea of genes, a soup of information ready to be uploaded. In part, this image mirrors the online, Internet techniques now used to scrutinize the sea. Many of the scientists of whom I write (DeLong, Chisholm, Venter) have lately been funded by the Moore Foundation, founded by Intel's billionaire CEO emeritus Gordon Moore, who believes that the ever-increasing processing and networking power of computers should be aimed at gathering data for planetary management.[119]

The floating genes of the information ocean described by Venter are animated by slippage and avoidance, modulated in the service of his panoptic view of the world as his oyster. At Sapelo, the marine institute slides away from island ecologies, carrying the ocean away in the form of microbial sequences, formatted to zip around the Internet. The place-based projects Bailey admires sit uneasily alongside the delocalized entity that is the Sapelo Island Microbial Observatory, most solidly instantiated in a website allowing web surfers to look at microbial genome data.[120]

Professional societies of marine biologists know better than most how the racial history of the United States ripples into the demographics of the discipline. Several institutions work to enlist underrepresented groups into marine science. Hampton University, a historically black college in Virginia, sponsors Minorities at Sea Together (MAST), inviting applicants to "spend three weeks under sail studying marine science, policy, and the heritage of African Americans and Native Americans on the Chesapeake; live aboard and operate a 53-foot sailing vessel; conduct scientific studies of the marine environment; and reclaim the heritage of African Americans under sail."[121] The program looks back to the days of black whalers but also brings marine biological questions into the present, asking participants to research the bay's water quality, in danger from supernutrification from industrial effluent.[122] Students (in 2004, six African Americans, three Puerto Ricans, two Native Americans, a Native Hawaiian, and a Chicano) gather data on oxygen depletion in the estuary, a proxy for rapid algal growth. If the digital, uploaded ocean can perform the purposes of an "exterritorial elite," it can be also be downloaded into novel remappings of space, place, and race. This is a polyvalent biosemiotics, though, with healthier waters having many possible outcomes: reduced environmental toxins, better fishing, but also out-of-reach private property values (abductions in the logical register, then, like abductions in the corporeal sense, retain the potential to do violence). If the work of MAST directs needed attention to environmental racism, it also reinforces epistemological hierarchies that have minority students doing community-minded work while others cruise around in yachts doing "global" science. In the twenty-first century, the color line has a baroque, nonlinear structure.

Diversity efforts often enlist role models to encourage careers in marine sciences.[123] Cornell-trained marine microbiologist Karen Nelson, whom I met at the extremophile workshop in Baltimore, is noted for her sequencing of *Thermatoga maritima* and work on gene transfer. She has been active in outreach to minority students, telling stories of her youth in Jamaica, her animal science studies at the University of the West Indies Trinidad and Tobago (she wanted to be a vet), and her path toward becoming a project leader at the Institute for Genomic Research, a story she has narrated for Black Entertainment Television.[124]

Weavings of sociology and biography here reveal a complex topography for racial formations and marine biology in the United States, one for which we need a metaphor besides the color line. Troy Duster has offered that race and racism are like water, taking many forms—solid, liquid, gas: durable like marginalized or gated neighborhoods, fluid like biographies and identity, ubiquitous but ephemeral like turns of phrase or tubes of sunscreen.[125] Boas knew that there was more to water than color: ice, rivers, rain, fog, and snow present humans with different possibilities and predicaments.[126] Marshlands like Sapelo's—fresh and salty, part soil, part water—point toward baroque conjunctions of land and sea, links that implicate social and political inequality. Venter's vision of a global genome looks different from this space. His Darwin costume, his dressing up as a great white sea captain, sits awkwardly next to those less globetrotting microbiologists concerned with the boundary spaces where nature and culture meet.

Eugene Thacker suggests that as genomic practice standardizes database formats, legal regimes about intellectual property in biomaterials, and even ideologies about the determinative power of DNA, we are witnessing the making of a form he calls "the global genome."[127] But, as I have suggested here, such a figure materializes only when people like Venter take it upon themselves to spin yarns about this outsized creature. Such stories can always be undone, unwound, and, indeed, always contain within themselves a swarm of other possible scales and tales. As John Law suggests, the global should be thought of as "something that is broken, poorly formed, and comes in patches."[128] These patches are woven of such baroque matters as the shifting color of seawater, the travel of gene-juggling marine viruses, the quality of drinking water, incongruent claims of property in biomatter, the nutrient load of rivers flowing to the sea, and the patterns of currents, eddies, flows, and turbulence that link bodies of water to one another. Comprehending such connections involves a mix of induction, deduction, and, yes, abduction—variously articulated and oriented amalgams of hope and reason. It also requires attentiveness to the unexpected scales at which seawater and human socialities can swirl together—can capture one another—a simultaneous caution about and openness to alien abductions.

6   Submarine Cyborgs

*Transductive Ethnography at the Seafloor,
Juan de Fuca Ridge*

I am preparing to sink into the sea, probably the first anthropologist to join the research submersible *Alvin* on a dive to the ocean floor. The three-person sub sits like a massive, oblong washing machine on the stern of the research vessel *Atlantis,* where a thick rope temporarily tethers it to an enormous metal A-frame rising from the ship's fantail. Clambering down a steep ladder into the submarine, I find pilot Bruce Strickrott already adjusting *Alvin*'s array of knobs, buttons, and computer screens. University of Washington geologist John Delaney descends next. Delivering a foul-mouthed oath, he wedges his tall frame into a tiny nook on the port side, a gap in the mesh of machinery that lines *Alvin*'s snug, 7-foot-wide interior sphere. He styles a red bandana into a headband and, as *Atlantis* lowers us into the open ocean, Bruce comments that the scarf makes the 62-year-old Delaney a dead ringer for country singer Willie Nelson. Outside, in the chilly waters of the northeastern Pacific, wetsuited escort swimmers survey the capsule exterior to make sure we do not go down gurgling. They snorkel past our individual 4-inch-thick acrylic viewports, each window just wide enough to fit the features of a face.

I have been able to join this dive courtesy of geologist Deborah Kelley of the School of Oceanography at the University of Washington in Seattle. A few weeks earlier, a berth on *Atlantis* opened up for this May–June 2004 NSF-funded trip to the Juan de Fuca Ridge, a site of undersea seismic activity about 200 miles off the Pacific Northwest coast. MBARI's Pete Girguis put in a word for me, and I soon found myself on the phone with Kelley, taking notes on steel-toed deck shoes and preparing for two weeks at sea with sixteen scientists, twenty-eight crew, and *Alvin*. The dive I join today will employ a high-resolution imaging sonar system called Imagenex to map portions of the Mothra Hydrothermal Vent Field, a region of "black smokers" on the Endeavour Segment, a narrow submarine volcano on the edge of the Juan de Fuca tectonic plate. The dive will be a standard eight-or-so hours long. I have been able to sign on because no groundbreaking research is slated for this routine excursion, Dive no. 4020—an indication of the safe and steady rhythm into which *Alvin* dives have settled since the Woods Hole Oceanographic Institution in Massachusetts began operating the sub in 1964.

As we begin our hour-long descent to the ocean bottom, 7,000 feet below, the easy image comes to me of *Alvin* as a ball of culture submerged in the domain of nature. After all, submarine settings take "to an extreme the displacement of the natural environment by a technological one."[1] As vent biologist Cindy Van Dover suggests more sensationally, "Descending the water column in a submarine is an unnatural act."[2] But natural and cultural dynamics develop dense interrelations as well, feeding back into one another in *Alvin*'s immersion. The assemblage of the sub and its encapsulated scientists is clearly a cyborg, a combination of the organic and machinic kept in tune and on track through visual, audio, and (human) metabolic feedback.

The word *cybernetics*, coined by mathematician Norbert Wiener in 1948 to describe processes of information-mediated self-regulation and self-correction, is formed, in a fitting maritime allusion, from the Greek for "steersman" or "governor."[3] Van Dover, who became the first scientist to pilot *Alvin* (and the first woman, after a line of forty-eight men), describes the cybernetic intimacy pilots often seek with the submersible: "When the sub was on deck, I would work inside her and, with my eyes shut, reach out to touch a specific one of the hundreds of toggle switches to learn their locations by heart."[4] Passengers can feel only a fragment of this almost erotic connection. As an anthropologist on *Alvin* I am anxious about my role in this circuit. Recalling an iconoclastic one-liner by Chris Kelty, another ethnographer of the hypertechnological, I ask myself, "What would Margaret Mead do?"[5] Delaney unwittingly offers a partial answer, wisecracking that my research will constitute a "recursive study of ourselves studying." Mead was fascinated by more than sex in Samoa and trance and dance in Bali; she was also a fan of feedback systems. In an article titled "Cybernetics of Cybernetics," she urged social scientists to locate their practice within the language of systems theory.[6]

In this chapter, I take up and tweak that charge. I examine how scientists employ the formulae of communication and control to direct *Alvin*, to describe how marine microbes operate, and even to articulate the relation of researchers to their instruments and social milieu. When I first got word that I would dive in *Alvin,* graduate students on *Atlantis* joked that I would now really immerse myself in the culture of deep-sea oceanographers, seeing their preferred medium with my own anthropological eyes. They spoke truer than they knew, for *immersion,* as a sense of unmediated presence, is indeed one effect of being embedded in a seamlessly operating cybernetic system. Manfred Clynes and Nathan Kline defined the cyborg in 1960 as an "organizational complex functioning as an integrated homeostatic system unconsciously."[7] Attending to the conditions that allow such automatic operation, I inquire into the technical and epistemological circumstances that produce the experience of immersion, multiply defined: as a descent into liquid, as absorption in some activity or interest (e.g., music), and as the all-encompassing entry of a person, like an anthropologist, into an unfamiliar cultural medium. I ask what forms of nature, culture, and science swim in this model of presence, and which get left out.

Donna Haraway in her paradigm-bending essay of 1985, "A Cyborg Manifesto," argued that cybernetics, far from reimagining organisms as rigid machines, opened up possibilities for recoding our bodies and selves, for short-circuiting the idea that a fixed "nature" dictated our personal and political destinies. Transporting Haraway's reoutfitted cyborg to the underwater realm, I develop the concept of the *submarine cyborg* to meditate upon how flows of feedback call forth a medium of interaction that blurs distinctions between inside and outside, artifice and environment. Such a medium is, like water around a sub, at once hyperpresent and invisible.

But my submarine cyborg also calls attention to the differences between media (e.g., seawater and air) that must first be bridged in order for boundaries to be blurred. I mean to make evident the translations across media— the transductions, to use the technical term—which, when hard-wired into machines and shuttled out of consciousness, produce the very sensation of being one with a fluid surround. I borrow the term *transduction* from the science of acoustics, where it refers to the conversion of sound signals from one medium to another—a borrowing I was inspired to by attending to how sounds from outside *Alvin* come to feel so present to those of us in the sub. I draw, too, on the more general, technical meaning of *transduction* as the transformation of energy from one kind to another, and I occasionally make use of its biological sense, referring to the relay of biological stimuli.

If the cyborg names an entity that exists through the maintenance of its equilibrium and boundaries—boundaries that shift and expand as they are networked to feedback dynamics across contexts (e.g., the coordination of submarines with surface ships) and scales (e.g., the modulation by microbial consortia of climate dynamics), submarine cyborgs make explicit the physical character of information translations necessary to maintain the integrity of self-regulating entities, to delineate as well as dissolve interiors and exteriors. I am less interested in the self-referential looping of a "cybernetics of cybernetics" than in the transduction of sounds, senses, and signals that support cybernetic sensibility and consciousness in the first place.

Submarine cyborgs steer me from my dive in *Alvin* to the microbiology lab in *Atlantis*, where microbes are described as "little living machines," and to an analysis of a networked undersea observatory NEPTUNE to be sited on the Juan de Fuca Ridge and controlled from cyberspace. At chapter's end, I call on cybernetics' resonance with governance (a meaning physicist André-Marie Ampère suggested for "cybernetique" in the nineteenth century) to listen to how scientists and policy makers at the United Nations hope to fine-tune informal and international mechanisms regulating the bioprospecting of vent sites outside national boundaries, locations beyond the circuits of the nation-state, zones that represent political varieties of the alien ocean. Transduction surfaces throughout the chapter to describe the textures of disjuncture that must be smoothed over in order for submarine cyborgs to function.

A CYBORG MANIFEST

Despite our deepening immersion, my immediate sense at the beginning of my *Alvin* dive is not of some submarine cyborg communion but of a crowding against other people's bodies. At one step of remove, passengers like Delaney and me, listed on the roster as observers, might be thought of as a cyborg manifest. As the pilot, Bruce has the strongest feeling for this cyborg.[8] We angle our legs around to find our most contained conformation. Jibes like "I don't want to get to know you *that* well" bespeak anxiety about this unaccustomed proximity. We are three straight men in a sub. Volumes could be written about how *Alvin*'s cyborg manifest has been marked by gender and sexuality. In *Water Baby*, journalist Victoria Kaharl reports the stories of women who found their presences in *Alvin* a source of worry for male sub mates who wondered how to urinate discreetly into their Human Element Range Extenders, more colloquially known as "pee bottles." Women worried, too; geologist Kathleen Crane wrote in her journal of 1977—six years after the first woman dove in *Alvin*—"I feel that to fit into this submersible operation I have to become completely sexless, so that nobody will notice that I am any different from the others."[9] Women like Crane often found themselves trying to be female men—a drag performance that did not always get them into the sub. Kaharl quotes *Alvin* electrician Bill Page: "The male graduate students would suddenly have their turn in the sub on the third or fourth day out, but if they were female graduate students, they might not get a turn at all."[10] This was a tale I heard repeated by prominent vent scientists; one did not dive until after her doctorate. The gender-neutral body was understood in the culture of mid-century American oceanography, as in early space travel, to be male.[11] In its initial incarnation, so was the cyborg, imagined as the ideal form for the astronaut integrated with the controls of his spaceship. David Mindell, commenting on the image of the steersman, reminds us that, "from sea captains and riverboat pilots to aviators and computer operators, these figures stood for a masculine ideal of control over two worlds, the natural and the technological."[12]

The comparison of *Alvin* to a spacecraft is easy in coming: "The spaceship has become the standard image of the megatechnic ideal of complete detachment from the organic habitat."[13] And like spaceships built by the United States, *Alvin* has done military duty, occasionally reeled in by one of its funding sources, the Office of Naval Research, to do such cyborgian tasks as searching for a mislaid H-bomb at the bottom of the Mediterranean.[14] But there are key differences from spaceships. With *Alvin*, the idea is that we are immersing ourselves, not achieving escape velocity. The sub's sensory prosthetics underwrite an aesthetics of *being there* in the sea.[15] As we drift down, Bruce calls to us to look out our portholes. Bioluminescent blobs stream upward like gelatinous gold. He flashes *Alvin's* lamps, dosing the critters with light. Neither John nor Bruce is a biologist, so when I ask what these thingies are, all I get is "Some bio." Bruce switches off our exterior lights to save power. The phone rings. Kelley on *Atlantis* has a speakerphone question for John about a grant proposal. Her voice, soaked with echo like a vocal track on a Jamaican dub recording, bounces around the sub as she and Delaney agree about an e-mail she will send. I scan our cybernetic senses, noting the personalized television screens on which we can channel surf through the kaleidoscopic video available from cameras and sensors arrayed on the exterior of the sub.

THE SOUND OF CYBORGS

We are well into our descent, some 400 meters down. We continue to sound—in the sense of sinking and measuring our distance into—the deep. Such sounding employs techniques, like sonar, which, in a confusing pun, transmit and capture sound: *sound* as fathoming has its etymological moorings in the Old English *sund*, "sea"; *sound* as vibration reaches back to Old English *swinn*, "melody." With the interior lights dimmed, a cycle of blips and bleeps captures my attention. Bruce identifies these as a 9-kilohertz tracking pulse sent out from *Alvin* to *Atlantis* every three seconds, a 9.5-kilohertz response from the ship, and a steady metronome of "pings" from transponders dispatched to the seafloor by *Atlantis* in advance of *Alvin* dives. Transponders are spheres the size of beach balls which, distributed across patches of ocean bottom, transmit signals that help the sub locate itself in three dimensions using triangulation. Bruce says he thinks of transponder pings as background noise. But they are not exactly the meaningless patter that Kaharl, who descended in *Alvin* in 1989, rendered in her dive narrative as occasional interruptions of "POP weewee wo WOP ka POP weewee wo."[16] For Bruce, the noises secure a sense that the sub is somewhere rather than nowhere, supported in a web of sound rather than lost in a featureless void. Even though he jokes that the prattle of pings can be an "acoustic will-o'-the-wisp" ("a thing that deludes or misleads by means of fugitive appearances" [OED]), these echoes are for him the warp and weft of a reassuring soundscape. "Without them, it'd be too quiet," he offers.

In "The Sounds of Science: Listening to Laboratory Practice," Cyrus Mody writes that "labs are full of sounds and noises, wanted and unwanted, many of which are coordinated with the bodily work of moving through space, looking at specimens, and manipulating instruments."[17] And so it is here in the oceanographic field; work in *Alvin* is coordinated by and through sound, even if we are not always fully tuned in to how. Indeed, such sound is crucial to making the undersea realm a "field" at all.

These sounds, I find, vitally contribute to a feeling of immersion. Submerging into the sea merges with a sense of submerging into a burbling, reverberating soundscape.[18] Such soundscapes, it is important to recognize, are not available to submariners without devices that permit hearing across media.[19] *Alvin*, maintained in its interior at one atmosphere of pressure (everyday sea-level pressure), can deliver to passengers a sense of an exterior soundscape only because of technologies that transduce underwater sound into signals audible in air.[20] Only in this way does it become possible to imagine ourselves emplaced in sound. Transduction, the alteration of "the physical nature or medium of (a signal)" and the conversion of "variations in (a medium) into corresponding variations in another medium" (OED), points to the material transformations involved in signal relays. If information directs us to questions of measurement as well as meaning, transduction adds the dimension of materiality.

Directional sound is a key information currency used by submarine cyborgs.[21] Submarines, at least for the United States, slipped into cybernetic waters in 1941, when oceanographers at Woods Hole published a report for the U.S. Navy titled *Sound Transmission in Sea Water*, which suggested ways military submarine pilots might hone their deadly games of hide-and-seek.[22] Directional sound made questions of feedback central to submarine warfare and demanded consideration of the properties of water. The speed of sound in water varies with temperature, and temperature with depth, so most of the time sound traveling obliquely through seawater does not move in a straight line but is bent like light through a prism. World War II submarine pilots confronting enemy vessels not equipped with local temperature profiles could fire upon them with crippling accuracy and then, predicting how the beams of their adversaries' targeting echo rangers would refract through the water column, take evasive action by hiding in sonic "shadow zones."[23] The result was that pilots imagined their adversary, as themselves, as constitutively oppositional, a vision that embedded in the cyborg bodies of submarines and submariners what Peter Galison calls "the ontology of the enemy."[24] In other words, submarines presumed an alien ocean even as they sought to secret themselves intimately within it. *Alvin*, as a research sub, travels through a less oppositional soundscape. Kaharl captures the inquisitive, even childlike exploratory ethos of most *Alvin* dives in the title of her book, *Water Baby*, a reference to Charles Kingsley's Victorian novel *The Water-Babies*, which tells the fanciful tale of a working-class boy who discovers Christian virtue and industry when he is reborn underwater to be tutored by sea creatures.[25]

Sound from outside (and, as we soon hear, inside) *Alvin* is an essential element in our submarine perception of immanent presence. According to Jonathan Sterne, the phenomenology inherited by Western science and religion tells us that "hearing is concerned with interiors, vision is concerned with surfaces. . . . hearing tends toward subjectivity, vision tends toward objectivity. . . . hearing is a sense that immerses us in the world, vision is a sense that removes us from it."[26] The sounds around *Alvin* reinforce the notion that we are in an interior, subjective space, sonically and wetly immersed in a world of objective, if dimly sensed, facts. Because sound travels far farther in water than in air, sonar pings and pongs create an echoing sense of being in a landscape that extends beyond the confines of the sphere, one reason few people become claustrophobic in the tight space of *Alvin*. In here, we are safe, water babies floating to the techno lullaby of transponders.

Importantly, we need not pay attention to such sounds; onboard computers process signals and transduce them automatically into visual data for navigation. "No one now wears headphones and a rapt, faraway look, attentive in ambient hush," observes James Hamilton-Paterson in *The Great Deep*. "For all that modern oceanography relies so much on acoustic techniques, it is the machines which do the listening."[27] In the early 1980s, when computers were first put on *Alvin*, they were divided into three kinds—named Collectors, Listeners, and Nodes—which gathered, stored, and displayed data and allowed a human interface.[28] "Listeners" were not strictly or only dedicated to sound processing but were so named because of their interpretative, sorting functions; they were programmed to make data presentable, worthy of attention. The word *listening* is important. Listening has been associated with active, often highly technical, efforts to interpret or discern, whereas hearing has been imagined as passive, letting sounds wash over the ear. Listening is work. If listening to sonar on *Alvin* has been delegated to machines, passengers now hear in a much more diffuse, less disciplined way than in earlier days. *Alvin* divers' *acoustemology*, their "sonic way of knowing and being in the world,"[29] has transformed from the attentive to the immersive. One result of this shift is that sound from outside *Alvin* becomes a just-out-of-consciousness buoy for our perception of floating presence. Because we do not need to work at the boundary between self and sound—because we do not have to be actively aware of transducing—the boundary becomes imperceptible, inaudible. The feeling is distinct from the largely visual experience of using a remotely operated vehicle like *Ventana* (see chapter 1), which does not deliver a three-dimensional soundscape and which, for all its immersive qualities, simultaneously reinforces a sense of experiential distance.

Immersion—in water, in sound—has come to suggest being submerged in a space as well as becoming one with it. Such a conception of immersion as a kind of communion achieved through dissolution first emerged historically when bathing in the sea became understood as healthy and appealing, in that late eighteenth-century age when Romantic poets sought through swimming to achieve a sublime union with the sea, a "merging with the elemental forces."[30] Earlier therapeutic immersions in the sea— in, for example, seashore Britain—delivered not a reassuring sense of fluid connection ("as if, being seven tenths water, one's body were transparent")[31] but a bracing shock. I take immersion these days to have lost that edge, to be a scaled-down version of what Freud in 1930 called the "oceanic feeling," a sensation of egoless unity with a fluid surround.

It is a cliché to say that anthropologists specialize in placing themselves in "the field" in order to immerse themselves in culture, whether these are social worlds distinct from their everyday lives or more finely inhabited versions of something they thought was familiar. Kirsten Hastrup invokes

Margaret Mead to argue that "immersing oneself in local life is good. . . . [F]ieldwork implies that the well-established opposition between subject and object dissolves."[32] An articulation in the old-school language of the discipline comes from Alexander Goldenweiser, pronouncing in a 1933 piece in *American Anthropologist* that "a field student who is also an ethnologist must . . . forget his own culture and immerse himself sympathetically (*Einfühlung*) into the primitive view-point."[33] Goldenweiser renders immersion—something like the participation pole of participant-observation—as a matter of seeing and sympathizing. Later formulations move into the register of sound, with language immersion the paradigmatic mode of such practice (this meaning enters English in 1965 with Berlitz's "total immersion" courses). Here, immersion has a person surrounded by a sonic medium in which words can be both vaguely heard and intently listened to, what immersion educators Swain and Lapkin refer to as a "language bath."[34] Language and culture become media analogized to water. No wonder diving in *Alvin* feels like perfect anthropological fieldwork. And no wonder, too, that the story of my dive often mimes scientists' own narrative deployment of *Alvin* dives to deliver a being-there "truth" to ground their own fieldwork.

It is not all hushed, ambient techno in the world of *Alvin*. There is a more familiar, interior, air-pocketed soundscape, too. As we continue our descent, a quiet classic rock soundtrack accompanies us from Bruce's MP3 player, plugged into the sub. Chandra Mukerji, in her analysis of videotapes from *Alvin* dives, suggests that music functions as a social and psychological means for "normalizing the process of working in a small sphere on the dark seafloor."[35] This contention in mind, it might come as no surprise that North Americans, the majority of users of *Alvin*, often compare it to a car. At the first conference on hydrothermal vents, in 1981, oceanographer Fred Spiess described an *Alvin* dive this way: "It's a little bit as if you've climbed inside of a Volkswagen bug and then wrapped a big blanket around the outside and punched three little holes in it so you could look out and then decided you were going to spend the day with the outside temperature running about a little above freezing."[36]

Since most passengers are academics with modest salaries, VW Bugs feature more often in analogies than, say, Hummers, the favored comparison of Bruce, once in the military. For scientist and pilot alike, however, making this environment comfortable involves turning on the tunes; the fact that Bruce is in charge of the music, especially on the way down, asserts his control. For him, this is a commute to work. We can listen to music because we need not listen to sonar. Playing music in automobiles, as

Michael Bull writes in "Soundscapes of the Car," serves to sever drivers from the outside world, creating a private, interior space.[37] In *Alvin,* such a severing keeps our sense of identity bathed in familiar melodies that shield us against the alien world outside. The musical soundscape creates a sense of absorption in the interior of the sub, but because it mingles with the transduced soundscape of the outside the effect is to feel at once inside a bubble and porously immersed in a wider world. The sounds of *Alvin*—echoing from outside, trickling from inside—reinforce the notion that we are in an interior space, sonically and wetly immersed. Having background music for looking outside the sub goes some distance, too, toward aestheticizing the environment. David Toop, in *Ocean of Sound,* writes that "the image of bathing in sound is a recurrent theme of the past hundred years: Debussy's *Images* and Ravel's *Jeux d'eau* ripple around the listener; Arnold Schoenberg's *The Changing Chord—Summer Morning by a Lake—Colours* wraps us in flickering submarine light."[38] *Alvin* divers may not favor such modernist compositions, but they do go for soundscapy music; Pink Floyd's *Dark Side of the Moon* is a favorite.

There is yet another soundscape in the sub, that of the fugitive speech of passengers. Not all speech is evanescent, though, for each passenger is provided with a cassette recorder to make verbal notes. Like anthropologists, *Alvin* divers are encouraged to chronicle their experience. Delaney dictates some impressions.

I ask, "If you're doing all this tape recording, does that mean you spend a lot of time back on land listening to your own voice?"

"Most scientists are very chatty with their machines, not each other," he replies.

"Yeah, their auxiliary brains," adds Bruce.

Recording automatic speech permits later listening, contributing to the notion that sound is immediate, a fleeting sign of reality itself. Staring at my tape recorder, I replay in my head a phrase from a conversation I had on board *Atlantis* with microbiologist Jim Holden, who hopes to culture extremophiles retrieved on our eighteen-day journey. "Microbes," he instructed me, "can be little tape recorders of their environment." So it is with *Alvin:* a large tape recorder of its environment. Subs in nature and microbes in culture have become kindred information-processing devices, their bodies inscribed by the places they have been.

And so, *Alvin* is a recording studio. Indeed, a previous chief engineer for *Alvin* had substantial audio experience: "Jim Akens . . . joined the Alvin group in 1977 after a decade in the rock-and-roll business; he built state-of-the-art sound systems for Joan Baez, Jeff Beck, Sonny Rollins, Steely

Dan, Joni Mitchell."[39] By the 1970s, recording studios had become stan-
dardized; they had become sites of signal routing, monitoring, controlled
feedback—control and communications systems, like *Alvin:* cybernetic
devices.[40]

Cyborgs have primarily been imagined in a visual, even textual, register—
made of inscribed surfaces, of information and codes. Cybernetics has been
a behaviorist science, insisting that the interior state of entities does not
matter to accounts of their equilibration. Observability is everything. But
observability is only part of the story in *Alvin.* In the sub's interior, our
sense of immersion, intimacy, a feeling for the cyborg, is accentuated by
our subliminal and subjective sense of the sounds that surround us, sounds
we are no longer encouraged to comprehend, let alone experience, as transduced.[41] What Hillel Schwartz names "the indefensible ear"—that organ
imagined as always vulnerable, always "on"—constitutes the channel we
imagine tunes us in to our innermost selves.[42] *Alvin* as cyborg, however,
draws attention to sonic dimensions of cyborg embodiment. As a submarine cyborg, *Alvin* can be used as a model for sounding the interiors of
cybernetic bodies, for calling into audibility the transductions that traverse
the boundaries of such entities.[43]

OTHERWORLDLY ASSOCIATIONS AND EARTHLY DOMESTICATIONS

We arrive at the seafloor. Bruce turns on the lights of the sub, illuminating
a rocky landscape. Spider crabs crawl lugubriously over brown boulders.
The 300 atmospheres weighing on the sub are impossible to imagine from
inside our titanium bubble. We scrabble around for words, images. As an
anthropologist, I have been inoculated against pronouncements about
hearts of darkness, threatening Others, forbidden fruit—so I am not likely
to repeat the associations made, for example, by Kathleen Crane, who in
her dramatic account of one *Alvin* dive wrote of "a darkness looming from
what seemed to be the darkest place on Earth. This alien blackness was
deepened by the fearful emotions generated by looking into a forbidden
world."[44] Such darkness may appear as such only to those with eyes like
ours; Van Dover has argued that some shrimp can see the infrared glow
emitted by vents' thermal radiation.[45]

Still, sorting through my store of symbols for making sense of this
event is like rummaging through my cultural unconscious. When Don
DeLillo writes in his novel *The Names* that "oceans are the subconscious
of the world," we read not some universal truth but a gloss of the

unimaginativeness in which I find myself floundering. My notepad scribbles are disappointing encounters with clichés about other planets—although the sheer fact of living through a sci-fi fantasy reminds me of Haraway's contention that "the boundary between science fiction and social reality is an optical illusion."[46] The erasure of the boundary between ethnographic science fiction and social reality is also an illusion, and perhaps a partly auditory one. My use of the ethnographic present tense has its own potentially immersive effects for you, reader, reading aloud or to yourself, and my calling attention to this device means to direct your awareness to how ethnographic experience is transduced into ethnographic text.[47]

We see a small yellow box tethered about a meter above the seafloor—a seismometer for detecting tremors. The landmark reassures Bruce and John. It even suggests a property claim: do not disturb, science under way. For oceanographers, devices such as this mark their ongoing labor. They stand apart from less wholesome artifacts periodically found on the seabed, like beer cans. In *The Universe Below*, William Broad offers a Reebok sneaker encountered on an *Alvin* dive as a reminder of how the deep is not immune from littering. Such icons configure the deep as a wilderness in jeopardy—though the collection of pingers, pumps, and scientific paraphernalia over this area reveals this as a working landscape. More, the sea has for at least fifty years been full of more deleterious stuff, like nuclear waste. Even when "untouched," these sites are not always hygienic; some vent sulfides contain barite, which emits radioactivity, not something in which one seeks to be immersed. When some samples from the Juan de Fuca Ridge were discovered in the 1980s to emit radon, "folks who kept sulfide paperweights on their desks and mantels got rid of them and stopped prying off pieces to give away as souvenirs."[48]

Having become oriented around the seismometer, Bruce and John embark on the day's mapping task. Before we set out, they take me on "an Eiffel Tower tour" of some local vent features. I dutifully write "16:38 GMT" in my notes, trying to be objective. We approach a complex of vent chimneys called "Faulty Towers," after the British television sitcom, and John says, "What you're going to see is what you see on the poster in the *Atlantis* dining room." This reference to the composite photo displayed in the mess hall of the ship gives me a template against which to judge my vision. Addicted to the recording capabilities of *Alvin*, I fiddle with one of the cameras provided in the sub. Delaney instructs me to look out the window: "Right now, if I were you, I'd be focusing exclusively on looking. Never mind the photography. I've got thousands of pictures. Just fill your eyes."

"What we're looking at is the Tower?" I ask.

"We're basically on the southern end of it," says John.

I take a picture anyway, of a silvery fish whistling by a hydrothermal black smoker. It comes out looking like a photo of a UFO hovering above an otherworldly spire. Such Lovecraftian images are never far away from prose about vents. Van Dover trades on a tradition of using analogies to primitive civilizations to describe the undersea realm: "Raw and powerful, black smokers look like cautionary totems of an inhospitable planet."[49]

For most vent scientists I know, this is a secular place. Even so, this deep is persistently spiritualized. In Fabian's analysis of the historical transition from religious to scientific travel among the learned classes of Europe, we read, "Religious travel had been *to* the centers of religion, or *to* the souls to be saved; now secular travel was *from* the centers of learning and power to places where man was to find nothing but himself."[50] Travel to vents, for scientists, has become a personal encounter with the sublime secrets of geology and evolution. Delaney has an obvious addiction to things profound. Before the dive, he showed me a poem about vents written by Michael Collier, poet laureate of Maryland. In "Fathom and League," composed after an *Alvin* dive, Collier wrote:

> Two miles down the sea floor is a skull,
> the wounded head of a monster—fractured,
> faulted, ridged.
> . . .
> if life begins
> in the ocean, it thrives on hot and cold,
>
> in the tumble and boil of the sea no longer silent.[51]
> In genesis there's no happiness, only awe
> and improbability. The hideous is beautiful:
> worms ten feet long, clams the size of Frisbees,
>
> and shrimp that swarm like insects. And something else:
> water burning inside of water, smoking spires
> and chimneys . . .

The associations here are with the deep as a place of origins and otherness. They also murmur with impending calamity; after all, this deep is a zone of volcanoes and seaquakes. Delaney avers that it would be great to die in such a disaster, to become part of the landscape he loves so well. He expounds upon undersea quakes and makes his first mention of microbes. With a nod to the rising profile of microbiology in vent science—a specialty competing with a previously dominant geology[52]—Delaney tells me seaquakes are not just geological events but, with the creation of hydrothermal sites, also "microbial blooms." I squint out the window. I want to see *some* sign of microbes, but my eyes are not properly trained. I have only words. And, like Collier's poem, they are rife with assonances and consonances—reminiscent of Coleridge's "sunless sea"—suggesting a commingling of breath and water, self and other. When will the science begin?

Two thousand meters down, at ten in the morning Pacific Standard Time, John and Bruce begin mapping segments of the Mothra Hydrothermal Vent Field, building on charts made by University of Washington geologists. Concerns with mapping put self and oceanic other back at a studied distance.

Mukerji writes that measurement and conceptualization keep *Alvin* scientists from freaking out beneath the waves: "By involving themselves in the use of their machinery, they protect themselves from the assault of sense impressions at the sites. . . . They do not see a strange world of wonder and awe outside their windows; they see a place to put down a pump to see how effective a signature can be in answering fundamental scientific questions about the vents."[53]

This strikes me as mostly correct, though I would adjust some of the theorization. Wonder and awe are in fact often taken for granted as the exciting hum beneath our cramped bodies. This recognition complicates another contention of Mukerji's—that "in both the extraction of specimens and their analysis, scientists alienate nature to serve the interests of science."[54] For Mukerji, there are four ways vent scientists alienate nature—that is, appropriate the physical world and estrange it from itself: by framing this realm within categories of science, by extracting samples, by transforming vents into representations, and by using samples and models to gain intellectual power over hydrothermal nature and social power over colleagues. Here, alienated nature becomes "part of the cultural capital that researchers can use to enhance their social standing."[55] The use of vent microbes to make biotechnological products such as Deep Vent might allow us to nominate a fifth, Marxist sense of alienation: the alienation from submarine pilots, research technicians, and postdocs of the material products of their labor to produce profit for biotech companies. These could all be aspects of an alien ocean. But what is left in place in Mukerji's account as well as in a classical Marxist framing is a romantic vision of nature itself as pristine, uncontaminated, possessed of a sublimely unknowable character. Nature is, even down here, thoroughly implicated with cultural activity—from radioactive waste dumping, to sonar pinging, to sewage, to deep-sea trawling, to tests of nuclear submarines. More, as

I saw in Monterey Bay, technological mediations often deliver a sense of immersion, not alienation.

More interesting to me is Mukerji's argument that vent scientists "use equipment to structure their encounters with the ocean in an interrogatory form."[56] In other words, the seafloor is littered with particular people's questions. Those questions are increasingly microbiological, posed in such forms as titanium incubators stuck into vent walls. Our charge for now, however, is to run lines up and down and back and forth along a defined area of vent field, an activity Bruce refers to as "mowing the grass." This task is one for which a human-occupied vehicle like *Alvin* is quite over- and underqualified; robots could do this job more exactly and for longer. Our cyborg manifest could be replaced with an inorganic system of communication and control. There is also no hope that this particular dive can be sociologically representative of *Alvin* dives in general. Of course, given that many scientists get only one *Alvin* dive in their career, there is little hope that any researcher's dive will generate data representative of any ocean ecology in general, either—even if people do use their dives as springboards into articles or dissertations. Questions of the representativeness of such immersive "fieldwork" bedevil both anthropology and oceanography.

If our vertical arrival at the sea floor was suffused with imagery of travel into space, across a threshold, into an alien world, horizontal motion takes us across a landscape to be surveyed, domesticated. In most narrations of *Alvin* dives, such movement is described as frontiering. As Van Dover puts it, "Deep-sea research . . . remains . . . a frontier science. The seafloor is the largest and least known wilderness on our planet."[57] William Broad, who also dove with Delaney, extends the American character of such imagery, offering that mid-ocean ridges are "like seams on a baseball" and that the Mid-Atlantic Ridge is like the Rocky Mountains and the Juan de Fuca Ridge is "akin to the gentle hills of the Appalachians."[58] Before I embarked on my dive, one scientist on *Atlantis* prepared me: "It makes you feel insignificant, being down there. If they were all visible, above water, these places would be national parks" (and, indeed, rocks from the Mothra vent field are now displayed in the next best thing, the American Museum of Natural History in New York City). The national park is a common image; Kaharl reports that the other observer on her dive said, "This looks like Bryce Canyon."[59] Bruce's summary of our day's work as "mowing the grass" domesticates such similes, casting us as doing the mundane work of keeping the space known and cultured, maintaining it as a sort of American subdivision.

Then again, careful scrutiny of our coordinates reveals that we are in fact in Canada—or more accurately, Canadian waters. In other words, we are not in a simply immersive space; rather, this is a zone in which our work is rigidly structured, even surveilled. Our submarine cyborg must move within spaces already configured by governance.

"WE ARE MERGING WITH OUR DATA"

As part of cruise planning Kelley had to get clearance from the Canadian navy to deploy *Alvin* in these seas, part of a Canadian marine protected area.[60] The science party had to work within a circumscribed zone, a circle of 5 nautical mile radius centered at 48°00′ N, 129°06′ W. The Canadian state department would receive a cruise report with a description of the expedition, including a list of samples collected. As this detail suggests, there are political as well as geological reasons why vent work is sometimes conducted outside national nautical boundaries. Crane reveals that the choice of the Galápagos area for the first explorations of vent sites in the late 1970s was largely political: "This particular site was chosen because it was located exactly 200 nautical miles from the Galápagos Islands and was legally out of the range of Ecuadorian gunboats, which were patrolling for illegal tuna fishing in their waters. At 200 nautical miles, we were in international waters and could carry out research without the fear of being hauled into port under the guidance of warships. It was in some ways political serendipity that led to one of the most interesting scientific discoveries of the twentieth century."[61]

The location of hydrothermal vent sites outside everyday territories has a distinct effect on how they are named. For Americans, vent names are up to the fancy of whoever first comes upon them. Unlike seamounts (larger formations about which there exist nomenclatural rules), vents are not central to military strategy or national nautical gerrymandering. Vent fields are given simultaneously awestruck and flip designations like the Garden of Eden, Hell, and Lost City (named by Deb Kelley, who told me she received earnest queries about whether she'd discovered Atlantis). Other fields have more idiosyncratic handles, like the Houston Astrodome, Clambake, or Broken Spur, named for objects scientists imagine they resemble. Scientists frequently give chimney formations whimsical names, like Edifice Rex. Unlike colonial explorers who often christened their discoveries with the names of benefactors or monarchs, vent scientists do not name structures after, say, NSF program managers. Nor do scientists name sites as resurrections of known areas of geological importance: no New Old

Faithful. These are spaces, scientists might say, divorced from grand politics, serious history. Thus, we have Godzilla and, of course, Mothra, Godzilla's adversary, a giant telepathic bug which, like vents, gives off toxic chemicals. In the close quarters of a sea journey, in-jokes dreamed up in the middle of the night become scientific designations. But they also become stern markers of territory for scientists staking careers on specific sites.

As Delaney and I look at *Alvin's* position displayed on one of the sub's screens, we pinpoint our location relative to mapped and unmapped portions of the seafloor. Delaney delivers a remarkable announcement as he watches the icon of *Alvin* move toward the already charted area of Mothra. Eyes fixed on the computer, he intones, "We are merging with our data." This idea of becoming one with the data, of the map becoming the territory—of culture folding into nature in a cybernetic one-to-one mapping—speaks to the intimacy Delaney feels with this terrain (*merge* derives from the Latin *mergere*, "to dip or plunge," the same root for *immersion*). His absorption in the sonar image and his more general immersion in the dive manifest a cyborg sensibility.[62]

A couple of days later, at a science meeting on *Atlantis,* Delaney enacts this sense corporeally. As he reviews the topography of Mothra, he directs a postdoc—the person who painstakingly created the final graphic—to pan and tilt a three-dimensional computer map, projected on a video screen. Delaney moves his body like a conductor and says, "music please," perfectly embodying the orchestrating, directing relation of professor to postdoc so characteristic of the natural sciences. In this dance, Delaney's body fuses with the map; he merges with the data. This relay of motion and energy can be interpreted as another genre of transduction, one akin to that described by Natasha Myers in her ethnographic exploration of how biologists who model proteins develop bodily intuitions about the movement of molecules, crafting a habitus that has their fingers, hands, and bodies responding to, miming, computer models of the protein structures they study.[63] In Delaney's dance, the transductive media of water, of the Imagenex sonar system, and of the body itself are all called into play. Many sorts of bodies—students, technicians, submarine pilots, computer scientists—are also part of the transductive chain through which, for Delaney, the alien ocean becomes the intimate, immersive, ocean. The presences produced through these transductions and immersions operate at scales beyond the individual, beyond the three passengers in the sub, to produce one version of the oceanographic field as such, a sense of ocean-space as a kind of virtual reality through which the appropriately cyborg subject might swim.

As Bruce maneuvers *Alvin*, often speaking to it by name, he scratches himself on the angled edge of a computer component. "There's got to be bits of my DNA on these machines, I hit them so often," he remarks. I nod and avail myself of the lunch packed for the dive by one of the *Atlantis* galley crew, Linda. The peanut butter sandwich cannot be eaten without recalling one of the most famous moments in *Alvin*'s history, when, in 1968, scientists had to bail out of a suddenly sinking sub, leaving lunch bags within, which were retrieved ten months later. In what has become a nugget of oceanographic folklore, *Alvin*'s baloney sandwiches survived their submarine exile intact, a sign that bacteria in that cold portion of the sea were metabolizing extremely slowly. Silently reviewing this story with a mouth full of white bread, I notice water droplets on the sub's window. Bruce reassures me that the sub is not leaking: "That's your breath. If you don't believe it, taste it. It's not salty."

Condensation is a sign of our life functions. In an earlier safety briefing in the sub, Bruce explained, "You breathe in O_2, metabolize it, and it comes out as CO_2—everybody knows that." He says now that the oxygen in the air is in lower concentration than we might be used to: "I like to keep the O_2 at 18 percent. If it's higher, it becomes a fire hazard and people get giddy. If it's lower, people come up tired." To be a sub pilot, Bruce must be able to gauge his passengers' moods, treating these as biofeedback signals. It is not the DNA all over *Alvin*'s interior that stands for the core of our vitality, then, but rather our metabolism. Just a few days previous, *Atlantis* microbiologist Jim Holden supplied me with a definition of life that revolved around metabolism: "Life is anything that transfers chemical energy from one compound to another for the purpose of replication. That's tricky, though, since that could include crystals. It would have to be a reaction that could not occur *abiotically* at the same rate. Biotic processes are those that happen through *metabolic* reactions, and enzymes that catalyze reactions that convert compound A to compound B facilitate those. And the reactions have to be catalyzed by organic molecules—even though they may contain inorganic components." On this view, humans and microbes share the process of metabolism, which unfolds for us in distinct media—the medium of air and peanut butter sandwiches versus the medium of water and hydrogen sulfide. Hyperlinking back to my discussion of the "message from the mud" in Monterey Bay, life is a media technology.

Or perhaps a cyborg technique. The biology of the past fifty years has been profoundly influenced by cybernetic terminology. We have become accustomed to describing organisms as information-processing and feedback devices. Vitality has been translated into a problem of coding, of systems management, whether at the level of genes, metabolic pathways, or ecosystems. Haraway asks that we be aware of what cyborg imagery occludes but also recognize that "machines can be prosthetic devices, intimate components, friendly selves,"[64] precisely because the logic of the cyborg can enable the liquefaction and denaturing of hierarchies and boundaries between such oppositions as human/machine, self/other, male/female, part/whole, and nature/culture. Scary and friendly both, *Alvin* lives just such a double cyborg life.

I have suggested here that immersion—which I have accessed less through cyborg imagery than cyborg sound—has become an experiential correlate of cyborg embodiment and that we should ask what the metaphor of immersion—cultural, linguistic, technological—obscures. Oceanographers do not just merge with their data. Subs do not just submerge anywhere they like. And anthropologists do not just soak up culture. Rather, to stay in the idiom of signal processing, transductions—material adjustments and translations—are necessary. Immersion functions as a rhetorical tool promising experiential truth through eliding the question of structure—whether of an ecosystem or a social order—positing a fluid osmosis of environment by a participant-observer/auditor. With submarine cyborgs as our guides, we can call into audibility the structures that permit such floating and absorption. I have in mind structures such as the programming of Imagenex, the university affiliations of *Alvin* divers (my joining the cruise was not hurt by my being an MIT professor), and the legal and political instruments that specify where to dive and what to do when one gets there. Instead of "recursive studies of ourselves studying"—old-time reflexive anthropology—I am offering a case for transductive ethnography, inquiry motivated not by the visual rhetoric of self-examination and self-correcting perspectivalism but by auditorily inspired, lateral attention to the modulating relations that produce insides and outsides, subjects and objects, sensation and sense data, that produce the very idea of presence itself. Rather than seeing from a point of view, then, we might tune in to surroundings, to circumstances that allow resonance, reverberation, echo—senses of presence and distance, at scales ranging from the individual to the collective. This is a model of anthropologists as transducers in circuits of social relations. A transductive ethnography would be a mode of attention that asks how subjects, objects, and field emerge in material relations that cannot be modeled in advance.[65] It would chart logical relations that could never be disentangled from particular cases. If the operation of abduction discussed in chapter 5 leaps to the future, transduction always travels across—athwart—the very present it is calling into being.

As we ascend, John takes a nap while Bruce and I talk about the making of *Volcanoes of the Deep Sea,* in which Bruce participated, and about how the political world above is headed to hell in a hurry. The conversation reminds me of another James Cameron IMAX movie, *Ghosts of the Abyss,* in which the crew of a submersible exploring the sunken *Titanic* surfaces on September 11, 2001, to hear news of the air attacks in the United States. In the cafeteria of the *Atlantis,* beside the photomontage of Faulty Towers hangs a framed, red-white-and-blue-beribboned picture of the Twin Towers before their smoking destruction. The national park and the national tragedy: mirror images of nature and culture, towers encountered, respectively, by submarine cyborgs and by airliners enacting a terminal ontology of the enemy.

Emerging from *Alvin,* I am approached by graduate students eager to drench me in cold water, the ritual greeting for first time divers. *Atlantis* chief mate Mitzi Crane points out over dinner that this is a baptism (yet another immersion), which, she jokes, affirms her sense that *Alvin* is a mother whose womb births new scientists.[66] The gender of *Alvin* is a live topic among the crew, who now see more women in the ranks of oilers, able-bodied seamen, engineers, and mates; is *Alvin* a "she," like most ships, or a "he," like an old drinking buddy named Alvin? That's the question making the rounds of the crew. *Alvin* is an object upon which we all project changing ideas about the nature of humans at sea, about scientists merging with their data, about cyborgs manifesting oceanographic and anthropological practices of immersion.

THE DEEP, ONCE OR TWICE REMOVED

During the two weeks I lived on *Atlantis,* I watched a scientific choreography that alternated between the careful cadence of ballet and the polite moshing of contact improv. Bustling up and down the steel stairs of the 274-foot-long ship, scientists and crew might be imagined on various levels of a metal scaffold, like actors clambering over an exposed iron armature in an avant-garde theater piece. More traditionally, their movements map a shipboard sociality that has long fused cultures of the merchant marine and lab science.[67] The senior scientists—four geologists, two chemists, one microbiologist—ranged between the top five levels of the ship, moving between the bridge, staterooms, galley/mess/library complex, and main lab, where they coordinated students and wrestled with recalcitrant instruments. Postdocs, graduate students, and lab techs—four geologists, two chemists, one microbiologist, one digital cartographer—moved between the galley/mess/library, lab, and windowless dormitory lodgings below deck. They could often be found in the computer lab, entering data. The crew, meanwhile, shuttled all over the ship, though deeper into the engine room than any scientist.

I was asked early on to take up a four-hour nightly watch, 8 P.M. to midnight, joining students as they searched by remote control for signs of new hydrothermal vents, a task they carried out in shifts till six in the morning. Though this duty sounded exciting, most students found it tedious. In the windowless computer room, staring at sluggishly updating screens of depth graphs, we coordinated the descent and ascent from *Atlantis* of a CTD-rosette package that detected anomalies in water temperature and acidity—possible signs of a rising plume of hydrothermal fluid. As the ship steamed slowly each night along a predetermined path between a pair of promising longitudes and latitudes, we remotely worked a pulley system linked to the CTD-rosette, raising and lowering the device like a yo-yo, a practice which, because it unfolded as the ship was moving, was called a "tow-yo." We worked in pairs, with one person checking a minimalist sonar display to make sure not to smack the package into the seafloor and the other plotting the ship's position on a maritime map.

My partner most evenings was a student I'll call Barb, a geologist. With other people coming by to chat, check e-mail, and eat microwave popcorn, the experience was pure dorm room.[68] Staying up all night at sea is a rite of passage for graduate students as well as a way to maximize data collected on NSF's dime. Senior scientists remarked to me that all-nighters made American oceanography distinct from Japanese marine research, which apparently stops whenever the ship's captain eats or goes to bed. Graveyard shifts, they also said, were harder to install on European ships, where wine and beer with dinner—not a feature on NSF-funded ships—make evenings more relaxed. No one noted, though, that *Alvin* was actually capable of diving around the clock; its eight to five schedule plots a time-clock workday for the pilot.

Barb and I glazed over gazing at screens for signs of a plume. The transduction of information from CTD to screen produced no sense of immediacy. If transduction might be imagined not only as the transformation of signals across media but also as the propagation of meaningful motion across bodies, Barb and I, at several removes from hydrothermal plumes, were unmoved, *not* merging with our data. When Deb peeked in, she remarked teasingly that we might show more excitement. When she was in school in the 1980s, Kelley was transfixed by screen traces. Barb was more eager to get into *Alvin*. I asked why.

It's cool as hell. There're some really bizarre structures down there. They're huge, enormous. It's really hard get a sense of them if you're just looking at footage from an ROV, because you're not in it, driving it—you don't get a sense of how things are arranged in space. In the sub, well, you're still not driving, the pilot is, but you're seeing things from your own point of view—port, starboard—you're *in* it. And you look at these huge structures and say: "How long have you been here? And how long have you been affecting the ocean?"

This account of immersion, pushing aside the transductions between passenger and landscape, is not just a declaration of epistemic preference. As David Mindell points out, "There's something more at stake here: scientists' image of themselves. Are they brave adventurers, risking all in pursuit of knowledge, or sober data analysts, poring over numbers and pictures from inside windowless laboratories? . . . Are you a real oceanographer if you don't descend to the seafloor?"[69] The question is similar to one I get as an anthropologist of science: are you a real anthropologist if you haven't suffered extreme climates, contracted dangerous diseases, if you study your academic colleagues and double-check your interview quotes over e-mail?

The answer to these questions is easy. What it means to do oceanography and ethnography is changing. In an age of remotely operated robots, Internet ocean observatories, multisited fieldwork, and online ethnography, presence in "the field" is increasingly simultaneously partial, fractionated, and prosthetic; it is not just distributed across spaces—multisited—but cobbled together from different genres of experience, apprehension, and data collection. It is multimodal. Indeed, scientific presence as such is not always necessary to produce representation. Against this background, *Alvin* is a throwback to an earlier mode of simply immersive fieldwork, one being eclipsed by multiply mediated attention to such phenomena as oceanographers' microbial assemblages and anthropologists' global assemblages.

Working tow-yo duty was a good way to learn how students made their way to marine science. When Kate, a geology student who taught me how to chart the ship's course, was growing up in Denver, her parents took her to hear Jean-Michel Cousteau, son of Jacques. She wrote him a letter as an assignment for her French class, and when he responded she was hooked. The one microbiology student on the ship, who had to attend to growth cycles of microbes in the ship's lab and couldn't do a shift, told me on a rare pass through the room: "I always knew I wanted to work at vents, ever since those *National Geographic* specials. When I was a kid, in Seattle, I always thought, 'there's the ocean, it's cold.' I had no idea that there were

all these hot chimneys and weird tubeworms. It's such a foreign environment." A chemistry student, Tiffany, had grown up as a "water person,"
spending her 1970s childhood at Lake Michigan where her grandparents
had a vacation home (Lake Erie, she editorialized, was frequently closed
because of contamination by *E. coli*). She attended Stanford, studying geology, then joined the Peace Corps, which convinced her she wanted to be a
doctor. She worked for Pfizer in San Diego, which she found unsatisfying.
While in La Jolla it occurred to her that going somewhere like Scripps
might allow her to combine scientific interests with her love of water. Her
studies of vent fluid at Washington had taken her, she lamented, "a long
way from doing something useful, like medicine." She glanced at a tow-yo
printout and sighed, "I'm not sure how this helps people in third world
countries." Wringing meaning out of vent fluid and microbes requires a lot
of steps, a lot of work to feel connected, immersed.

MICROBES ARE LITTLE LIVING MACHINES

When I met the microbiologist on *Atlantis*, Jim Holden, professor at the
University of Massachusetts at Amherst, he had just acquired fragments of
vent chimneys from an *Alvin* dive. My conversations with Holden wound
up centering on the place of his microbial work in his moral imagination,
though when I first found him in the *Atlantis* biolab he was focused on
more mundane matters, filling test tubes with chemical media, hoping to
isolate novel hyperthermophiles. He planned to submerge granulated vent
rock in tubes filled with iron, nitrogen, sulfur, and carbon dioxide to determine which creatures grew on which substrates. The tubes were incubated
at 95°C, high enough to keep the microbes in their preferred temperature
zone but not so hot they would boil, a danger at 100°C without appropriate high pressure.

Holden was less interested in DNA than in how microbes worked in
the wild. He was curious about the lifestyles of hyperthermophilic
chemolithoautotrophs, heat-loving organisms that use chemical energy to
fix inorganic carbon from the environment using inorganic electron donors
like hydrogen. Chewing on such an exact mouthful of Greek prefixes, suf-
fixes, and infixes is required to articulate the metabolic milieu within which
organisms exist. In the *Atlantis* library I found a concise classification in a
textbook by Van Dover:

> Metabolic processes of living organisms have three basic requirements:
> a source of energy, a source of carbon, and a source of electrons. The
> energy source may be light (photo-) or chemical (chemo-); the source

of carbon may be inorganic (auto-) or organic (hetero-); the electron donors may also be inorganic (litho-) or organic (organo-). By this nomenclature, photosynthesis is more thoroughly described as photoautolithotrophy (where H_2O is the inorganic electron donor); chemosynthesis is chemoautolithotrophy (where, for example, inorganic sulfide and oxygen yield the energy, and sulfide serves as the electron donor). Animals are chemoheteroorganotrophs.[70]

Humans, Holden said, are chemosynthesizers who eat other chemosynthesizers, like cows, which are in turn fueled by photosynthesizers (vegans' and vegetarians' mileage may differ). Chemolithoautotophy, I commented, seemed to place geology into the heart of a biological classification. Holden said yes, telling me that some anaerobic creatures "breathe (i.e., respire) iron like you or I breathe oxygen. They reduce rust, converting ferric iron into magnetic iron." He went on to explain why gene sequences alone were not enough to explain such processes:

> A genome sequence by itself has limited utility. Without physiology and biochemistry, the genome is useless. The real utility of the genome depends on manipulating the organism; it's what you're able to *do* with genome information once you have it. Now, seeing genes in their environment—that's environmental genomics, sequencing genes from the environment, metagenomics, to see what genes are there. But just because a microbe has a gene doesn't mean it's using it. You'll miss things with the environmental genomic approach. You'll miss the fact that one organism can use both sulfur and iron as an electron acceptor, and you won't know which it is actually doing. This is the tricky part about linking the lab to nature. The hope is to study organisms in their natural environment.
>
> So, a complementary approach is to do physiological ecology. You think of an environment as a collection of genomes. How does organism A turn on or off organism B? Metabolic pathways don't have to happen in just one organism. Organisms live in communities, like cities; there are bakers, doctors, lawyers. But we don't understand this at all. We have a tendency to focus on individuals rather than communities. Some scientists study monocultures; others, genes out of context. The question is: how can we use physiology and genomics to understand how organisms interact with their environment and one another?

"Context" appears as the marker of relevant relations, in this case of what "genome information" *does* in an environment. When I met up with Holden later, he had moved on to another experiment: exposing vent samples to gas mixtures so he could monitor changes in these mixtures' compositions in response to different metabolisms. It was a simulation of possible environments. He summarized his research questions: "Which form of metabolism is preferred by hyperthermophilic autotrophs? And when and where in nature will you find various metabolisms and under what conditions? How do hydrothermal vent systems and the organisms that reside at these vents affect one another?" The idea was to "look for chemical cues that life exists. The environment can record the presence of life. For example, we can analyze the composition of the atmosphere for oxygen and methane content. If life is altering a planet, maybe we can sniff life even if we can't see it."

This definition of life as based in part on its "smell," on the signature of its alteration of an environment—a biofeedback trace—moves away from the code metaphor often animating genetic portraits of vitality. But the search for messages between organisms and environments remains central. It reappeared in Holden's explanation that, while one could look at environmental processes to infer the presence of particular metabolisms, one could also do the reverse: look at microbes to learn about context.

> Microbes can be little tape recorders of their environment. By understanding their physiology, we can understand the microbe-environment interface. Coming out on this cruise is an attempt to learn about this interface in the field, to find out what happens in nature, to ground-truth what we know from the lab. Microbes are incredibly versatile. They can do what other organisms can't: they can grow in many compounds. Iron and sulfur, for example. We just live off potato chips and hamburgers and produce water and CO_2, which is not very interesting. What's more, even a single kind of microbe can radically switch its form of metabolism, processing nitrate as a heterotroph and then iron as an autotroph. They can respond to changes, and what's interesting is to understand how their genes are turned on and off to adjust to these changes. Microbes are little living machines that are not *consciously* able to adjust to their environment but have sensors that respond to environmental cues. The fact that they can change so rapidly, and in a way we can follow and manipulate—it's not so easy to grind up humans and find out what they do—makes them really interesting. They're full of surprises.

What should capture our attention is the cybernetic language of "switching," "turning on and off," and "adjusting" using "sensors"— actions that produce messages, what Holden calls "tape recordings" of the environment, a metaphor that suggests that such records can be played back, erased, recorded over. This is a vision of life forms as self-regulating systems of communication and control. Holden's observation that "metabolic pathways don't have to happen in just one organism" also speaks to the power of the cybernetic model to reorganize boundaries across scales, in this case from single organism to microbial consortium. But the materiality of such transformation was key for Holden. One mechanism that triggers changes in metabolism is called *signal transduction,* in which proteins on a cell membrane react to environmental cues to modulate gene expression and enzyme production. The operative metaphor here is still informatic, but, as with sound transduction, the physical structure of the substances that facilitate meaningful transformations are key for these little submarine cyborgs.[71]

I was keen to learn how Holden came to be captivated by marine microbes. He explained:

> I grew up around water—spent my whole life on or near water. This environment has always attracted me, from watching Jacques Cousteau, going to the beach, learning about the otherworldliness of marine life. I took Oceanography 101 and I was hooked. I studied hydrothermal vents in the classroom—this was 1984—a short while after they were discovered. Why is ocean life otherworldly? There are life forms living in marine environments so completely different than what you see in Iowa—there are fishes, eels, octopi, things living in three dimensions and in so many different colors and forms. The most extreme example of otherworldliness is at hydrothermal vents. You get down to a vent site in a sub and you wonder: "Am I still on Earth? This is so completely different." That's the attraction.

He placed his interest in a larger frame: "The draw is discovery—and that stems from my personal philosophy. I find beauty in discovery. I consider creation to be like a house at night with its lights on and we're standing some distance from it. Every time you make a discovery, you're closer to the house."

"And now," said Holden with a quiet pause, "I'm going to say something that will throw your anthropology right out the window, and that is that I'm a Christian. For me to make discoveries is like seeing the handiwork of the Creator. There are natural processes in the world, but I don't think it's limited to that. There is something beyond the laws. I don't like the deist, blind watchmaker argument. I think that God is intimately involved in the creation and can intercede in it; whether he goes through natural or supernatural laws, I'm open to both."

Holden told me he was not persuaded by Stephen Jay Gould's argument that religion and science are "nonoverlapping magisteria," answering questions of morality and fact, respectively.[72] Holden believed in a God both immanent (present, knowable, and working within the world) and transcendent (beyond the limits of our experience). Investigations into the evolution of hyperthermophiles are perfectly consistent with views of the world as a Creation in which God is continually present, coeval with the evolutionary becoming of reality itself. Holden's metaphor of a house at night—especially in association with hydrothermal vents—is powerful. The mythic space of the underworld becomes reassuring, safe. The superheated alien ocean becomes a hearth.

I asked Holden whether he considered microbes fellow creatures. He replied: "I'm just an observer and manipulator. I don't feel any oneness with them. Obviously, we share metabolic capabilities, but I don't have an attachment to them. That makes it easier to take a kilogram of them and grind them up. It's like being a farmer who raises pigs or cows. They are what they are. I am what I am. But I do see them as part of the handiwork of the Creator. We both have metabolism, which makes us alive. But we humans have something beyond metabolism, something that makes us special."

This position on the link between life forms and forms of life is consistent with Holden's vision of microbes as little living machines—a position similar to Descartes' explanation of the ensoulment of people as against animals. Though microbes and people share metabolism, embedding us in the world, for Holden humans have something beyond the observable; the language of observation he uses to describe his relation to microbes operates in a different register than immersion, in the idiom of objectification rather than subjectification of cells. It will be intriguing to learn whether recent attempts to amplify the sounds of cellular processes transform microbiologists' sense of connection to the world of the very small. The "sonocytology" of UCLA chemist Jim Gimzewski, dedicated to bringing the vibrations of cells into human audibility, may attune scientists to new worlds of cellular flourishing and distress. One of Gimzewski's collaborators reports that the "frequency of the yeast cells the researchers tested has always been in the same high range, 'about a C-sharp to D above middle C in terms of music.' . . . Sprinkling alcohol on a yeast cell to kill it raises the pitch."[73] In the transduction of cellular quivering into sound we may hear the music that Ernest Everett Just once imagined vibrating through cells. As Sophia Roosth's analysis of sonocytology makes clear, however, such transduction works on human sentiment to the extent that its own operation is occluded, that auditors are encouraged to treat as transparent—or, better, transsonant—the techniques that access and create the immersive subcellular soundscape Gimzewski conjures as echoing with "a kind of music."[74]

Holden was interested to discover how microbial machines became what they were. It was this becoming, I surmised, that made him interested in culturing extremophiles rather than sequencing them. But there were practical reasons as well; his lab in Amherst does not have the expertise, equipment, or time to create DNA libraries. He was curious, though, about what one might learn from such libraries. He had arranged for New England Biolabs—the company behind DeepVent—to sequence and archive the DNA of hyperthermophiles he gathered. In exchange for giving Holden information derived from genetic libraries, the company would retain the right to keep these libraries and explore their contents for biotech leads. Holden gave me examples of hyperthermophilic biotech, like stonewashed jeans (which, he joked, might better be called bacterially eaten jeans) and laundry detergent.

Since much of our work had been in the Canadian Exclusive Economic Zone, the question of international sampling agreements arose. Jim said the bureaucracy behind collecting samples with *Alvin* had been taken care of by Deb's request for permission to use the sub in these waters.[75] When I contacted Holden a year later, he told me the samples were still in his freezer; he probably would not seek clearance from Canada until he got around to looking at them and only then if he had sent them to New England Biolabs.

It is not only scientists who see microbes anew these days. Micro-organisms materialize for the *Atlantis* crew in novel ways. On an engine room tour—de rigueur for these kinds of trips—I joined Barb, Kate, and Tiffany to learn about the ship's motors and the controls for the A-frame that lifts *Alvin*. I asked the engineer about dumping ballast water, and he told me it had become more regulated in his seventeen-year career. "There's more and more paperwork," he said. "We're much more careful about ballast water nowadays—and with good cause, I suppose. There are a lot of strange things—barnacles, sealife—that you don't want to move from one ocean to another." The bridge, he said, keeps a record of where the ship takes on ballast water, so engineering crewmembers do not have to keep track of it or know the regulations of every port.

Many things have changed on ships like *Atlantis* over the past decade or so, not only the ways microbes have crawled into the consciousness of scientists and crew. In *The Universe Below*, Broad described the scene on *Atlantis* in 1993: "The milling scientists aboard the ship tended to be soft and blurry, like a tribe of aging hippies, a mélange of beards and sandals and T-shirts. . . . An anthropologist from another world would no doubt judge oceanographers a distinct race of *Homo sapiens* by virtue of their physical similarities and dress."[76] Since Broad's trip, the crowd has become younger and includes more women; eight of the sixteen science people on my trip were female, though there are fewer in senior ranks. As for the question of a "distinct race" on the ship, it must be said that all the scientists on my journey were white (the crew was more ethnically mixed). Holden, who spent much of his career in Georgia, observed that even if international and minority students were in his classroom, "when you get out to a ship like this, you look around and say, 'Wow, it's all white'" (see chapter 5). In terms of the hippie vibe Broad detected, soft and blurry beards were still in evidence, as, indeed, were antiauthoritarian, ecological attitudes.

A heated shipboard discussion on the question of whether Christianity in America had become a force against environmental consciousness brought to the surface some of today's marine scientists' worries about the public reception of their work. During a science meeting, one researcher made an offhand remark conflating creationism with Christianity and added a complaint about the backward beliefs of "farmers in Iowa"—a stereotype that stood for a land-locked consciousness, a way of thinking characteristic of the "flyover states" between the coasts that are home to so many oceanographers. Holden made a valiant attempt to complicate the discussion, but the Bush-era clichés of the red state–blue state divide in the United States—faith-based versus reality-based, Republican versus Democrat—were in full effect; reds did not comprehend scientific discussions of life forms, blues did, and so on. Circling back to how marine scientists might communicate the reality of climate change, Delaney wryly worked this rhetorical landscape: "The ocean is the environmental flywheel of climatic stability, of the planet. It is midwife to life. To paraphrase Hillary Clinton, I think 'it takes an ocean' to support life."

The discussion was symptomatic of wider cultural tensions about the roles of science and religion in framing matters of social concern. Delaney's vision was a cybernetic one, with scientific communication ideally guiding social forms of life. Science generates knowledge generates care generates stewardship in a feedback loop that tunes the social order into the holistic complexity revealed by science.

NEPTUNE

One of Delaney's passions is a multimillion-dollar project called NEPTUNE, an underwater "plate-scale observatory" in the planning stages that would cover the Juan de Fuca tectonic plate with a 3,000-kilometer-long network of fiber-optic and power cables, enabling remote computer observation and control of undersea sampling and data collection devices.[77] NEPTUNE, the name of the Roman god of the sea and of earthquakes, speaks to the ambitious scale of this venture (the acronym expands as "NorthEast Pacific Time-series Undersea Networked Experiments"). Thinking of Neptune alongside Gaia, the cybernetically imagined planet this project will be enlisted to monitor, calls to mind Haraway's pronouncement that not just "'god' is dead; so is the goddess"—and her immediate second thought: "Or both are revivified in the worlds charged with microelectronic and biotechnological politics."[78]

Delaney explained the vision one rainy afternoon in a presentation to the science party and crew of *Atlantis*, which he gave the provocative title "Will Research Ships Become Obsolete?"[79] NEPTUNE is to be a webwork of underwater sensors "built to look in real time, through online tools, at fluid flow and microbial communities. It will look at three things: the dynamics of plate tectonics, how these generate fluid flux, and how these connect to populations of microbes in the water." The project has financing from the W. M. Keck Foundation for astronomical research, startup monies from NSF, and funding from Canada, a partner in this binational enterprise. An executive team includes members from the University of Washington, Woods Hole, and MBARI. The vision is huge. Submarine cyborgs such as *Alvin* and robotic vehicles such as *Ventana* are to be stitched into a project of communication and control that will ask engineers, programmers, oceanographers, and others to fashion an infrastructure that realizes and represents a vision of the sea as a space that can be remotely monitored and managed. Ocean space and cyberspace will be woven together.[80]

Delaney hopes to bring in industrial partners like Cray, Microsoft, and Intel as well as ExxonMobil, which may be interested in deep-sea oil. Delaney's pitch to *Atlantis*, however, was in the register of interdisciplinary science, an endeavor that would take scientists from "mantle-to-microbe":[81] "Microbial production at the seafloor may rival that of phytoplankton. Plate tectonics enables microbial productivity at the base of the food chain. When the rocks crack, the bugs bloom—and this might even be proportional to earthquakes. We can learn how planets support life."

A deep-sea Internet represents a shift in the way marine science might be done, with in situ remote-controlled experiments complementing (even replacing) expeditions (see figure 26).[82] Many of the scientific activities I observed might be pressed into chips, wired into semiautonomous networks. Indeed, my *Alvin* dive had gathered data for a map to be plaited into NEPTUNE.

NEPTUNE Essential Elements

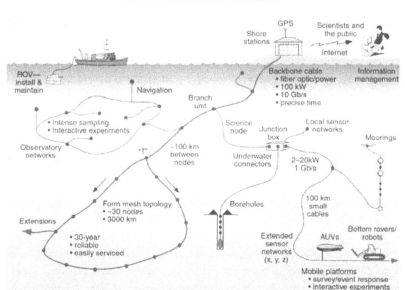

FIGURE 26. Envisioned NEPTUNE network and sensors. Image provided courtesy of the NEPTUNE Project and Center for Environmental Visualization. Reproduced with permission.

Delaney envisioned a democratization of science through online links between NEPTUNE and classrooms, permitting students to play with parts of a large experimental system. A skeptical scientist toward the back of the room heckled, "So they can learn how to crash an autonomous underwater vehicle into a black smoker?" Undeterred, Delaney showed a computer animated fly-through of a wired underwater world, with seismometers on tethers swaying in sync with commands from land. The visuals were hypnotic, like the tendrils on the computer-graphicked dancing microbes in *Volcanoes of the Deep Sea.*

What sort of seascape will NEPTUNE occupy? The NEPTUNE website reports that "because some portions of the NEPTUNE cable will be laid on public submerged lands, the program will be subject to several legal and regulatory requirements of state and federal agencies. Other entities with vested interests in NEPTUNE are the U.S. Navy, Native American tribes, fishing organizations, environmental groups, coastal communities near NEPTUNE shore landings, and telecommunication companies."

This networked ocean, if it is realized, will circuit through the resistances and capacitances of landed politics. Geographic information systems used to map such processes as the spawning of salmon in the Pacific Northwest already entangle knowledge communities—Native American fishers, the U.S. Environmental Protection Agency, timber companies, computer programmers—with different priorities about and powers over the flow of water and fishes.[83] Insofar as scientists immerse themselves in the media ecology of online NEPTUNE, it will result from a chain of transductions that permits scales and contexts of action—microbial, technological, tectonic, political—to be calibrated to one another. The mythically named NEPTUNE may turn out to underwrite an underwater "god-trick," "seeing everything from nowhere"[84]—squeezing an "objective" picture of the sea out of the Internet by damping down dissenting epistemologies—or it could, as Delaney hopes, broker a more participatory, distributed mode of ocean research, incubating hybrids of scientific and social practice. NEPTUNE, then, should not be understood only abductively, through the forward-looking logic of hope, but transductively. Transduction, writes philosopher Gilbert Simondon, "maps out the actual course that invention follows, which is neither inductive nor deductive but rather transductive, meaning that it corresponds to a discovery of the dimensions according to which a problematic can be defined."[85] Transduction is a mode of engagement that like the laterality described by Bill Maurer might "refigure the practices delineating the interior and exterior of inquiry—the observer and the observed, the sensorium and the sensed."[86] Transduction pays attention to impedance and resistance in cyborg circuits, to the work that needs to be done so signals can link machines and people together, at a range of scales, from private to public.

Still, such transductions may be difficult to track because of the dispersed and submerged character of the submarine cyborg that is NEPTUNE. During the 1960s and 70s, outer space was represented by such agencies as NASA through the broadcast medium of network television, presented for collective witness to "citizen-spectators."[87] But NEPTUNE, promising that the secret sea of the cold war will be opened up to the eyes of an auditing environmentalist public, offers a more fragmented, individuated, virtual view of a remote realm of scientific exploration. The notion of surfing NEPTUNE's cyberspace draws on conceptions of the sea as a place for individual immersion.[88] This ocean, broken down to be built up again on websites, fractures its publics into unsteady constellations of distracted web surfers, tuning in and out of an ocean both intimate and alien.

ALIEN BIODIVERSITY

After my journey on *Atlantis,* I learned that of the twenty-one vent sites in the Juan de Fuca Ridge, twelve are in Canadian jurisdiction and nine fall outside any national territory.[89] There is another alien ocean here to consider. Many vent organisms reside in legally ambiguous areas, being "primarily found in the deep sea in areas beyond national jurisdiction, typically occurring in conjunction with volcanically active undersea mountains in water 3,000–4,000 m deep."[90] Not just alien in a fanciful sense, such microbes are alien as citizen creatures—and easy to alienate as property, since no nation owns them. Their stateless location has become attractive to those who see spaces outside national jurisdiction as ripe for first-come, first-served appropriation.[91] According to legal officer Lyle Glowka of the Environmental Law Centre, World Conservation Union, there may also be nontrivial money at stake: "The potential market for industrial uses of hyperthermophilic bacteria has been estimated at $3 billion per year."[92]

But if plans for accessing extranational biodiversity have until recently depended on a gold rush attitude—with U.S. government agencies, universities, and companies leading the way toward mining unclaimed biotic wealth, aided by declassified military maps—this approach has come under scrutiny. The United Nations has asked whether deep-sea creatures, microbes included, might need international representation and stewardship, since "approximately 60 percent of the ocean is in that jurisdictional never-never land, the 'global commons,' where policies for ecosystem protection are largely in the discussion stage."[93] In June 2005, a year after my *Alvin* dive, I learned more about such discussions at a public event at the United Nations in coordination with the Ad Hoc Open-Ended Informal Consultative Process on Oceans and the Law of the Sea.

Discussion about the disposition of deep-sea resources began in 1967, when Maltese ambassador Arvid Pardo addressed members of the UN General Assembly, drawing attention to research in the seabed that revealed deposits of industrially important minerals. "Pardo urged the assembly to declare the deep ocean floor the 'common heritage of mankind' and to see that its mineral wealth was distributed preferentially to the poorer countries of the global community."[94] Adherents of free-market ideology demurred, notably the United States, which continued to mine the deep seabed, claiming freedom of the high seas and taking no heed of governance being set in place by the nascent Law of the Sea Convention. The ideological battle continues today. What has changed since the 1960s are the resources at stake. The manganese nodules that prompted debate proved prohibitively expensive to extract. Deep-sea biodiversity has reopened debates about access, ideology, and money. Broad writes, "It is no small irony that the greatest excitement to date in undersea mining centers not on deep minerals" but on "the mining of life. . . . By weight, these single-cell organisms are worth far more than gold."[95]

In 1995, the United Nations commissioned a study of the state of the oceans. The Independent World Commission on the Ocean—chaired by Mário Soares, former president of Portugal, with members drawn primarily from second and third world countries—in 1998 produced a report contesting the notion that ecologies of the deep were up for grabs for the first nation or company to exploit them.[96] The commission hoped to organize a cooperative international project toward equity around vent ecologies called, optimistically, abductively, HOPE: Hydrothermal Ocean Processes and Ecosystems.[97] This vision linked vent organisms to projects of rethinking property, equity, and the commons in a contested seascape: life forms to forms of life.

Writing against the appropriative frontier logic of American ocean behavior, the report's authors took oceanspace as a blank slate on which to rethink the distribution of resources: "Contrary to what occurs with terrestrial resources, which can be individually possessed and appropriated in forms developed and consecrated over the centuries, marine resources are by their own *nature* common, and are generally considered as such."[98] For these writers, marine resources have a life of their own resistant to national politics and the free market—a utopian view that a glance at anthropologies of contemporary fishing fights would immediately destabilize.[99] "Turning to the Sea: America's Ocean Future," the document that came out of the Monterey National Ocean conference in 1998, offered a contrary vision of what the necessary machinery of care should look like. Observing that "there is no mechanism currently in place to ensure that profits derived from publicly owned resources will be shared with the public and used appropriately," the United States located the solution in partnerships between government, universities, and industry rather than at the level of UN negotiations.[100] Where the Independent Commission argued that "greater equity in the oceans would contribute to reducing poverty and underdevelopment in general,"[101] the United States defended the Lockean view that mixing one's labor with nature turned it into property; on this English model, whoever first plants a tree or raises a garden in fallow soil—or turns an extremophile into an industrial product—makes a legitimate property claim.[102]

The deadlock between these two views of resources continues, as I saw at the UN meeting. The hour-and-a-half event, organized by United

Nations University (UNU)—a think tank dedicated to bringing scholarly advice to bear on issues of international concern—was convened to present to UN delegates a UNU report released that day, June 9, 2005: *Bioprospecting of the Genetic Resources in the Deep Seabed: Scientific, Legal and Policy Aspects.*[103] The gathering was opened by two speakers who informed the audience of seventy about deep-sea biology and biotech. Cindy Van Dover gave a compressed account of vent science. Diversa's Eric Mathur—dressed in a trim black suit rather than the aloha shirt I had seen him wearing in Hawai'i—repeated his call to "protect not pillage." Delegates were concerned to think about which legal instruments might govern the bioprospecting of the extraterritorial ocean.

Genetic resources collected from the deep sea could potentially fit within one of two distinct oceanographic zones described by the UN Convention on the Law of the Sea. They might fall within the "Area," defined as the seabed and ocean floor beyond the limits of national jurisdiction. Resources from this domain, following Pardo, are considered "the common heritage of humankind," and economic value resulting from their development is to be managed by the International Seabed Authority (formed in 1982 to deal with mining claims). Alternatively, such resources could be considered as floating within the "High Seas," the world's seawater beyond the limits of Exclusive Economic Zones (see figure 27). Resources in this zone—think migratory tuna—are considered "open-access," unregulated and available to anyone to use or capitalize. "Under the regime of the High Seas," the report summarized, "hydrothermal vent species would . . . be openly available for all to access and sample."[104]

But there is a complication. "Living resources" from the High Seas are defined as "non-sedentary." This might exempt vent fauna like tubeworms and their associated microbes, which are planted in places, making them a part of the Area and therefore "the common heritage of humankind." The question is whether the context of the Area or of the High Seas should apply. The report summarized: "It is unclear whether the living resources, more particularly genetic resources, of oceanic ridges and the seabed in general would fall under the regime of the Area or that of the High Seas."[105] Discussing this point, a delegate from India, lawyer Manimuthu Gandhi, suggested that chemosynthetic microbes, drawing nutrients from the seabed but living just above it, might "by nature" confound boundaries between the Area and the High Seas, between the living and the nonliving. But he also suggested that science might be able to sort out which direction the law should lean. In my view, though, such microbes, blurring the boundaries between the organic and inorganic (like proper cyborgs), make

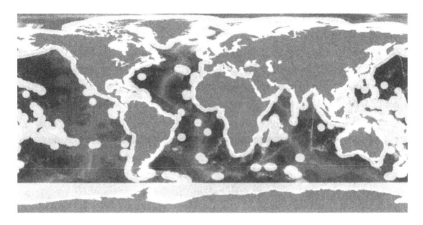

FIGURE 27. World Exclusive Economic Zones, in lightest gray. Black represents the High Seas/the Area, with mid-ocean ridges represented as light gray lines in this space. Map by Mathieu Belbeoch. Includes data supplied by General Dynamics Advanced Information Systems (GDAIS) ©1999–present. These data and this information are provided on a best-efforts basis, and GDAIS does not guarantee their accuracy or warrant their fitness for any particular purpose. Such data or information has been reprinted with the permission of GDAIS.

clear the baroque character of divisions between nature and culture, common-heritage and open-access, upon which international laws are based in the first place.

Brazil's delegate, Marcos Almeida, saw not a question of science but one of equity: "I think it's not fair . . . that developing states become dependent on the developed states—because, if you consider that all the property rights are taken now because the developed states, they have the technology to do that. In thirty years, fifty years, when the developing states would be raised to a position to have this technology—I hope—would there be any microorganisms that could be taken to some new use that was not already patented? That is the question."[106] Almeida brought up the question of benefit sharing, which raised the issue of whether rules for deep-sea bioprospecting should fall under the Convention on Biological Diversity.

Vent scientists have themselves attempted to fashion new norms for investigating the deep, creating their own rules. Since 1992, an international consortium of marine scientists called the InterRidge group, mostly from the United States, the United Kingdom, France, Germany, and Japan, has been devoted to collaborative, international research along mid-ocean ridges, which extend some 46,000 miles around the planet. In the 1990s, they pronounced upon the propriety of accessing genetic resources outside national jurisdiction. In a piece of gray literature titled "Management and Conservation of Hydrothermal Vent Ecosystems," InterRidge expressed concern about the effects of bioprospecting in jurisdictional gray areas, charging that parties to the UN Convention on the Law of the Sea and the Convention on Biological Diversity should be obliged to protect and preserve marine biodiversity.[107] Realizing that key players such as the United States do not recognize the Convention on the Law of the Sea, InterRidge argued for informal mechanisms among scientists themselves, suggesting zoning vents for different uses, including mining and biological sampling. In their formula, mining would happen on the periphery and bioprospecting closer to the center, leaving the absolute center untouched—a literalization of a Lakotosian view of science that keeps pure nature/science at the core. Noting, "At this time there is no agency with a mandate to oversee marine scientific research activities or biological resources on the seabed,"[108] InterRidge proposed that it act as a "research reserve system regulated entirely by consensus."[109] This plan for a self-correcting system of communication and control—for governance—remains vague; the UNU report referred to it, but only in passing.

What is notable in the InterRidge document is scientists' move into the realm of do-it-yourself policy, a mode of action seeking to place scientific work in social context (and therefore to place social context into science) *before* scientific research is undertaken. In this approach, to borrow Marilyn Strathern's analysis from another context, "science incorporates society into its aims in order to pre-empt society's verdict."[110] This is a scientific form of life abducting anticipated feedback from society into the investigation of life forms. But this subsumption of social context within science operates with a model of society which, while recognizing different interests, elides the question of whether there might not be irresolvable conflicts—for example, around whether capitalist enterprise should be mixed with science at all. These differences cannot always be brought into equilibrium through communication and control, cannot be dissolved, immersed, into the solution of science. A "recursive study of ourselves studying" (to page back to Delaney's words), an enactment of what Ulrich Beck calls "reflexive modernity"—in which knowledge operates on itself, immersed back into that which it describes—cannot fully predict the transductions that will emerge from the overlapping of forces and energies that bring a zone like "the deep sea" into consequence.[111]

Not quite tuned to these transductions, InterRidge still grants science a privileged position. The final frame in which InterRidge places its project is biology itself. Noting that "vent organisms may prove to have a large biotechnological potential, in terms of both enzymes and specialised compounds," and reporting on a case in which "an Australian-led expedition to the Bismarck Sea north of Papua New Guinea dredged seafloor vents, prospecting for minerals and micro-organisms of potential economic value," the InterRidge report argues that the biodiversity that supplied this potential must be protected "for evolution and for maintaining life-sustaining systems in the biosphere."[112] Here, vent sites are important not only as locales for science and commerce but for evolution itself, pitched as an agential stakeholder in ecological policy, an entity that needs diversity to maintain the feedback systems of its dynamic equilibrium. Here B and B'—biology and biocapital—once again collapse. In this cybernetic formulation, social regulation can be harnessed to sustain the natural systems of which scientists are a part, systems both familiar and in need of further exploration, alien. The alien ocean is a zone suspended between places— between steady ships and submerging subs, between sensors at the ocean floor and laptops on land, and between national jurisdictions and areas outside governance.[113] It is a medium, I maintain, to which humans connect only through chains of negotiation, translation, and transduction.

7 Extraterrestrial Seas

Astrobiology and the Nature of Alien Life

The salt marshes of Sippewissett, Massachusetts, a few miles north of Woods Hole's Marine Biological Laboratory (MBL), host sheets of microbes that detail a history of early Earth. This history is not so much archaeological—revealed by peeling back ever-older layers—as it is analogical: microbial mats similar to these have likely existed on Earth for more than three billion years. I am visiting these squishy structures today with Lynn Margulis, one of the architects of the Gaia hypothesis and an evolutionist known for her theory of symbiogenesis, the idea that evolutionary novelty emerges from the symbiotic fusion of different sorts of cells and organisms. According to this increasingly accepted view, all of today's nucleated, eukaryotic cells evolved through incorporating once free-living prokaryotes—like the oxygen-respiring bacteria that became mitochondria and the cyanobacteria that became chloroplasts, entities that now constitute indispensable organelles in the cells of animals and plants. Margulis has been working for forty years to investigate the possibility that structures such as the tiny hairs on the edges of cells, the filamentous threads in mitotic cell division, and the tails of sperm—all known to have common descent—come from an earlier incorporated ancestor called a spirochete, a swimming, corkscrew-shaped bacterium. Margulis is leading us to the marsh to collect mats of organisms she knows host similar spirals stirring from a dormant, rolled-up form; she wants to detect signs of emergent symbiogenesis, evidence that these wiggly creatures are insinuating themselves into their neighbors' cellular structures.

There are seven of us on this modest expedition into Earth's microbial past: Margulis, some students from Woods Hole, a journalist from *Discover,* my anthropologist spouse, and me. We have trailed Margulis's red compact car over dirt roads that string together gray-shingled Cape

Cod vacation homes along the marshlands. "Private Property" and "No Trespassing" signs do not deter Margulis; she moves in a different time and space continuum than most New England humans. As a key collaborator with James Lovelock on Gaia since the 1970s, her mind tunes in at once to the microscopic and macroscopic, to the subvisible and superorganismic, to baroque and romantic complexity.

After parking our caravan of cars—fueled by oil, itself the result of ancient microbial processes we nowadays encounter as a brand of biocapital (what Margulis calls "unearned resources"[1] and I'd call "necrocapital")—we hike out to the intertidal zone, the sort of threshold region Rachel Carson celebrated in her ecological Ur-text, *The Edge of the Sea*.[2] Margulis wears knee-length waders and a fleecy hooded pullover and walks determinedly into the mush, grasping a spatula and large spoon, ready to hunt microbes. She is the only one dressed appropriately, though sympathetic to those of us freezing in the drizzle on this unseasonally cold day in May 2005. Waving us into the marsh, she points out grasses that are signs of nearby mats: *Salicornia* and *Spartina* (I have seen that one before, in Georgia). She directs us to shallow pools bottomed by multicolored mud. Pointing out that the alternately exposed and submerged character of this site is ideal for mat growth, she digs up a lichen-like sample of microbial mat and instructs us to look at its dripping wet cross section, half an inch thick. Its rainbow layer cake of orange, green, pink, and black is composed of diatoms, cyanobacteria, purple sulfur bacteria, sulfate-reducing anaerobes, and other microbes living in complex interdependency. The mats are ecosystems, with cyanobacteria and purple anoxygenic photosynthetic bacteria as primary producers—"the acme of evolution," she remarks. "Cyanobacteria can do *everything*," Margulis says, "except talk." "Everything" means they gather their necessaries from vastly different media: electrons from water, energy from sunlight, and carbon dioxide from the air. They are expertly suited to the earthly elements, the proportions of which, according to the Gaian model, they have themselves had a role in determining.

Margulis is inviting us to look anew at planet Earth, to peer through these mat systems all the way back to the Archean eon, 2,500 million years ago. The mats are like time machines, media that transmit messages from an age when the ocean was otherwise, when Earth was just becoming the planet it is now. Cyanobacterial photosynthesis long ago filled the world with oxygen, pushing into marginal zones such anaerobic creatures as methanotrophs, underdogs in the Lovecraftian parable that MBARI's Steven Hallam told me about the ancient ocean (in chapter 1). Ancestors of blue-green bacteria like the ones in microbial mats consigned methanotrophs to environments we now consider extreme—the reason scientists call them and other such nonstandard life forms *extremophiles*.[3] Standing in the marsh with Margulis, we are searching not for a message from the mud but for embodied, living analogs of ancient life. There is ample evidence for mats in Earth's fossil record in the form of stromatolites, rocks left behind by the trapping, binding, and deposition of carbonate (and other sediment, such as sand) by cyanobacterial mats. Some biologists suggest that if bacteria evolved on other planets they would leave behind stromatolite-like formations, which might then be considered signatures of extraterrestrial life.

The extraterrestrial analogy is not far from our minds because our trip into the marsh comes at the end of a weeklong workshop on astrobiology at MBL. Astrobiology is an area of inquiry devoted to thinking about biological systems—actual ones on Earth and possible ones elsewhere—in a cosmic context. Astrobiologists want to know how life emerges on worlds in general and in particular; they are curious, for example, whether Mars might host microorganisms akin to those found in extreme environments on Earth. In this scientific venture, microbes in such locations as salt marshes become meaningful as proxies for extraterrestrial life. Scientists read them not just for clues about Earth's past, present, and future but as a means for considering the category of life itself in a more ample, universalistic frame.

Margulis's work on comprehending earthly life as transformative of planetary biogeochemistry has been centrally important to astrobiology—a field of inquiry institutionally established by NASA in the late 1990s though in existence in a slightly different version since 1960 as "exobiology." Woods Hole's MBL is one of sixteen national centers of NASA's Astrobiology Institute, a loose consortium of university and college affiliates chartered in 1998. Whereas some centers focus on planetary geology or the detection of planets outside our solar system, MBL Institute faculty use their expertise in marine microbiology to model life on other worlds. For example, in an ongoing research project titled "Terrestrial Analogues for Early Mars," one team examines metabolism in microbes found in metal sulfide deposits at the Juan de Fuca Ridge, locations they believe "represent potential analogues for an early, wetter Mars."[4] Our workshop on life on other worlds, then, fits neatly into the mission of contemporary marine microbiology, even as it expands the ambit of such biology to embrace research and speculation into extraterrestrial seas.

In this chapter, I examine how marine microbes figure in research in astrobiology, an interdisciplinary field that includes microbiologists, geologists, planetary scientists, astronomers, and other natural scientists. I am interested in how earthly oceans have become analogs for extraterrestrial seas (i.e., alien oceans in the most cosmic sense)—and how, in turn, marine microbiologists have reimagined Earth's oceans as potentially similar to seas on other planets. My point of entry is the Woods Hole workshop, a gathering that included biologists, historians, and philosophers and centered on getting attendees like myself up to speed on the latest research in astrobiology.[5] My discussion is organized less ethnographically—around the conference—than thematically—around research into oceans and life, past, present, and possible on Earth, on Mars, and on Jupiter's moon Europa, those sites most spoken of by biologists of the extraterrestrial. Visiting these three worlds not only allows me to meditate upon specific alien oceans but also serves as a vehicle for revisiting themes and arguments of this book. Transposing this text's concerns into the key of the extraterrestrial makes explicit the limits of life as they have been encountered and defined in this volume. Astrobiological inquiry into the prospect of extraterrestrial microbes is animated by stories of origins and kinship, questions of classification, sentiments of optimism about the promises of new life, definitions and redefinitions of locality, and arguments about the scale and context within which to think about the subject matter of biology. In motion are the conceptual and spatial limits of oceans and, indeed, the definition of life itself.

What are life forms? As I have sought to demonstrate, what Foucault called "life itself" is today in flux. Life itself no longer resides solely in individual organisms. When it becomes possible to speak of "environmental genomics," life disperses into distributed, deterritorialized webs, a net that for microbial oceanographers sometimes gathers up the planet itself. In molecular biological practice, life also fractionates into such entities as genes in DNA libraries, information traces in databases, compounds in test tubes, products for sale, and circuits of communication and control. What materializes, then, is a meshwork of only occasionally intersecting networks of transferable plural lives—what Franklin and Lock call not "life itself" but "lives themselves."[6] More, none of the various *life form* grids into which microbes might be slotted (phylotypes, ecotypes, guilds, assemblages) is fully stable—to say nothing of *species*, which, because that concept, at least etymologically, asks after visible forms, may never have been workable for subvisible creatures.[7] The trouble is not only microbial; once metazoans (at least) are understood as what Richard Doyle framed to me as "conscious polyps of a chattering bacterial multitude," finding one-size-fits-all forms for describing living things may become impossible (recall Doolittle's scenario in which "a species [or phylum] can 'belong' to many genera [or kingdoms] at the same time"). The *form* in "life forms" is deliquescing.[8]

Alien oceans as seas on other worlds help us sound the distributed, networked, emergent, and divergent character of life these days. The *extra-* in "extraterrestrial" points to categories outside and beyond, contexts stretched from the earthly to as yet unknown limits. As scientists seek to make microbes meaningful in the orbits of outer space, we see a stark instance of the ways life may be becoming other to its historical origins and to itself. To draw on an obscure meaning of *alien*, life as a material and semiotic relation is being aliened—transferred across contexts, leaving whatever it is or was transmuted.

After calling on Earth, Mars, and Europa, I light in this chapter upon a final marine planet—a literary example of an alien ocean: Solaris, a world imagined by Stanislaw Lem in his 1961 novel of the same name. Solaris is a globe swathed in a sentient sea. A watery world that endlessly perplexes its observers, Solaris provides a figure for considering the ways alien oceans are spaces that mix the familiar and strange, the real and ideal, the utopic and dystopic. They are what the inescapable Michel Foucault calls "heterotopias."

Before setting out on this hop through the solar system and to Solaris, let me provision our ship with a quick history of astrobiology and an accounting of the theory and methods of this science that is at once inductive, deductive, and abductive.

LIFE IN SPACE

Astrobiology's search for extraterrestrial life might bring to mind the Search for Extraterrestrial Intelligence (SETI) with its chorus lines of radio telescopes gyrating in the desert. But there are major differences. SETI came of age in the early days of UFO sightings and amid the alternating currents of cold war paranoia and countercultural optimism about the possibility of meaningful signals from the sky. In 1993, the more fringe associations of SETI caught up with it and NASA funding was cut from the enterprise. Astrobiology, advocates hope, will have a different profile—connected to the latest microbiology, remote sensing, and computational analysis and drawing from established research in planetary science and origins-of-life research.[9] SETI listened for complex communiqués from the stars; astrobiologists look to other planets for stripped-down signs of life. The object of scientific study and yearning is no longer intelligence, but life. No longer culture, but nature. Not extraterrestrial messages, but otherworldly organisms. Astrobiologists busy themselves with such tasks as scrutinizing Martian meteorites fallen to Earth for traces of biogenic activity and scanning the atmospheres of other worlds, searching for indicators of oxygen and methane, which may signal the presence of vital materials and metabolism. Scientists once speculated about canals on Mars, signs of a spent civilization; astrobiologists now look for outlines of evaporated rivers and indices of subsurface seas.

Methodologically, astrobiology has brokered a shift from listening to looking, from auditory attentiveness to active reading. Whether scanning Martian meteorites or atmospheres, the new object of scientific pursuit is something astrobiologists call "the signature of life" or sometimes a "biosignature," defined as "any measurable property of a planetary object, its atmosphere, its oceans, its geologic formations, or its samples that *suggests* that life was or is present. A short definition is a 'fingerprint of life.'"[10] The biosignature returns us to the realm of semiotics, the making and transmission of meaning through signs and symbols. If SETI sought signals in an ocean of noise, looking for the arbitrary and organized surprise—what scientists have come to call *information*—astrobiology searches in a more impressionistic mood, scouting primarily for what Charles Peirce called *indices*—indirect representations, traces, of its object, life (fingerprints and smoke are canonical Peircean indices—of fingers and fire, respectively). Astrobiologists cannot look to DNA libraries for signs of the life they seek. Their readings are not of genetic text already gathered but of signatures suggesting that such texts *might* exist—or might have been there just a moment ago. A founding challenge presents itself, which, according to David Des Marais, is that astrobiologists face the difficulty that "our definitions are based upon life on Earth" and that, "accordingly, we must distinguish between attributes of life that are truly universal versus those that solely reflect the particular history of our own biosphere."[11] This is no simple task, since what counts as universal is precisely what is to be discovered. This puzzle—of how to extract the general from the specific—has long animated biological theory, but it takes center stage in astrobiology.

The shift in attention from alien intelligence to alien nature has suggested a novel methodological strategy to those who would scout for extraterrestrial life. Astrobiologists treat unusual environments on Earth, such as methane seeps and hydrothermal vents, as models for extraterrestrial ecologies. Framing these environments as surrogates for alternative worlds has made marine microbes like hyperthermophiles attractive understudies—what scientists call *analogs*—for aliens. It would be difficult to guess from its staid exterior that the New England town of Woods Hole is now buzzing with scientists thinking about space aliens, even sifting for analogs in their own backyard, in places like the Sippewissett marsh.

Such sifting has been made possible in large part by the category of the extremophile. Lynn Rothschild, an evolutionary biologist at the NASA Ames Research Center in California and one of the featured speakers at the MBL workshop, has written informatively on this topic.[12] *Extremophile*, coined in 1974 as a scientific-sounding hybrid of the Latin *extremus* and the Greek *philos*, gathered together organisms—thermophiles, halophiles, and the like—that previously had little to do with one another.[13] Before, if one spoke of these creatures in the same breath it was in the realm of food preservation; pasteurization, freezing, salting, drying, and irradiating are all methods of killing the hardy bugs that live in food. Extremophiles became tightly bound together in the scientific imagination when Carl Woese named the Archaea, noting that many methanogens, halophiles, and thermophiles fit into this new category. The characterization of heat-loving microbes in vents coupled with the claim that hyperthermophiles might be aboriginal life forms made extremophiles attractive to scientists thinking about the origins of life on Earth and, maybe, elsewhere.[14] Rothschild's own curiosity has centered on halophiles, salt-loving microbes that can survive extreme desiccation in suspended animation between waterings. With bacteriologist Rocco Mancinelli, she helped design an experiment for the European Space Agency in which halophiles were exposed to the extreme cold and unfiltered solar radiation of space; during a stint on a recoverable satellite in 1994, these microbes survived for two weeks, a result that supports the possibility that living things could be transited to Earth from such sites as Mars, if indeed Mars sports such life.[15] Extremophiles thus become proxy aliens—a framing employed by marine microbiologists, NSF administrators, science museum curators, and Hollywood directors to prompt public interest in these creatures.

For all their differences from one another and from most earthly life, extremophiles share one thing: they require liquid water. So, in spite of the often outlandish features of extremophiles, they all fit nicely within scientific and symbolic associations of water with life. Astrobiologist David Des Marais argues that "all life requires complex organic compounds that interact in a liquid water solvent."[16] And NASA's motto for looking for life on Mars has been "Follow the Water."[17]

As this book went to press, astrobiologists stepped away from assuming this elixir to be a biological necessity, offering that ammonia, sulfuric acid, and methane could also work as solvents.[18] What remains important in this repositioning, however, is an emphasis on liquid. For astrobiologists, life, extremophilic or no, will exist in a liquid medium. It is for this reason that extraterrestrial seas—alien oceans—are such objects of fascination.[19]

In what follows, I follow the astrobiologists, looking for liquid, beginning with Gaia and Mars.

A TALE OF TWO PLANETS

At the astrobiology workshop, Margulis gave us a compressed, autobiographically organized chronicle of the Gaia hypothesis, which she joined Lovelock in developing in 1970. As she told us, this model originally emerged to look not at Earth but at Mars, to determine whether it might be possible to discern from a distance whether the red planet supported life. In 1965, Lovelock had been invited by NASA to design an experiment for detecting life on Mars. Using his expertise in gas chromatography, he wagered that the best way to look for life remotely would be to search for signs of metabolism in planetary atmospheres. As our workshop organizers, historians Steven Dick and James Strick, put it, Lovelock suggested that "the most obvious activity of living things which offsets entropy [is] that they keep the gas composition of a planetary atmosphere far from equilibrium. For example, if a planet's atmosphere contain[s] significant amounts of both methane and oxygen simultaneously, for any length of time, Lovelock argued, this is so far from the equilibrium condition that it is strong presumptive evidence of life."[20]

According to Dick and Strick, the insight Lovelock gained from meditating on Mars was "that the gases that living organisms most actively affect, especially carbon dioxide, methane, oxygen, and water vapor, are just those gases that most dramatically shape the climate of the planet."[21] Such disequilibrium mixtures of gases constitute a biosignature, "a feature whose presence or abundance requires a biological origin. Biosignatures are created during the acquisition of the energy or the chemical ingredients that are necessary for biosynthesis or both."[22] Biosignatures of this kind, then, are semiotic traces that can be read, that require literacy in organic chemistry.

Lovelock next transposed his thinking about what he was coming to call "geophysiology" back to Earth. Employing models from cybernetics, he speculated that living things might use feedback to regulate Earth's temperature, oxygen level, and the acidity of the ocean. By the late 1960s, a view of the geophysiology of the planet had acquired an iconic representation in photos of Earth coming back from astronauts' cameras. Haraway suggests that, "as Lovelock realized, the cybernetic Gaia is . . . what the earth looks like from the only vantage point from which she could be seen—from the outside, from above. Gaia is not a figure of the whole earth's self-knowledge, but of her discovery, indeed, her literal constitution, in a great travel epic."[23] In other words, Earth *as* Gaia is one result of extraterrestrial travel. It is the planet rediscovered from outer space. In this reimagining, humans become passengers on what Buckminster Fuller famously called Spaceship Earth.[24] The out-of-body experience of looking at Earth allows the possibility of seeing the planet as an alien would see it and invites an uncanny sense of being alien to one's own consciousness. Susan Lepselter, in her analysis of satellite portraits of Earth, "From the Earth Native's Point of View," writes, "The self is felt in its moments of disorientation, when the imagined spaces it inhabits—spaces as vast as the earth and as intimate as the body—veer suddenly into strangeness, perceived at once from the inside, and from the objectifying vision of beyond."[25] To look at Gaia is to be invited to discover and incorporate the alien, to extraterrestrialize a Thoreauvian transcendentalism.

Lovelock and Margulis's research, funded in large measure by NASA, has been key to this disorienting discovery. One entailment of Gaia, according to biologist Harold Morowitz, was that scientists might now consider that "life is a property of planets rather than of individual organisms."[26] Gaia posits that living systems must be understood in an ecological webwork that scales all the way up to the planet as a whole—a view incorporated into the recent genetic and biogeochemical models of people like Ed DeLong, Penny Chisholm, and David Karl. On Earth, oceans are key nodes in the Gaian web, cycling nutrients.

In her workshop talk, Margulis emphasized that the Gaia hypothesis did not suggest that the planet was some perfect Eden, as many critics assert. She pointed this out again during our jaunt into the salt marsh: "The ocean is too salty. Does anyone know the pH of the ocean? Most biologists will say 7 because that's neutral and they want to be neutral. But it's not. It's 8 or so. The ocean is *too* salty." Fishes are often happier, she remarked, in the lower salt concentrations of water provided in aquariums. "Gaia is not God and didn't do anything perfectly." Like Steven at MBARI, Margulis emphasized that Gaia could care less about humans. She dismissed, as well, the idea that Gaia demanded that Earth be considered an organism: "No organism can consistently eat and live on its own waste." If Gaia sometimes veers toward the romantic—a holistic vision of harmony—it also includes an attention to such baroque complexity as the never fully equilibrated relation between chemistry and biology, or the weave between life forms and forms of life. Rather than only gaiasociality, then, symbiopolitics.

Lovelock's view that living systems can shape their home planets has gathered more adherents over the years, but the model remains controversial. In astrobiology, one of the more visible challenges to the biocentric character of Gaia comes from a hypothesis that goes by the name of Snowball Earth.[27] According to this model, Earth has experienced three distinct climate states: nonglacial, glacial, and pan-glacial (Snowball Earth), when virtually all continents and the oceans were ice-covered. The last of perhaps three snowball events ended 635 million years ago, some 3.8 billion years into Earth's history, when ice began to spread from the poles toward the equator, eventually covering the entire planet in a deep freeze. Ice-covered Earth reflected light off its surface and grew colder and colder, leading to more ice, and to colder temperatures—and so on, in a runaway feedback effect that led to the extinction of much Earth life. What eventually yanked the planet out of this snowball state were earthquakes and volcanoes. When volcanic carbon dioxide accumulated in the atmosphere, an extreme greenhouse effect melted the snow; the few surviving organisms—autotrophs and heterotrophs that persisted in crack systems—eventually adapted to the new world by becoming multicellular in the evolutionary efflorescence known as the Cambrian explosion.

When I spoke with the primary advocate of this theory, Paul Hoffman, a geologist at Harvard, he remarked that Snowball Earth may be in tension with the Gaia hypothesis: "life" is not sufficient to regulate Earth out of snowball feedback; plate tectonics seem to have been necessary. In other words, Snowball Earth questions whether microbes are always central to transforming the planet. Against a gradualist Gaia model, the Snowball picture posits a planet moving through catastrophic changes. Hoffman told me that thinking about icy moments in Earth's history was like "doing planetary science, except it's on Earth. It's the Earth behaving like an alien planet."

The Gaia versus Snowball discussion is reminiscent of the nineteenth-century debate between geological gradualists like Lyell, who saw Earth changing slowly, and catastrophists like Cuvier, who envisioned Earth's history as marked by successive disasters (mostly engineered, in Cuvier's view, by God; the Noachian flood was iconic). At issue at that time was the age of Earth—not a matter adherents of Gaia and Snowball Earth debate. The question today is about the rate of and impetus for evolutionary change. This can also be understood as a dispute between biologists and geologists, with each side granting their object of study a key role in earthly change. Biologists seeking messages from the mud and analogies to early Earth meet geologists reading the record of the rocks. Whereas messages require decoding and analogies elaboration, the record of the rocks requires only replaying with the proper device. As Bowker has it for geology, "Life itself writes its history into the earth"[28] (even as such writing is always under erasure and erosion by the ocean).

Moreover, two different sentiments about life on Earth circulate here. Gaia has Earth as a biological system, one that is fragile and can be thrown into disequilibrium by forces that include human enterprise, forces that might radically transform the nature of life. Snowball Earth has the planet as a durable if unruly geological object to which life must adapt to survive. In the first model, the ocean is a significant force shaping life on Earth; in the second, the ocean depends upon the geological substrate on which it exists. The models offer distinct senses and sensibilities about how life forms come to be and, perhaps, of how forms of life might respond to these geneses (do we cut down fossil fuel use? think differently about quakes?).

Let me move to Mars, inspiration for Gaia. Mars has long attracted speculation about otherworldly life, and water has been central to such conjecture. But if life has scaled up on Gaia, the quest for living things on Mars has, over the past century or so, tracked ever-smaller signs of life.

In 1878, Italian astronomer Giovanni Schiaparelli announced that he had seen through his 8-inch telescope what looked to be a network of canals crosshatching the planet. Subsequent observers seconded this claim, suggesting that canals were made visible by adjacent swaths of vegetation. For Schiaparelli, Mars was a planet with a water vapor atmosphere, two polar ice caps, and a few sparse seas. He believed that Martian channels carried water from seasonally melting poles to the rest of the planet, with this "network of canals . . . perhaps constituting the principal mechanism (if not the only one) by which water (and with it organic life) may be diffused over the arid surface of the planet."[29] Harvard-educated astronomer Percival Lowell in 1894 speculated that the canals were artificial, signs not of natural providence but of intelligent design, an architected response by an advanced Martian civilization to a dehydrating world. Lowell pressed his hypothesis vigorously over the next decade or so, mapping some 183 canals and offering in his 1908 *Mars as an Abode for Life* a defense of his belief that a dying Martian civilization had engineered an irrigation system. H. G. Wells's 1898 *The War of the Worlds* delivered its own Lowellian scenarios, scripting Mars as the dystopic destiny of an industrial-age Earth exhausting its natural resources. In Wells's novel, Martians travel to Earth searching for fresh territories but are brought low by terrestrial microbes. Wells, putting a Panglossian spin on Pasteur, concludes with a vision of humans and microbes as conjoined allies against the Martians. Earthly Nature wins against alien invaders:

> Germs of disease have taken toll of humanity since the beginning of things. . . . But by virtue of this natural selection of our kind we have developed resisting power. . . . there are no bacteria in Mars, and

directly these invaders arrived, directly they drank and fed, our microscopic allies began to work their overthrow. . . . By the toll of a billion deaths man has bought his birthright of the earth, and it is his against all comers; it would still be his were the Martians ten times as mighty as they are.[30]

In this tale, microbes do not make Earth more hospitable for humans in a completely benign way; insofar as they are our partners, we exist in a choreography of life and death. At the beginning of the twenty-first century, any Martians, if they do exist, are imagined by astrobiologists quite differently; they may *be* microbes.

By the 1930s, Lowell's canals were shown to be optical illusions created by low-resolution telescopes. But the idea of vegetation on Mars persisted into the 1950s. Astronomers supplemented visual observations with spectrographic analyses, used to infer the presence of carbon dioxide in the atmosphere and of water in the ice caps. Now, it was believed, conditions were suitable for photosynthesis, though, because nothing akin to a spectral signature of chlorophyll could be found, dark spots on Mars were explained as moss or lichen. Soviet astronomer Gavriil Adrianovich Tikhov pronounced in 1955 that "the problem itself indicated the necessity of combining astronomy and botany. And so the new science of astrobotany was born,"[31] and in 1959 Harvard astronomer William Sinton signed on with a high-profile argument in *Science* supporting the case for Martian vegetation. But this claim too was eventually revealed to derive from artifacts of observation; Sinton himself determined that his lush readings were caused by the interference of heavy water in Earth's own atmosphere. The idea of liquid water on Mars was evaporating, and possible Martian inhabitants were being scaled down from cultured irrigation engineers to weeds to, next, microorganisms.

The 1971 flyby of the NASA *Mariner 9* orbiter showed gullies resembling dried riverbeds, resurrecting the question of water and galvanizing the development of life-detection experiments to be delivered to the Martian surface by the next round of spacecraft christened with maritime names, the Viking landers, which touched down in 1976. Sanitary engineer Gilbert Levin designed one of the experiments that made it to Mars. His experiment required the lander to submerge soil samples in an aqueous solution dosed with organic compounds to see whether carbon dioxide or methane would emerge, signs that microbes were decomposing the substances as food. Many biologists found the experiment problematic since it assumed an abundance of water uncharacteristic of the Martian surface, whose atmospheric pressure is too low to allow liquid water to exist at the surface.[32] Any water that

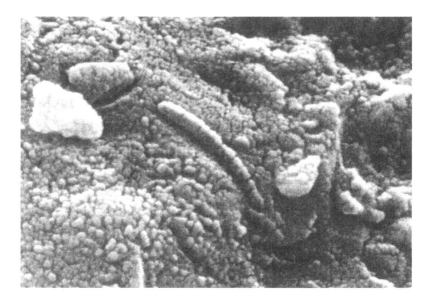

FIGURE 28.   Ovoid forms inside Martian meteorite ALH84001, as depicted through scanning electron microscopy. The elongated shape in the center is some several hundred nanometers in length. Image by NASA.

may have been responsible for the gullies captured by *Mariner* would have to have existed, they claimed, at a time when the Martian atmosphere was denser than today. None of the life-detection experiments delivered a positive result, and the question of life on Mars was once again put to bed.

The slumber lasted until 1996, when fossil-like traces were discerned through electron microscopy in a Martian meteorite discovered in Antarctica in 1984. Mars meteorite ALH84001, which had been blasted off Mars 16 million years previous by the impact of another meteorite and arrived on Earth 13,000 years ago, harbored elliptical shapes some believed to be outlines of ancient microbial life (figure 28). David McKay and colleagues in a 1996 *Science* article wrote, "Ovoid features . . . are similar in size and shape to nanobacteria in travertine and limestone. The elongate forms resemble some forms of fossilized filamentous bacteria in the terrestrial fossil record."[33] The argument went beyond visual pattern matching; there were also traces of organic carbon compounds—though either biotic or abiotic processes could have produced these—as well as magnetite and sulfides similar to those made by some earthly bacteria. McKay and company concluded that they had found "evidence for primitive life on early Mars" with an age of about 1.3–3.6 billion years.[34]

Other scientists were skeptical, accepting that the tomato-sized sample was from Mars but disputing the story told by the teensy shapes and compounds within, seeing them as artifacts of abiotic chemistry, not signs of life. In his 1997 *Wittgenstein: On Mars,* playwright George Coates put his finger on the uneasiness researchers may have felt. Coates offered a dramatic comparison of the ovoid in ALH84001 to the famous drawing in Wittgenstein's *Philosophical Investigations,* a scribble that suggests the outline of either a duck or a rabbit, depending on how one looks at it (and that might put some readers in mind of the evocative geological form described by tabloids as the "face on Mars").[35] Another objection pointed out that traces on ALH84001 indicated formations two orders of magnitude smaller than any microbe, too cramped to fit within their boundaries all of the apparati cells need to function. Supporters of life signs in ALH84001—including vent scientist John Baross—countered that these were not microbes but nanobes, or maybe components of a larger cell.[36] Though many scientists see the search for life in ALH84001 as a microscopic rerun of the hunt for canals on Mars, the case is far from closed.

THE MESSAGE FROM MARS

When the European Space Agency's *Mars Express* polar orbiter spectrometer detected methane in 2004, physicist Vladimir A. Krasnopolsky of Catholic University argued at a meeting of the American Astronomical Society's Division of Planetary Sciences that methanogenic bacteria might be "a plausible source of methane on Mars."[37] Enter a potential message from Martian mud.

Any message from Martian mud, like any from earthly cold seeps, will depend on a variety of media—spaceships, spectrographs, computers, and, that old standby, water. Water set the stage for investigation and was the first scientific grail for a pair of predecessors to the *Mars Express:* the NASA Mars exploration rovers *Spirit* and *Opportunity,* remote-control wheeled vehicles that began to tool around on the Martian surface in 2004, directed from Pasadena, California. NASA's website explains why *Spirit* and *Opportunity* would look for water: "On Earth, all forms of life need water to survive. It is likely, though not certain, that if life ever evolved on Mars, it did so in the presence of a long-standing supply of water. On Mars, we will therefore search for evidence of life in areas where liquid water was once stable, and below the surface where it still might exist today. Perhaps there might also be some current 'hot spots' on Mars where hydrothermal pools (like those at Yellowstone) provide places for life."[38]

The *Spirit* and *Opportunity* rovers, like MBARI's ROV *Ventana*, are camera-toting robots that can be manipulated from afar. As Zara Mirmalek documents in "A Martian Ethnography," researchers in Pasadena operating the rovers began to "live on Martian time" (phasing into odd working schedules in order to stay in sync with the 24-hour, 39.6-minute Martian day), bringing a faraway world close, becoming intimate, like MBARIans on *Point Lobos*, with a realm in every other way inaccessible to them.[39]

It was the orbiting *Mars Express*, however, that first offered a potential message from Martian methane. *Mars Express* has less in common with *Ventana* than with the satellite components of MBARI's projected ocean observatory, MARS (Monterey Accelerated Research System). Any *Express* message from Mars would come not from dredged up DNA but from spectrographic readings. In place of the genetic library, researchers examine that other once-removed, textually imagined, sign of life-made-legible: the biosignature—a ghostly cousin to such semiotic concepts as the genetic code. According to petrologist Monica Grady, there are both direct and remote "signatures of extraterrestrial life."[40] Direct signatures include measurements that show evidence (e.g., in rock samples) of the production of organic molecules through biological process; ALH84001 was believed by some to contain such direct signatures. Remote signatures of extraterrestrial life, meanwhile, include such items as the spectral signature of the atmosphere, which can point toward such bioproducts as ozone or methane.[41] This is the sort of biosignature scientists like Krasnopolsky believe is contained in transmissions from *Mars Express*.

How can we understand a potential methane-infused biosignature emanating from Mars? Literary scholar Robert Markley suggests that "methane on Mars functions as both a material trace and a 'message' in search of a context yet to be defined by systems of representation and knowledge-production."[42] The context begged by the message from the methane on Mars is a suitably ecumenical definition of life, translated into an operational—that is, working and measurable—definition of the biosignature. For methane on Mars to become meaningful astrobiologically, in other words, it must come from life. Otherwise, so what? In terms of social sentiment, the context, like the one offered by Steven Hallam in his Lovelock-Lovecraft parable, will be one in which alternative life forms are revealed to be both other to us and part of a more encompassing ecology. Mars will no longer be Wells's dystopia of canal-carving capitalists but a world of humble methanogenic microbes tuned in, like contemporary environmentalists, to a biosphere. Indeed, at the Woods Hole workshop it was notable that no one worried that extraterrestrial creatures might cause us any harm; rather, in presentations and informal discussion, the solar system (at least) was mostly figured as a unified ecology, full of possible life to be considered innocent until proven guilty. Over drinks at Woods Hole's Captain Kidd bar, several participants revealed to one another their active hopes to one day encounter extraterrestrials.

Such extraterrestrial life, if it exists, is still ghostly. Biosignatures are a strange kind of biomedia. They are not recontextualizations of biotic matter, like DNA libraries, but are themselves contexts in search of life, which, to stay with the signature metaphor, threatens to appear as a forgery, a sign of something other than itself, an artifact, like *Urschleim*. In "Signature Event Context," Jacques Derrida writes that taking the signature as a trace of the authentic, of presence, depends on the absence of the signer, resulting in the error of attributing presence to signature itself: "In order to function, that is, in order to be legible, a signature must have a repeatable, iterable, imitable form; it must be able to detach itself from the present and singular intention of its production. It is its sameness."[43] But Derrida's claim is too singular; a signature might rather be thought of as a family of differences rendered related by witnesses who attest to their similarity, not their sameness.[44] The outcome, as astrobiologists are more than aware, is that the search for extraterrestrial life is strongly constrained by what scientists believe they have witnessed of life on Earth: "Our concepts of life and biosignatures are inextricably linked."[45] And the concept of life is itself unsettled.

But if the definition of life has been a frustratingly evasive object for theory in biology, this elusiveness has been the foundation for astrobiology as a research program. Markley's analysis of signs of methane on Mars in his *Dying Planet* provides a helpful clue as to why: "The value of methane on Mars lies precisely in the fact that the evidence for its sources, if not its existence, is ambiguous, and consequently it provokes dialectical responses—skepticism and belief—that in turn will promote new research programs, new technological innovations, greater economic investments in interplanetary exploration, and new incentives for the continuing exploration of Mars."[46]

That exploration—that search for methane, life, or both—continues to look for water, even oceans. NASA announced in late 2006 that photos of Martian gullies taken years apart by the Mars Global Surveyor suggested that water might occasionally and evanescently flow across the surface.[47] And results from the orbiter *Odyssey* in 2002 suggested the presence of a large amount of subsurface water ice near the north pole of the planet. The MARSIS (Mars Advanced Radar for Subsurface and Ionosphere Sounding) antenna, which became operational in 2005, is designed to look remotely, transductively, for this ice, using radar, as NASA puts it (abductively) to "search for underground features much the way an ultrasound device looks at an unborn child inside a mother's womb."[48] In May 2008, NASA's *Phoenix* Mars Lander set down and started digging into the soils of this region (NASA's Mars-as-mother metaphor suggests life waiting to be birthed, perhaps by cosmic C-section). Some researchers believe that Mars once maintained a substantial ocean in its northern lowlands, perhaps as deep as 4,000 feet. Vic Baker of the University of Arizona proposed in 1991 that such an ocean might come and go over geological time if the planet has extreme heating and freezing cycles.[49]

DISPLACING THE TREE OF LIFE

If Mars had or has some sort of ocean, an antique or sun-shy sea, the possibility arises that life originated on Mars and was conveyed to Earth in meteorites. One advocate of this view, physicist Paul Davies, asks readers in an article in *Astrobiology,* "Does Life's Rapid Appearance Imply a Martian Origin?" He asks that we "suppose that . . . life is very hard to start (i.e., that the expectation time for life to emerge spontaneously on a suitable Earth-like planet is very much longer than the habitability duration of that planet)."[50] Davies argues that Mars was habitable earlier than Earth, that Mars's small size may have attracted fewer disruptive meteors and cooled more quickly than Earth, "permitting the early establishment of a deep subsurface zone in which hyperthermophilic organisms could take refuge from the bombardment."[51] It is possible, he claims, that life originated on Mars and subsequently traveled to Earth in a process he calls *transpermia* (to distinguish it from more general theories of *panspermia* associated with such figures as astronomer Fred Hoyle, which imagine life to be widespread throughout the universe).[52] As Davies puts it, "If life emerged from a series of highly improbable chemical and physical steps, as is widely assumed by biologists, then a Martian origin for terrestrial life is probable."[53] A genealogical alienness for life on Earth appears here not as the laterally transferred alien gene of extremophilic microbes but as the planetarily transferred alien gene that seeds the tree of life on Earth in the first place. Rothschild and Mancinelli's experimental suggestion that microbes can survive transit through space may lend credence to this view—depending, perhaps, on what one thinks of experimental archaeology and on whether one regards the retrodiction of the paths of possible meteorites as being as compelling as the reconstruction of ancient

Polynesian voyaging trajectories—another form of travel that has set contexts within which organisms might count as native or alien species.

When I talked with Paul Hoffman about the idea that life originated on Mars, he said that the story is now so often told that if it turned out to be true he would find it "interesting, but not surprising." Other scientists, however, find such a possibility frankly thrilling. Gentry Lee, chief engineer for *Spirit* and *Opportunity*, exclaimed on a 2004 *NOVA* television special that "maybe life evolved first on Mars and was knocked off the surface and carried to Earth. Maybe we are all Martians!"[54] This story of Martian origins is a tale of kinship through descent, of identity secured by ancestry. With transpermia's conception of a space-traveling seed (the Greek *sperma*), this might also be read as a narrative of obscured and revealed paternity. It is a story that promises a truth that once disclosed cannot be revoked, like the identity of a long-lost father in a space-age soap opera. This futuristic tracing back of the tree of life to Martian soil is rather nostalgic, finding not the confusing alien scrambles of gene transfer but a newly grounded narrative of origin. Part of the reason the story is so neat is that there is as yet no trail of information, genetic or otherwise, to follow or to complicate this romantic possibility; this interplanetary tree is at the moment strictly hypothetical.[55]

Just as with models that site life's origins in deep-sea vents, the hypothesis of a Martian origin unsettles creationists. Charles Carlson, the man who led a Bible study group on intelligent design at the George W. Bush White House, warns those keeping up with the search for life on Mars to "be aware that there is another agenda that some folks have here. It's not just a fascinating space exploration. It's making a case for how life began without a Creator."[56] Other believers argue that the fact that scientists are entertaining a Martian origin for Earth life demonstrates the outlandish lengths to which they will go to disprove a divine genesis on Earth and shore up evolution; a posting on a Christian millenarian website offers, "It takes far more faith to believe that we accidentally evolved from Martian debris, than it does to believe that we were created in the image of a loving God."[57] If the unrooting of the tree of life from sites like vents pleased intelligent-design folk, creationists are less than happy about the rerouting of life's root to Mars.

Hoffman's "interesting, but not surprising" interpretation, meanwhile, suggests that Earth and Mars are connected across the gulf of space, now configured not as a blank space but as a topology with its own gravitational currents and turbulences, through which items such as meteorites might travel in some directions and not others. Interplanetary space becomes akin to an ocean, a macrocosmic sea. No wonder spacecraft sent to Mars are christened with names like *Mariner.* In a relation in which gravitational differences and dynamics have meteorites moving from Mars to Earth, but not vice versa, while spaceships travel the other way, Mars becomes nature to Earth's culture—and even, in Martian origin-of-life scenarios, the microbial spark of life transferred to Earth's founding nature, the ocean. This interpretation, just as for hyperthermophiles in *Volcanoes of the Deep Sea,* moves these putative microbes from being what that film termed "alien to us" to "most certainly related." Red planet and blue planet become kin, even twins.

ULTRAVIOLET CAPITALISM?

The parameters of life's possible history expand. Sociologist Melinda Cooper suggests that life's newfound extraterrestrial elasticity is not just a function of work in the biosciences: "The notion of life itself is undergoing a dramatic destandardization such that the life sciences are increasingly looking to the extremes rather than the norms of biological existence. Importantly, these new ways of theorizing life are never far removed from a concern with new ways of mobilizing life as a technological resource."[58]

Cooper argues that attention to extreme life forms is coincident with a capitalism seeking to overcome its own ecological limits. In the wake of the Club of Rome's "Limits to Growth" report of 1972, which predicted environmental collapse if world industry and population continued to grow exponentially, capitalists began looking for new modes of capital accumulation. Rejecting the geochemical finitude of Earth as the last word on limits, Reagan-era futurologists chided the Club of Rome for a failure of imagination, for not anticipating the promise of biotechnology. Cooper detects in contemporary interest in extreme life forms—in researches into how biological systems continually redefine the limits of life—raw ideological material for fresh kinds of capital that burrow into the generativity of living things to create new fantasies of endless frontiers of surplus. Call it *ultraviolet capital*—a capital beyond the sky-high, ocean-deep blue-green capital of chapter 3, a capital beyond the visible spectrum, where extremophiles that survive ionizing radiation become intriguing candidates for cosmic biotechnology. Though I have spent too much time with biologists of extreme life to boil their work down to a delirious emanation of capitalism, I agree with Cooper that scientific and economic speculation often emerge at the stressed boundaries of existing forms of life.

Back, then, to life on Earth. And Mars.

ALIEN SPECIES, ALIEN POLITICS

A view of Earth and Mars as part of an expanded zone of local nature represents a shift from earlier interest in space biology, which worried much more about boundaries, difference, and contamination. Exobiology, the previous incarnation of astrobiology, was founded around 1960 by Nobel Laureate Joshua Lederberg, a bacterial geneticist at Stanford, who was anxious about the possibility that spacecraft returning to Earth might infect the planet with extraterrestrial microbes. Lederberg's finger was on the popular pulse. In "Danger from Space," *Time* magazine in 1961 pronounced: "The invaders most to be feared will not be little green Venusians riding in flying saucers or any of the other intelligent monsters imagined by science fictioneers. Less spectacular but more insidious, the invaders may be alien microorganisms riding unnoticed on homebound, earth-built spacecrafts."[59]

Exobiology began, then, as the biology of invasive alien species, a defense against alien nature. Although Lederberg was also concerned with the contamination of other planets—and particularly the Moon—by earth biota, public and funding focus centered on the "back contamination" of Earth by alien life. In "Germs in Space," historian Audra Wolfe notes that these concerns were saturated with cold war imagery: "The American duty to protect freedom, through interplanetary settlement if necessary, might be challenged by invisible internal enemies."[60]

One of the complicating features of exobiology during the cold war was that to be comprehensively protective would require cooperation between the United States and Soviet Union. Such collaboration was not in the cards, and outer space, like the high seas, was treated by the United States as a zone at once outside politics and constantly in need of defending as "free." This belief that there existed realms beyond politics might make sense of the fact that one of the major lapses of NASA's exobiology program was its treatment of Earth's oceans as spaces outside human geography. Although the Apollo astronauts who went to the Moon were sequestered after their return to Earth, their capsule landed in the middle of the Pacific—a less than careful biocontainment effort, as some scientists at the Centers for Disease Control complained at the time.

The biological defense of the planet from extraterrestrial contaminants is still a going concern at NASA, though it has nowadays taken on a more relaxed aspect—so low key, in fact, that NASA was able to spare the chief planetary protection officer, John Rummel, as a speaker at the

Woods Hole astrobiology workshop, prompting one attendee to ask, "Shouldn't you be out protecting the Earth?" Rummel wore his title a bit ironically, pointing out that he was not trying to deal with all the meteorites raining down on Earth—and that his office had only one other staff person. "We're the Planetary Protection Office, not the Department of Homeland Security," he joked. And, though he was once head of NASA exobiology, he remarked that his job could come into conflict with astrobiologists who would love to bring back alien rocks. I had known of the Planetary Protection Office from an interview with vent biologist Colleen Cavanaugh, an advisor to this body. She told me that everyone who hears of the Office thinks of the film *Men in Black*, in which anonymous, besuited spies rummage for aliens hiding in humanity's midst. The Planetary Protection Office liked the connection and even purchased the movie's signature Ray-Ban sunglasses and took on the *Men in Black* tagline: "Protecting the Earth from the Scum of the Universe," tacking on "and Vice Versa."

This vice versa is where much of the attention of the Office lies these days; most of what Rummel oversees is the sterilization of spaceships sent to Mars. American frontiering narratives morph into sentiments of interplanetary ecological stewardship. In the next decade, NASA plans to send spacecraft to Jupiter's icy moons, which astrobiologists also consider promising destinations in need of protection. NASA: "Planetary protection requirements for this mission will focus on ensuring that the spacecraft will not inadvertently crash into one of the target moons and thereby risk melting ice and possibly contaminating liquid water. Planetary protection standards recommended by the National Research Council's Space Studies Board for missions to Europa (Preventing the Forward Contamination of Europa) would limit the probability of contaminating liquid water on Europa with a viable terrestrial organism to less than 10^{-4} per mission."[61]

Post–cold war environmentalism balloons up to include the entire solar system, with planets akin to islands in a macrocosmic sea. As with alien species in the Hawaiian Islands, invasive organisms travel with humans. Alien species are conveyed through culture and technology. The Planetary Protection Office acknowledges that, if the jury is still out on whether alien life forms travel on such natural vehicles as meteorites, they most certainly travel on cultural vehicles such as our own spacecraft. Like "alien algae," they come on ships. Just as in discourse about alien invasive species, nature is not alien to itself but rather becomes so when culture brings separated "natures" into juxtaposition.

ALIEN ABDUCTION

The line between nature and culture, of course—apparent in the debate about alien algae in Hawai'i—relies on how contexts are adjusted to frame the distinction. Rhetorical work positions Mars and Earth as hosting separate or similar "natures." Markley comments on the shifting, analogical relations of Mars and Earth: "Because it works by induction, a point-by-point comparison of observed characteristics or phenomena, analogical thinking has the potential to call into question the very principles that allow such comparisons to be made. . . . Not surprisingly, even as Mars has been perceived and described scientifically as an earthlike planet, Earth has become, in effect, a Mars-like planet."[62]

This is what gives some of us an uncanny sense that our home is not what we thought it was; this is what makes us aliens to ourselves. By extraterrestrializing Earth, scientists redraw boundaries of what can be considered natural on our planet. On this interpretation, however, life on Mars does not offer any grand repositioning of how we might think of life in the universe more generally. By bringing Mars and Earth into ecological embrace, scientists learn little about how life might originate in completely independent places.

But if induction animates analogies between Earth and Mars, abductive reasoning is not far behind, working as the anticipatory, hopeful end of the equation between the two worlds. Astrobiologists are themselves well aware of the many styles of reasoning that go into their enterprise. They are acutely tuned to the instability of their definitions of life and hope to be awake to surprising multiplicity, one reason researcher Baruch Blumberg suggests in his account of astrobiology that the field is open to a promiscuity of evidentiary regimes:

> Astrobiology is an interesting mixture of scientific processes. One emerges from the historical sciences that make up a large part of the astrobiology enterprise: astronomy, ecology, field biology, geology, oceanography, paleontology, and others. The events being investigated have happened, and it is the task of the scientist to tell the explanatory story. It is *inductive* science in that the data are collected first and then the hypothesis is formulated. . . . A second scientific approach emerges from the ethos of contemporary medical/biological research. It is *deductive* in the sense that it is hypothesis driven. . . . There is a strong emphasis on experimentation, in which the scientist creates his or her own universe that is, or is assumed to be, a simulacrum of the real world beyond the laboratory bench.[63]

Astrobiologists disagree, then, with nineteenth-century Nantucket astronomer Maria Mitchell, who wrote, "There is nothing from which to reason.

The planets may or may not be inhabited."[64] Astrobiologists hold that there are many sites and logics from which to reason about extraterrestrial life—from Earth as one planet among others, from organic chemistry, from optics. Astrobiologists are even open to the idea that they might not yet know what to reason from. As Blumberg suggests, "Life has the characteristic, using philosophical terminology, of 'being' and 'becoming.' It exists in a particular form now, but has the potential, because of the diversity in its off-spring, of becoming something related, but also different."[65]

Within this awareness—phrased though it is in terms of descent and inheritance—is a sense that astrobiology is in part an enterprise that depends upon abduction, the argument from the future. Abductive reasoning appears on NASA's Mars Program website: "The challenge is to be able to differentiate life from nonlife no matter where one finds it, no matter what its varying chemistry, structure, and other characteristics might be. Life detection technologies under development will help us define life in non-Earth-centric terms so that we are able to detect it in all the forms it might take."[66] Abductive reasoning appears again in the NASA Astrobiology Roadmap: "Catalogs of biosignatures must be developed that reflect fundamental and universal characteristics of life, and are thus not restricted solely to those attributes that represent local solutions to the challenges of survival."[67] This quest for the universal, looking up and out into space, is a quest for romantic complexity. In this abduction, life becomes deterritorialized into what it might yet become, but with an implicit commitment to the idea that it will always be "itself"—and that it will therefore always take "forms"; a National Research Council report on astrobiology is "Dedicated to Non-Human-Like Life Forms, Wherever They Are" (a strange way of putting it, since so many nonhumans are already on Earth).[68]

Just as in the abduction of Sargasso sealife into narratives of the genome of Mother Earth, the abductive reason at work here requires skipping steps and scales. It skips over the fact that scientists do not yet know what all life on Earth consists in. It skips over the possibility that *life itself* might be a historical rather than universal category. Just as *man* was represented by a slim slice of humanity in the Apollo Moon landings—that is, by white men (what Gil Scott-Heron in 1972 called "Whitey on the Moon")—so *life*, modeled by what we know of life on Earth, could turn out to be a parochial construct. *Life*, like *race*, may be revealed not only not to be a natural kind but also to be a constitutively heterogeneous—baroque—category all the way down.[69] Or across, athwart. *Life* may be a local, epistemologically contingent category, not even destined for theoretical immortality on Earth.

FIGURE 29. Photomosaic of fractured and smooth regions on the surface of Jupiter's icy moon Europa, imaged by robot spacecraft *Galileo* during a 1996 flyby. Image by Galileo Project, JPL, NASA.

AMPHIBIOUS CYBORGS ON EUROPA

Astrobiological witching for water as a possible pointer to life has found no more attractive site than Europa, one of the moons of Jupiter. Discovered by Galileo in 1610 (though named by rival Simon Marius), Europa is noted for its ice-covered surface, believed to be perhaps 5 kilometers thick and first seen in high resolution in 1979, with the *Voyager 1* flyby. Planetary scientists believe there is a liquid layer underneath, as much as 50 kilometers deep, kept in fluid form by tidal flows generated by Jupiter's immense gravitational pull. The motion of this water leaves spectacular cracks on the moon's surface (figure 29). Some scientists suggest that beneath its cracked crystal shine Europa hosts structures akin to hydrothermal vents, which may support chemosynthetic life forms.

My dive partner in *Alvin*, geologist John Delaney, has advanced just such a view; he has long been an advocate of extrapolating from extreme earthly environments like the Mothra Hydrothermal Vent Field to think about extraterrestrial life. In his contribution to the *National Geographic Atlas of the Ocean,* he speculates about the possibility of life on Europa, writing that "this seemingly hostile moon may hold the highest probability of encountering viable life elsewhere in our solar system, for tidally driven heating may stoke the volcanic fires within Europa. Scientists hypothesize that underwater volcanoes on Europa may support life-forms analogous to those found in seafloor hydrothermal vent systems on Earth."[70]

Although it is not clear that Europa has a plate tectonic system and its attendant vulcanism, it could be that tidal torque is sufficient to open up ventlike rifts in the Europan seabed. Scientists also do not know if present beneath the ice are the reduction-oxidation gradients characteristic of Earth's oceans, those chemical boundaries at which, for example, many deep-sea creatures make their oxidizing or reducing livings. And the radiation that blankets the moon—the product of Jupiter's mammoth magnetosphere—could either power a biosphere or quash it. It depends on whether this radiation is fed into a feedback loop that supports self-organizing systems or short-circuits them.

The most ambitious project in astrobiology would see an autonomous robot sent through space to land on Europa, drill through the ice, and then submerge itself, à la a self-guided *Ventana,* into the Europan ocean, transmitting signals back to Earth through a complex chain of transductions. Such an amphibious cyborg, crossing media—space, land, water—would be hugely expensive and NASA does not have enough money for such a project. At a talk about Europa I heard in 2003 in Moss Landing, Jere Lipps from the Department of Integrative Biology at UC Berkeley exhorted marine biology students in the audience to begin thinking about plans for such future exploration: "I encourage you to join in the search for life in Europa. Because it's you young people in the back who will do it, not us old folks up front, unless they invent a new elixir of life!"[71] Lipps's enthusiasm was such that an older professor in the audience asked him: "You're obviously excited about this. Have you considered having yourself cryogenically preserved?" Lipps brushed off the joke. But it was clear that for him Europa is a utopic destination, an alien ocean holding the promise of new life.

The pull of the possibility of life on Europa has popular power. Representations of this satellite as an abode for life are unfailingly optimistic, and Europans are a far cry from Wells's dyspeptic Martians. Europa features in Arthur C. Clarke's *2010* as an aqueous cradle for new life in the solar system. And Europa makes a spectacular appearance at the end of James Cameron's 2005 three-dimensional IMAX film, *Aliens of the Deep.* Cameron in this production leads the viewer on a dramatic tour of environments like hydrothermal vents, posed as analogues for extraterrestrial ecologies.[72] Unlike *Volcanoes of the Deep Sea,* this film is played primarily for grandeur. Whereas *Volcanoes* follows senior research scientists, for *Aliens* Cameron cast a crew of fit and trim graduate students, whose wonder as they descend in submersibles exudes youthful exuberance. Uniformed in jumpsuits emblazoned with the words "Extreme Life," they spend much of the movie saying things like, "That is absolutely unreal!" or "That's like another planet!" They seem so overwhelmed by their presence in the deep, they have a hard time speaking in the present tense;

looking out of the high-tech sub is "like a glimpse back in time" and some-times leaves them "feeling . . . transported into the future."

After a string of shots of students ogling sulfide chimneys from the bubble-like windows of submersibles, viewers are transported at the end of the film through the magic of computer graphics to the waters below the ice of Europa, to what the voiceover calls "an alien ocean." One of the film's stars is Dijanna Figueroa, a young African-American grad student at UC Santa Barbara who studies vent invertebrates. In a dramatic final sequence, she is digitally pasted into an interplanetary submarine plying the seas of Europa, readying to meet extraterrestrials. From within her sub, Figueroa spies a digitally rendered squidlike alien who meets her gaze by extending a friendly tentacle toward her outstretched hand.[73] The camera pulls back to reveal our friendly E.T.'s home, a shining city that resembles a glowing Atlantis, a Neptunian Jerusalem, a bioluminescent underwater City upon a Hill, populated by what look like aquatic angels. This is the alien ocean as supernatural—in two senses: supernatural because this alien ocean is located in a nature bigger than one pertaining solely to Earth; supernatu-ral, too, because this alien ocean is a transcendent paradise; the natural, encountered elsewhere, spirals up to intimate the existence of a higher order or system. The dizzy rhetorical work functions to shore up a partic-ular version of nature—a cosmic, harmoniously ordered nature. This alien ocean is a utopia, a place where creatures commune, touch, across inter-planetary difference.

SOLARISTICS

As I have been at pains to argue in this book, alien oceans are neither dystopias nor utopias, neither chaotic others nor warm and wet macro-cosms for our 71-percent-water individual selves. Rather, they are what Foucault might call "heterotopias." Unlike utopias—perfect places for-ever out of reach—or dystopias—perfectly awful worlds, like Orwell's Oceania in *1984*—heterotopias are constantly shimmering into solid existence; they are sites where the social order is "simultaneously repre-sented, contested, and inverted."[74] They are real places, neither fully per-fect nor imperfect. They are "capable of juxtaposing in a single real place several spaces, several sites that are in themselves incompatible."[75] Foucault names the world of the sailing ship—with its overstated instal-lations of landed social hierarchies (think Royal Navy) and its inversions of everyday life (think pirates)—a heterotopia par excellence. *Alien Ocean* has tipped Foucault's microcosmic heterotopia into the wider sea, finding today's heterogeneous microbial ocean a fitting example and symbol of the representations, contestations, and inversions that criss-cross the very possibility of making sense of life forms and forms of life, microscopic and macrocosmic.

One of the most fantastical representations of an alien ocean appears in Stanislaw Lem's sci-fi novel *Solaris*. In this tale, astronauts living in a space station hovering above a mysterious planet named Solaris discover that the sea embracing this world is a sentient entity able to dip into their dreams in the most disquieting way. Solaris can conjure up material copies of the persons haunting the astronauts' sleep. When we join the narrative, the space station is populated with doppelgangers of our characters' loved ones and elusive objects of desire—people left behind, even dead, on Earth.[76] But these ghosts are not mere Freudian Frankensteins; their behavior is too erratic to put down to wish fulfillment. No one in the novel knows how or why Solaris causes such apparitions to appear or, indeed, whether Solaris has any intelligible agency, intentionality, or motivation at all. We learn that the interdisciplinary field of "Solaristics" has for over a century been unable to comprehend the fluid alterity of the planet. Early research journeys—which left measuring instruments physically trans-formed—presented relatively benign paradoxes compared to the latest psychological entanglements, which are mostly incomprehensible and cause the main characters variously to conduct scientific experiments obsessively, seek speechless communion with Solaris, drink, or commit sui-cide. Lem gives no aid in sorting out which response makes most sense. The point of his story, rather, is to deliver a portrait of humans encountering an alien entity that none of their empirical or interpretive frames can capture. Solaris, an alien ocean, is a heterotopia. It is not a space that invites an easy gaiasociality, a sentiment of rational responsibility to an ocean planet, for it will not yield to the dreams of reason.

Solaris might be compared with the alien ocean of this book. It is a medium that reflects and refracts human self-conceptions. It is a space of familiar and unfamiliar kinships, of tangled pasts and futures.[77] It is a ter-rifying other, brought to us courtesy of our own obsessions, fantasies, mis-takes. It is a realm that scientists, managers, and policy makers seek to domesticate, to tame. It is a space for hopeful, if uncertain, speculation. It is an object of expert scrutiny, though also out of sight and mind of landlub-bers and laypeople. The alien oceans I describe in this volume are not all alien in the same way, not all analogous or homologous, though they are sometimes—to borrow a term from lateral gene transfer—paralogous, similar through borrowing.

In his analysis of *Solaris,* "The Book Is the Alien," Istvan Csicsery-Ronay suggests that the coexistence of multiplicities of incongruous inter-pretation is itself both a sign of and productive of the alien: "In the face of that-which-does-not-correspond, the most diverse and contradictory ways of making sense become a single self-reflecting set of correspondences, an amorphous mythoscience thrashing in its inability to articulate the alien."[78] In other words, an excess of uncertainty often promotes the sight-ing of aliens. Attending to all possible uncertainties at once, however, may lead to reading too much into alien objects, landing the analyst into a mad library of mirrors, where everything makes sense precisely because we have no coordinates for situating our knowledge. DNA libraries—or, indeed, all the library materials I poured into this book—can never fully capture the sea, the alien ocean, as such—even as we may try constantly to represent it. It may be that the best we can do—as oceanographers, as anthropologists—is construct good enough representations, mindful that the ocean and our portraits of it produce realities that require a constant recalibration of our interpretative activities.[79]

OTHER OCEANS

In reporting on the search for planets outside our solar system, one pre-senter at the MBL astrobiology workshop, Philip Crane, described worlds that might support life as "other Earths." The phrase frustrated many par-ticipants, who protested that *our* Earth is the only one there is; "Earth-*like* planets" might be a better term, they offered, though even that would frame the search for life too narrowly. Crane's use of *Earth* as a general term leads me to consider whether the extension of the concept of life to cover as yet undiscovered instances might entail a similar interpretative overreaching. How far can Earth or life be translated into theoretical terms that can float free of particular embodiments?

If we follow Margulis on her travels in the salt marsh, the answer might be "not too far." Margulis periodically totes mud from Sippewissett back to her lab in Amherst, where she places mats in nutritive media to see if spiro-chetes materialize and attach themselves to other cells. She writes, "I believe that with much help from colleagues and students, we will soon be able to show that certain free-swimming spirochetes contributed their lithe, snaky, sneaky bodies to become both the ubiquitous mitotic appara-tus and the familiar cilia of all cells that make such 'moving hairs.'"[80]

For Margulis, living things are forever incorporating one another, engag-ing not just in lateral gene transfer but also in lateral genome transfer.[81]

But Margulis will not know how such incorporation works unless she actually runs an experiment, tries to jostle spirochetes awake to see what they do next. In the language of Richard Doyle, she is in the realm here of "wetwares," "encounter[s] with flesh as a refrain, a repetition of algorithms or recipes of sufficient complexity that only through instantiation can they be experienced."[82] In other words, Margulis's spirochetes produce a signature of life that needs to be fed to be read. The semiotics of life needs living things to signify, and those things cannot exist except in contingent, real time—coming into liveliness through such material activities as eating, which always happens in a web whose coordinates are never fully in place prior to their habitation and creation.[83] There is no Platonic world of *life,* or of *form.*

As I have documented, *life* these days is being distributed into material and semiotic networks that scale from the fidgeting and floating gene to the Gaian globe, with lots of baroque curlicues in between. As the marine biologists with whom I have interacted argue, this does not mean that life therefore achieves escape velocity from materiality— even if it sometimes becomes reticulated into physicalities (like genetic or ecological networks) that are not bounded in the way individual organisms are. Consider, for example, geobiologists Mark and Dianna McMenamin's theory of "hypersea," which argues that life on land exists in and creates a kind of rhizomatic terrestrial sea: "In a way, the land biota has had to find ways to carry the sea within it and, moreover, to construct watery conduits from 'node' to 'node.'"[84] This is romantic complexity downloaded into baroque webworks of life forms like fungus and their entangled ecological relations with soil and trees. What provides the anchor for "life" here—and for most scientists I interviewed for this book—is "the ocean," the ocean, as Hawai'i's Dave Karl described it to me, as "a flowing medium." The *the* in "the ocean" importantly designates an entity apprehended as global—a framing that is a historical, scientific, and social accomplishment, a scalar narrative; after all, there are also oceans. And some of those other oceans exist as other descriptions or versions of the sea—just as Gaia or Snowball Earth are the closest we have to "other Earths"—versions that are right here, on this planet. Or were. Or might be: it is possible that Earth's oceans may be traveling toward a state that is alien to the ocean we have known, more acidic, for example, or deeper, owing to climate change. The question is: What is our relation to those alien oceans? Are they allies, friends, strangers, remixed selves? What symbiopolitics are in the making? What oceanicities?

BIO, THAT ALIEN VISITOR

I want to close by asking a question that has animated biology since Treviranus asked it in 1802, a question Erwin Schrödinger took as the title for his famous 1944 book: What is life?

For Schrödinger, "life" pointed to the structure of an aperiodic crystal capable of containing and transmitting the logic of heredity, which he predicted would be legible as a kind of "code-script." The description of the double helix of deoxyribonucleic acid and the subsequent deciphering of the "genetic code" both fulfilled and were dependent upon Schrödinger's metaphorical armature; life was understood to unfold from the unified, formal structure of DNA. Fifty-one years after Schrödinger, Lynn Margulis and her son, Dorian Sagan, offered a less unitary definition in their book *What Is Life?* Margulis and Sagan delivered a distinct answer to the question for each of life's five kingdoms: bacteria, protists, animals, fungi, and plants—emphasizing not some underlying logic or an overarching metaphysics but rather the situated particulars of microbial, fungal, plant, and animal embodiment (one reason Margulis resists the three-domain model of Woese, which is based on a genetic reduction). In *What Is Life?* life is not something to be compressed into the linear logic of a code but is always emerging into an assortment of bodied manifestations.[85]

The question of what life is continues to visit us and, indeed, posing the question—and, more, asking why we have been obsessed with asking it—has become a growth industry, and not only in the natural sciences. Historians, philosophers, literary theorists, and anthropologists increasingly offer their views on the topic, and my analyses in this book are diffracted through such contemplations. The emerging consensus is that a funny thing happened to life as it traveled through the twentieth century. On the one hand, as Hannah Landecker argues in *Culturing Life,* her history of cell biology, the material components of living things, cells and genes, can now be rearranged and dispersed, made to live in distributed laboratory choreographies that have them frozen, amplified, multiplied, and exchanged.[86] The substances of biology are no longer only the property of discrete organisms. On the other hand, as Doyle argues, the representation of life itself has been overtaken by ambiguity and multiplicity: "'Life,' as a scientific object, has been *stealthed,* rendered indiscernible by our installed systems of representation. No longer the attribute of a sovereign in battle with its evolutionary problem set, the organism its sign of ongoing but always temporary victory, life now resounds not so much within sturdy boundaries, as between them."[87]

In other words, life has been materially and semiotically detached from its previous address in bounded bodies. In this world of biology unbound, the biological is less simply reproductive than transductive and abductive. As I have argued here, life forms are being fractionated, rendered age-nealogically related, turned into networks, shored up as property, deterritorialized, and scaled down, up, and over. Twenty-first-century biology places scientists and their interlocutors increasingly amid this flotsam, this domain of "life at sea." Like ethnographers or biological oceanographers who can never be fully and everywhere present to study the lives of people or plankton, so too life slips between presence and absence as it forms and deforms, as it evanesces between bodies and ecologies. Life is spaced out.

Oceans, as I have sought to demonstrate, are good to think with about life. In his 1962 *Totemism*, from which I borrow the "good to think" locution, Claude Lévi-Strauss argued that anthropologists' desire to form a universal theory of totemism—the use of animal or plant figures to represent kin groups—rests on the fundamental error of believing that this is a universal form to begin with: "Heterogeneous . . . beliefs and customs [have been] too hastily lumped together under the label of totemism."[88] For Lévi-Strauss, totemism was an illusion. More basic was how human action grappled with the relation between nature and culture; the selection of a clan animal was an emanation of that struggle, not an object in itself. These days, of course, nature and culture themselves are in upheaval, revealing this pair, too, as a parochial parsing of the world.[89]

Resting my argument on this eroding foundation, then, I suggest that "life itself" is presently in the process of being revealed as a theoretical abstraction. If life was once believed to reside firmly in the territory of nature, such practices as biotechnology now place it in a riptide created by the shifting currents of nature and culture—a dissolving binary I have sought to mark in this book by using the suggestively overlapping terms *life forms* and *forms of life*. I have argued here that emergent scientific accounts of the ocean are making explicit the shifting boundaries of life— as well as uncertainties in the meaning of this term. "Life" migrates into quotation marks—not just in cultural studies of science but in theoretical biology—because it is so definitionally unstable.[90] We have entered the age of what Sandra Bamford calls, in a felicitous phrase that harkens back to the preface of *Alien Ocean*, "biology unmoored."[91] Life cannot be captured through a general *theory* of vitality—something philosopher Carol Cleland advocated at the MBL workshop—but is better apprehended by working athwart theory, zigzagging from abstraction to tangibility, or moving aft, aloft, and adrift amid explanation and substantiation.[92]

Let me return to the distinction between *zoë*, "the simple fact of living" (Agamben's "bare life") and *bios*, "the form or way of living proper to an individual or a group," in order to revisit the proliferation of the prefix *bio-* in many of the practices of which I write.[93] In the marine biology I have surveyed, we are visited by many *bio-* words: biogeochemistry, bioinformatics, biotechnology, biodiversity, bioprospecting, bioinvasion. Each neologism argues that *bio-* adds something distinct to the prefixed word. *Biogeochemistry* tells us that planetary ecology involves substances and processes ontologically entangled with the geological and chemical. *Bioinformatics* promises that there is a logic to the biological that can be abstracted and moved across substrates. *Biotechnology* offers that that logic as well as any accompanying biotic substance can be instrumentalized. Each of these three terms might properly be understood as working in the territory of *zoë*. A movement toward *bios* becomes evident in the remaining compounds on this list. *Biodiversity* has latent within it a conjunction between *zoë* and *bios*; bare biotic matter can be seen as organized into a form that has value—both for "evolution" (e.g., as the InterRidge scientists have it) and for social forms like genetic engineering. *Bioprospecting*, as Cori Hayden suggests, employs *bio-* to "launder"[94] prospecting of its dirtier, more exploitative history in mining—one reason that its rebranding by opponents as "biopiracy" is so morally condemnatory; *bio-* is *bios* in both formulations. Finally, *bioinvasion* is nothing but value judgment, with *zoë* hijacked into a perverse *bios* whose origin is human. The unsteadiness in these six *bio-* terms—flagged by the fact that it is quite difficult to describe them in exact parallel structure—is a signal that the link between life forms and forms of life is uncertain. Adding *bio-* to these words brings up as many questions as it answers. Neither nature nor life can speak for itself.

The attempt to make explicit the scientific and cultural meaning of life is not restricted to the natural sciences. *Bio-* words from the humanities and social sciences also file through this book: biopolitics, biosociality, biocapital, biomedia. *Biopolitics* is the inspiration for all the others, centering as it does on the fashion in which "life and its mechanisms" have been turned into subjects for governance and management.[95] *Biosociality* distributes biopolitics into the lives and decision making of everyday people, asking how, for instance, patient advocacy groups think about turning manipulable *zoë* (like the chromosomes of future children) into *bios* (say, a biography free of an undesired polymorphism). *Biocapital* seeks to investigate whether the use of *zoë* for the production of surplus value and profit describes a new permutation of capitalism. Unlike the terms *biodiversity, bioprospecting,* or *biotechnology,* these formulations seek to identify new conjunctions of life forms and forms of life, not to invent them or infuse them with value. This is not to say that the terms do not map out fields of contest and disagreement. Certainly, many biologists are uneasy with the commoditization of biotic stuff, a dynamic named by biocapitalism.

There *is* a novel empirical claim in these terms, however, which can be illustrated though Thacker's concept of *biomedia.* His coinage offers a way to think about how social contexts are enlisted and created to reformat what will count as the biological at all. Thacker writes, "Biomedia returns to the biological in a spiral, in which the biological is not effaced (the dream of immortality), but in which the biological is optimized, impelled to realize, to rematerialize, a biology beyond itself."[96] This "biology beyond itself" makes explicit the ways *zoë* is increasingly under reconstruction by *bios.* Call these media processes "mash-ups" of *zoë* and *bios.* As with "a piece of popular music created by merging the elements of two or more existing songs" (OED), life forms are being remixed with and within our forms of life—one reason that, in operating athwart theory, scouting for unexpected approaches to today's biopolitics, I often mash up anthropological and marine microbiological theory (about, e.g., relatedness, definitions of locality, and, indeed, life).

In the age of molecularization, when life is broken down to be built up again, what emerges is not always again "life itself."[97] Life is no longer itself but rather its partibility, its relationality. If SETI imagined its work would have ethical and cultural implications—and maybe even extraterrestrial content providers for these categories—astrobiologists, in their search for biosignatures, imagine the bare life of the cosmos as something that cannot speak for itself, that needs *us* to interpret its ethical content. The extraterrestrial arena makes explicit what is happening to life today, here on Earth. Life forms are no longer stably able, if ever they were, to underwrite a *bios* on which everyone will agree.

This is one reason that the alien has surfaced in this book; it has appeared when the relation of life forms to forms of life, of *bio-* to whatever word it is prefixing, is unsteady, when flux crosses form. A sampling: Biogeochemistry: are all life forms that could stabilize the biosphere familiar to us? Bioinformatics: will life forms provide information consistent with linear origin stories? Bioinvasion: what does it mean for a life form to be native to an environment? Biocapitalism: to what extent can life forms become alienable property?

The uncertainty culminates in the *biosignature,* in which the *bio-* stands for life forms scientists have yet to encounter, and indeed for possible alien life forms. One take-home message of this book—if one can return home after so many alien encounters—is this: *life* as a scientific concept has become alien to its origins. *Bio-* is an alien visitor.

At the conclusion of *The Order of Things*, Foucault, in a phrasing that evokes Rachel Carson's description of the seashore world, suggested that *man* may someday "be erased, like a face drawn in sand at the edge of the sea."[98] He did not mean that humanity might be wiped out by oceanic inundation—though such a literal reading is freshly thinkable and not altogether irrelevant in the wake of the Indian Ocean tsunami of 2004, 2005's Hurricane Katrina, and growing evidence of global warming.[99] Rather, Foucault speculated that the *human*—that biological, language-bearing, laboring figure theorized by human sciences ranging from anatomy to anthropology to political economy—might not endure forever, just as archangels, warlocks, and savages are no longer so thick on the ground of our social imagination as once they were, and just as *race* as a biological category now wobbles between phantom and Frankenstein as it has been set afloat in a sea of genes. In a seascape swimming with ambiguous ancestors, organisms out of place, and submarine cyborgs, it is worth asking whether the microbial ocean, as it becomes a site in which links between life forms and forms of life are redrawn, might also key us in to transformations around the boundaries of the *human* and the biotic substances and connections through which this being is imagined. We see in this book inklings of such symbiopolitical reconfigurations, from microbes indexing climate dynamics inhospitable to humans, to lateral-gene-transfer-tweaking tales of origin and deep genealogy, to neurotoxic marine bacteria flowing through water supplies into human brains, to biotech promises to enlist microbes into a globalized capitalist labor force, and to the emplacement of biological embodiment on Earth into zones of extremes and extraterrestriality.

By the time I finished this book, descriptions of human bodies as 70 percent salty water were being edged out by pronouncements like this one, delivered by microbiologist Jo Handelsman: "We have ten times more bacterial cells in our bodies than human cells, so we're 90 percent bacteria."[100] Once upon a time, the *human*, plunged into the sea (as blood, sweat, tears, milk), was baptized into communion with the planet. But plunged into the sea as a swirl of microbial genes, something more unsettling happens. Microbes are not simple echoes of a left-behind origin for humans, orphaned from all evolutionary association. Microbes are historical and contemporary partners, part of our bodies' "microbiomes."[101] "The" human genome is full of their stories, revealing that *all* genomes are metagenomes.

The links between the scale of human bodies and ecologies become baroque, spatially and temporally. The bacteria that inhabit our bodies do not simply mirror the bacteria that inhabit the sea—as might brine in our blood. This is not human nature reflecting ocean nature. It is an entanglement of natures, an intimacy with the alien.

Such dynamics shift the grounds upon which *anthropos* might be figured, perhaps transforming humanity into *Homo alienus*.[102] The microbial sea, imagined in a molecular, microbial, genomic idiom, may not usher in what philosopher Martin Weiss sees as a "dissolution of human nature" coincident with the rise of biotechnology.[103] Instead, we might take Ed DeLong's PowerPoint juxtaposition of Leonardo's Vitruvian Man against the whole Earth (back to figure 1) as an icon foretokening not the dissolution of humanity, "like a face drawn in sand at the edge of the sea," but rather this: the saturation of human nature by other natures.

Notes

1. DeLong and Karl (2005), Proctor and Karl (2007).

2. "Linking genomes to biomes" is the motto of the Center for Microbial Oceanography: Research and Education, coordinated by the University of Hawai'i.

3. See Hamblin (2005).

4. Steinbeck, in Ricketts and Calvin (1952: vi).

5. Latour (1988: 39).

6. Capitan, quoted in Latour (1988: 37).

7. According to the OED, *life-form*—meaning either "a living creature; any kind of living thing" or "a habit or vegetative form exhibited by any particular plant or which characterizes a group of plants"—enters English in 1899 from German *Lebensform*, referring to "groups of similar adaptational form [that] by no means coincide with natural families or groups of species" (e.g., annuals, rhizomatics). The manifestation of life in forms more generally was an obsession of nineteenth-century biologist Ernst Haeckel, noted for his geometrical lithographs of marine protozoa; see *Art Forms from the Ocean: The Radiolarian Atlas of 1862* (Haeckel 2005). Alexis Rockman, whose "Biosphere: Microorganisms" graces this book's cover, borrows from Haeckel (see Brody 2002). *Life form* appears in its general sense in English earlier than the OED documents, in *American Naturalist*, which states that "all life forms have come into being by a process of evolution from primitive organic germs" (Morgan 1878: 676).

8. Wittgenstein (1958). Wittgenstein's German was *Lebensform*, though he distanced the word from natural science meanings.

9. Compare Shapin and Schaffer (1985), Collins (1985), Lynch (1992), Lenoir (1997), Fischer (2003), Lynch (2005).

10. See Franklin (2005) and Rose (2007: 80).

11. Foucault (1970: 127–28).

12. Quoted in Coleman (1977: 2).

13. Foucault (1990: 143).

14. Yanagisako and Delaney (1995).

15. Fischer (2003: 37).

16. Sheridan (2005).

17. Keller (2000).

18. When I began this research, I was completing an ethnographic study of "Artificial Life," a brand of theoretical biology devoted to computer simulations of living things. In *Silicon Second Nature* (Helmreich 2000), I argued that Artificial Life takes to logical conclusion a view of life as a program coded by genes. Researchers at New Mexico's Santa Fe Institute for the Sciences of Complexity were so compelled by the genetic program metaphor that they considered self-replicating computer programs convincing stand-ins for life, which they believed could manifest in cyberspace. If Artificial Life theories script life as detachable from particular substrates, the latest marine biology similarly renders life as possessed of an as yet unmapped elasticity—though one always anchored in organic chemistry.

19. Earle (1995: xii).

20. Quoted in U.S. Cabinet (1999: 6). Clinton is not alone in this comparison. Marine microbiological pioneer Claude ZoBell: "Nearly 3/4 of the earth's surface is covered with water. . . . Nearly 3/4 of the human body is water" (1959: 34). And see Safina (1997: 435).

21. Leonardo, edging a kindred comparison into the language of anatomy, wrote, "As man has within him a pool of blood wherein the lungs as he breathes expand and contract, so the body of the earth has its ocean, which also rises and falls every six hours with the breathing of the world" (quoted in Ball 2001: 22).

22. Earle (1995: 15). Philip Ball reprimands those who make too much of the saltiness of blood: "The composition of blood plasma and the watery cell liquid called cytoplasm does have similarities with seawater: all contain sodium, potassium, and chloride ions (which is why blood tastes salty) as well as bicarbonate. But the resemblance is largely incidental: cells simply need these ions to function" (2001: 242–43). Penny Chisholm (personal communication) objects that, since cells evolved in the sea, more than coincidence is at work.

23. Haraway (1997: 174).

24. Davidson (1998: 4). In *The Sea around Us*, Rachel Carson (1951) refers to the ocean as "Mother Sea."

25. Rabinow (1992). See Rapp (1999) for amplification.

26. Earle (1995: 201).

27. U.S. Cabinet (1999: 22).

28. COMB (Center of Marine Biotechnology) main page, www.umbi.umd. edu/~comb/, accessed April 15, 2002.

29. Haraway (1997: 41).

30. National Oceanic and Atmospheric Administration (1998).

31. Thomson et al. (2003). See also Drake et al. (1999), Doblin et al. (2003).

32. De Kruif (1996: 9).

33. Langewiesche (2004: 3, 39).

34. Jackson (2006). Cf. Woodard (2000).

35. Ruggieri and Rosenberg's 1969 *The Healing Sea: A Voyage into the Alien World Offshore* speaks this double tongue.

36. See Needham et al. (2000), Wakeford (2001).

37. Shapiro (2000: 3).

38. Latour (1988: 44).

39. Viewing the ocean as a network does not require microbes. See Crawford (1997).

40. Paxson (2008: 17).

41. See National Research Council (1999).

42. Margulis and Sagan (2002).

43. Theologian Catherine Keller (2003: 25) argues that gendering the sea *she* emerges from patriarchal interpretations of biblical creation stories in which a "feminization of the flux" warrants an ordering of the primordial, watery chaos by the Word of a masculine God.

44. On early modern seafarers, see Rediker (1987); on the seagoing sublime, Raban (1993); on seascapes and the American imagination, Stein (1975); on slime/sublime, Giblett (1996).

45. Microbeworld: Discover Unseen Life on Earth, www.microbe.org.

46. Munn (2004: xiv).

47. Sea creatures have been compared to aliens before. Dolphins were imagined in the 1960s as analogs for alien intelligences—though emphasis was on their language, not their biology, their "life"; see Lilly (1961).

48. Dean (1998: 175).

49. Stewart and Harding (1999: 294).

50. Eva Hayward describes a phasing between alien and other in her analysis of sea creatures at Monterey Bay Aquarium. Jellyfish exhibits are introduced with "'alien' allusions. All propose that you are entering a different place, a place of difference, a place of first encounters. . . . Yes, the organisms are 'other,' but their difference is woven back into the culture through the display technologies, and away from timeless, ahistorical, mythologies" (2004).

51. Hall (1990: 229).

52. Comaroff and Comaroff (2001).

53. See Dean (1998: 167), Marciniak (2006).

54. For reviews, see Acheson (1981), Pálsson (1991), Johnson (1996).

55. Pálsson (1991: xvii).

56. Malinowski (1984: 4).

57. Kuper (1983: 12).

58. See Lenček and Bosker (1998) on the history of beaches.

59. For sociology of fisheries biologists, see Finlayson (1994); on shipboard interaction, Bernard and Killworth (1973), Bernard (1976), Rozwadowski (1996). Historians of science have researched the genesis and development of oceanography and marine biology. I rely on their accounts here.

60. Cf. Philbrick (2003) on the U.S. Exploring Expedition, 1838–1842. See Rozwadowski (2005) for a history of oceanography.

61. Rehbock (1992).

62. Conferences I attended: Sept. 29–Oct. 4, 2000: "International Marine Biotechnology Conference," Townsville, Queensland, Australia; July 23, 2001: "Extremophiles: Theory and Techniques," Center of Marine Biotechnology, University of Maryland, Baltimore; April 23–26, 2002: "Fourth Asia-Pacific Marine Biotechnology Conference," East-West Center, University of Hawai'i, Honolulu; May 30–June 1, 2002: "Assembling the Tree of Life," American Museum of Natural History, New York City; March 17–19, 2003: "Third International Marine Bioinvasions Conference," Scripps Institution of Oceanography, La Jolla, California; March 7–9, 2004: "WAS . . . IS . . . MIGHT BE . . .: Perspectives on the Evolution of the Earth System," Massachusetts Institute of Technology, Cambridge; April 17, 2004: "Microbial Sciences: A Symposium," Harvard University, Cambridge; January 19–21, 2005: "InterRidge Workshop: Tectonic and Oceanic Processes Along the Indian Ocean Ridge System," National Institute of Oceanography, Goa, India; May 15–22, 2005: "Cosmic Evolution and Astrobiology," MBL-Dibner Seminar in the History of Biology, Cosponsored by the NASA History Office, Woods Hole, Massachusetts; June 9, 2005: "What Are the Stakes of Deep Seabed Bioprospecting?" United Nations University Institute of Advanced Studies, United Nations, New York; November, 7–8, 2005. "Vital Signs: The Diagnostics of Earth System Evolution," Massachusetts Institute of Technology, Cambridge.

63. Research expeditions I joined: March 7, 2003: R/V *Point Lobos*, Monterey Bay and Canyon; March 26, 2003: R/V *Point Lobos*, Monterey Bay and Canyon; June 18–22, 2003: R/V *Roger Revelle*, Hawaii Ocean Time-Series, cruise 149, to Station ALOHA; May 11–21, 2004: R/V *Endeavor*, cruise 393, Sargasso Sea; May 23–June 9, 2004: R/V *Atlantis*, cruise AT 11–13, Juan de Fuca Ridge; June 1, 2004: DSV *Alvin* dive 4020, Mothra Hydrothermal Field.

64. I often use *oceanography* and *marine biology* interchangeably, though distinct histories can be teased out. In Europe, oceanography developed out of fisheries science, whereas in the United States the world wars gave American oceanography a more physical cast (Shor 1978). My interlocutors used the two fluidly, to draw distinctions between physical and biological oceanography or to assert that marine microbiology should be part of oceanography.

65. Cf. Rabinow (2003). See Riles (2000) on how anthropologists studying professional communities find they often share analytic languages.

66. Cf. Hacking (1983).

67. Latour (1999: 64).

68. Maurer (2005: 15).

69. Azam and Worden (2004: 1622).

70. Feyerabend (1975). Cf. Kaiser (2006), which traces how objects of theory in physics may have histories underdetermined by theoretical considerations (conditioned, e.g., by pedagogical practice) even as such objects become legible only because of their articulation in terms of theory.

71. Knapp and Michaels (1982).

72. Strathern (1991: 75).

73. Powdermaker (1966: 14).

74. Haraway (2008). See Callon (1986) on associations between fishermen and scallops.

75. See Latour (1988: 38–40).

76. Nelson (2003: 250).

77. Conceptions of the sea as politically uncontested can be traced back to 1609, when Dutch jurist Hugo Grotius designated seas beyond political dominion as the "High Seas"; what is new is the American frontiering frame as well as a view of the ocean as simultaneously generative and in need of stewardship. Following David Nye, I consider such a model to be very American, embracing "three interlocking stories that together define . . . a sequence: the wilderness tale, the second creation story, and the recovery narrative" (2003: 299–300).

78. Venter, on Voice of America's "Our World," March 17, 2007, transcript at www.voanews.com/english/archive/2007-03/2007-03-16-voa3.cfm?CFID= 84736433&CFTOKEN=35525614. Fernando de la Cruz and Julian Davies (2000: 128) see things similarly: "It is clear that genes have flowed through the biosphere, as in a global organism."

79. Geertz (1973: 5); Marcus and Fischer (1986: 29).

80. Fabian (1983: 130).

81. Morrison (1992). Trinidadian writer C. L. R. James, composing *Mariners, Renegades, and Castaways* while detained on Ellis Island in 1953 as an "illegal alien" and "foreign subversive," offered a reading of *Moby-Dick* that detected a "parallel between Ahab's illegal change of the contract and the emergency powers claimed by the Cold War state" (Donald Pease in James 2001: xviii).

82. Lilly (1961).

83. Melville (2001: 7).

1. THE MESSAGE FROM THE MUD

1. Fuller (2005) provides a history of *media ecology,* finding early use of the term in a 1971 issue of *Radical Software.*

2. See Monterey Bay Aquarium Research Institute, Where Are the MBARI Research Vessels? www.mbari.org/cruises/both.asp.

3. Höhler (2002).

4. Quoted in Earle (1995: 138).

5. ROVs also have mundane industrial origins, in undersea oilrig maintenance. International Submarine Engineering built *Ventana* as a science-centered device.

6. Hamilton-Paterson (1992: 168).

7. Anaerobic methane oxidation occurs mostly at continental margins, though it has been detected in the Black Sea and Mono Lake (D'Hondt, Rutherford, and Spivack 2002). Hoehler and Alperin (1996) first hypothesized reverse methanogenesis.

8. Orphan et al. (2002), DeLong (2005).

9. See Reeburgh (2003).

10. McLuhan (1964: 7).

11. McLuhan (1964: 18).

12. See Stocking (1982).

13. Woodward (1983).

14. Nathan Sawyer, Moss Landing antique dealer, found the issue of *Game and Gossip* (Mann 1972) from which I adapt this history.

15. See Strayer (2002) for a sociology of Moss Landing fishermen.

16. Kant (1952).

17. The association of flying with underwater vehicles has not always taken the airplane as its model. In the 1950s, Auguste and Jacques Piccard built the bathyscaph *Trieste* on the principles of ballooning. Where hot air balloons used hydrogen, *Trieste* used a tank of gasoline—lighter than water—to raise and lower its human-occupied submarine.

18. See Fink (2005) on the media sublime.

19. Goodwin (1995: 257, 254, 256).

20. Goodwin (1995: 240). See Latour and Woolgar (1986) on inscription.

21. E.g., Bernard and Killworth (1973).

22. Hutchins (1995). On "cascades of representations," see Latour (1987).

23. Mukerji (1989: 153).

24. Bolter and Grusin (1999: 5).

25. Haraway (2008). See Thurtle and Mitchell (2004).

26. Hillel Schwartz suggested "intimate sensing" to me.

27. On Dutch seascape paintings, see Corbin (1994).

28. Hayward (2004). Contrast the aesthetic of San Diego's Sea World (Davis 1997).

29. Cf. Jenkins and Tulloch (1995).

30. Later, it returns with me, a jarred keepsake: "Like the Oriental ornaments collected in European homes in the eighteenth and nineteenth centuries, like the exotic flowers nurtured in the gardens of the same period . . . , rubbery grey specimens floating in laboratory jars are to trained eyes exotic marvels from distant worlds" (Mukerji 1989: 146).

31. See Hayward (2005).

32. Jennifer Paduan, "Ode to the soboL tnioP."

33. This exchange with Knute is spliced in from a voyage I took on *Lobos* with chief scientist Shana Goffredi on March 26, 2003.

34. Manovich (2001).

35. Though compare the well-funded *National Geographic* "Crittercam" project, in which sea creatures carry lightweight cameras into circumstances impossible for humans. For another angle into "saving the whales," see Cipriano and Palumbi (1999) on travels to Tokyo's fish market to flush out illegal whale meat using a suitcase-sized gene sequencer. On Tokyo's market, see Bestor (2004).

36. Schrader (2006) discusses the "phantomatic" character of dinoflagellates, whose toxicity is revealed only after the fact, through traces like dead fish.

37. McLuhan and Fiore (1967).

38. See Ellis (2003).

39. DeLong (2002).

40. Pace et al. (1986) first proposed such approaches. See Handelsman (2004).

41. See ZoBell (1949). Interest in marine microbes begins earlier. In 1894, Fischer wrote in *Die Bakterien des Meeres*, "There is no doubt whatsoever that our admiration and lively interest is aroused not only by the magnificent properties of the oceans, but also by the number, variety, and interesting characteristics of these tiniest of organisms, the marine bacteria" (quoted in Karl 1978). And see ZoBell (1946).

42. "An example of a widely used medium for routine culture of many marine bacteria is Zobell's 2216E medium, which contains low concentrations of peptone and yeast extract plus a mixture of various salts" (Munn 2004: 26).

43. "Probably less than 1% of prokaryotic species have been cultured" (Munn 2004: 27).

44. Or *potentially* viable marine forms, since genes could come from dead organisms. Ghosts sometimes flitter through the medium of the gene library. Sommerlund (2006) notes that the distinction between living and dead is elided in gene-sequencing approaches to microbial ecology. The distinction between lab and field also transforms. Bringing the seascape into the lab in environmental genomics turns the lab into a field in miniature; experimental work is often overtaken by a kind of natural history of genomic collections (cf. Kohler 2002).

45. Quoted in Pollack (2003: D1).

46. Colgan (2006).

47. The basic analytic unit in a library is one well (i.e., a plate coordinate). Following the text metaphor, a plate is like a page of phrases. Libraries consist of a portion of a plate or of one or more plates.

48. New England Biolabs, Deep Vent$_R$™ (exo-) DNA Polymerase, www.neb.com/nebecomm/products/productM0259.asp.

49. No one is worried the microbes might infect us. Our vaccinated consciousness makes it easier to consider them friends.

50. Thacker (2004: 11). Cf. Knorr-Cetina (1999: 157).

51. See Redfield (2000: 15).

52. Laporte (2000) argues that privatization of human waste led to distrust of the olfactory, to the idea that what is true cannot smell.

53. Leach (1964: 37–38).

54. Pauly (2000).

55. Knorr-Cetina (1999: 99–100).

56. Knorr-Cetina (1999: 95). Cf. Myers (2007).

57. See Kay (2000). The library metaphor is optimistic, in view of the "formless and chaotic nature of almost all . . . books" (Borges 1962: 53).

58. Flanagan (2001).

59. In *The Log of the Sea of Cortez*, Steinbeck reports the fanciful tale of a "decomposed body of a sea-serpent . . . washed up on the beach at Moss

Landing" (1995: 27). If the remains of this serpent are anywhere, they are in the Necropiscatorium.

60. Benjamin (1968: 61).

61. See Béjà et al. (2000).

62. See Masco (2004) on "mutant ecologies."

63. Taylor (1995) offers a history of Mexican oceanography (based at the Escuela Superior de Ciencias Marinas of the Universidad Autónoma de Baja California in Ensenada), centered on fisheries and seaquake research.

64. Steinbeck (1995: 7).

65. Lovecraft (2001: 283).

66. Steven's tale recalls an analogy between coral reefs and culture offered by a Boas student, Alfred Kroeber. Kroeber used the image of the reef and its polyps to describe the layering of culture: "The firm, solid part of the reef consists of calcium carbonate produced by the secretions of these animals over thousands of years—a product at once cumulative and communal and therefore social" (1952: 135). See Helmreich (forthcoming) on reef imagery.

67. Hallam et al. (2004: 1457).

68. In the film *Alien Nation* (Baker 1988), extraterrestrials come to Earth to escape slavery. They can withstand methane, an ability humans exploit to employ them as methane-mining wage laborers.

69. Consult Sconce (2000).

70. Haraway (1997: 8).

71. Fabian (1983: ix).

2. DISSOLVING THE TREE OF LIFE

1. Low (2003).

2. See Cavanaugh et al. (1981) on tubeworm endosymbionts.

3. Darwin (1964: 490).

4. Quoted in Schlee (1973: 84).

5. Broad (1997: 31).

6. Hamilton-Paterson (1992: 191).

7. Beer (2000: 118).

8. Huxley (1868). See Rehbock (1975).

9. Gordon (2002); cf. McGraw (2002).

10. Oreskes (1999).

11. See Broad (1997), Mukerji (1989).

12. "The discovery of hydrothermal activity on the seafloor has a longer history, dating back to the 1880s when hot water was discovered in the Red Sea by the Russian research vessel *Vityaz*. The Swedish ship *Albatross* confirmed this astonishing finding in the 1940s. . . . However, due to increasing political unrest and war, the Red Sea soon was designated off limits to research vessels and further exploration" (Crane 2003: 54–55).

13. Van Dover (1996: 55–56).

14. "Microbiologists first observed chemosynthesis more than 100 years ago . . . the biogeochemical significance of chemosynthesis emerged only upon discovery of deep-sea hydrothermal vent systems" (Van Dover 2000: 117; see also Jannasch 1984).

15. Binns and Decker (1998: 96).

16. Broad (1997: 109).

17. Fabian (1983: 1).

18. Darwin (1964: 421).

19. Darwin (1964: 433). And "Our classifications will come to be, as far as they can be so made, genealogies; and will then truly give what may be called the plan of creation" (486).

20. See Ritvo (1998).

21. Darwin (1964: 433).

22. Klapisch-Zuber (1991) traces the history of family tree diagrams in Europe from the 9th to the 16th century. Medieval genealogical schemas placed ancestral figures at the top of graphic representations, with descendants subordinate to kingly predecessors. Such schemes were pictured as downward-flowing streams, as houses filled with inhabitants paired down the length of a hallway, and, occasionally, as trees. When the church adopted the concept of *arbor juris successionum* (tree of legal succession) from Roman civil law to describe tables of consanguinity, tree representations became common. But tree imagery presented an ideological problem, since planting the roots of family trees in soil, at the bottom of a page, meant situating illustrious ancestors in a position inferior to descendents—indeed, in dirt. The difficulty was solved by a detour through genealogies of Christ, which, placing the roots of the Tree of Life from the Garden of Eden at the bottom of illustrations, allowed branches to rise up into lineages leading to the perfection of Jesus. Such images suggested that future generations might, like Christ, aspire toward Heaven. The ascending tree was secularized after the Renaissance, so that by the time upper-class families in Europe employed it to map their earthly histories and hopes, Darwin could appropriate it to chart a grand history of all life.

23. Beer (2000: 86). Cf. Wertheim (2007) on the thicket-like character of Darwin's early (1837–38) sketches of the tree.

24. See Zuckerkandl and Pauling (1965) for an early articulation.

25. Recall the usual chain of taxonomic levels: species, genus, family, order, class, phylum, and kingdom.

26. Woese, Kandler, and Wheelis (1990: 4578). In 1977, Woese and Fox introduced "archaebacteria" to name a lineage they tentatively designated as one of three "*urkingdoms*" into which life might divide.

27. Woese, Kandler, and Wheelis (1990) divided Archaea further, into Euryarchaeota (including methanogens, halophiles, and some thermophiles) and Crenarchaeota (mostly thermophiles). They argued that, "since thermophily is the only general phenotype that occurs in both major branches of the Archaea, it is presumably the ancestral phenotype of the Archaea." Crenarchaeota, then, are so named—joining the Greek for "spring, fount" to

archaea—"for the ostensible resemblance of this phenotype to the ancestor (source) of the domain Archaea" (4579).

28. See Doolittle (1996: 8797).

29. Woese (2000: 8393). In designating rRNA as genealogy's backbone, Woese follows Pace et al. (1986).

30. Doolittle (1999: 2127).

31. Baross (1998: 3).

32. Ellis (2001).

33. See Corliss, Baross, and Hoffman (1981).

34. Baross (1998: 9).

35. Van Dover (1996: back cover).

36. Taylor (1999).

37. Bock and Goode (1996).

38. Van Dover (1996: 82).

39. Davies (2001).

40. Water appears in some origin stories for *Homo*, too, as in the aquatic ape hypothesis, proposed by marine biologist Sir Alister Hardy in 1960 and elaborated by Elaine Morgan (1984). This scheme has humans develop from swimming anthropoids, a model offered as an answer to the apparently pressing puzzle of why we are so hairless.

41. As, Ts, Cs, and Gs are not, as *Volcanoes* would have it, "the four base chemicals of life's universal alphabet." Just because nucleotide bases can be labeled with letters does not mean they constitute an alphabet; "'letter' frequency analyses of amino acids yield only random statistical distributions" (Kay 2000: 2). No one could speak or write "the genetic language" as they could English. Kay (2000) argues that the textual metaphor for DNA represents an inheritance from Judeo-Christian origin stories that have God creating the world with a "word," as author of the "book of nature."

42. In more technical language, the question is whether such similarities—often called homologs—are orthologs rather than paralogs. "For molecular phylogenies to reflect organismal phylogenies, the compared genes must be *orthologous*—that is, differences must have arisen from speciation events. The inference of organismal phylogeny will be obscured if instead *paralogous* genes, which have arisen from duplication events, are compared" (DeLong 1998b: 5). See Fujimura (1999) on how *homology*, which biologists have used to refer to similarity by common descent, has become synonymous with simple similarity, necessitating these new words.

43. The use of parsimony in such software as PAUP is less a demand of the phenomenal world than the need to prune the space of possible solutions.

44. Borges et al. (1996).

45. Doolittle and Brown (1994: 6724), emphasis added.

46. See Bowker (2005: 193).

47. Mackenzie (2003: 321).

48. Moreover, such "information" can be lost over time: "When producing computer models, we assume that time is unidirectional: species cannot lose

characteristics once acquired. And yet we know empirically that some species do just that" (Bowker 2005: 214).

49. Lateral gene transfer was first detected in pathogen evolution in the 1950s, "when multidrug resistance patterns emerged on a worldwide scale" (Ochman, Lawrence, and Groisman 2000: 299).

50. Hilario and Gogarten (1993).

51. Doolittle (1999: 2127, 2128). See Doolittle and Longsdon (1998) on the "mixed heritage" of Archaea; also Doolittle (2000). One critique of this position is that it assumes the vertical model it seeks to explode (Gupta and Soltys 1999).

52. Paul (1999: 48, 46).

53. Deleuze and Guattari (1987: 11).

54. Doolittle (1999: 2127).

55. Kay (2000). See Lenoir (2002: 125) on how information science has "changed the picture of biological theory itself."

56. See Oyama (2000). Woese applied information theory to biology in earlier work, in his 1967 *The Genetic Code,* in which he called DNA and RNA "informational molecules," comparing them to "tapes" and "tape readers," respectively (Kay 2000: 16).

57. Thacker (2004: 44).

58. Keller (2000: 7–8). Cf. Keller (1992), Doyle (1997).

59. Turner (1967: 19).

60. Gogarten, Doolittle, and Lawrence (2002: 2227).

61. Schneider (1968:23).

62. Doolittle (1997: 12751).

63. See Franklin and McKinnon (2000). Indeed, Schneider wrote, "It does not follow that every fact of nature as established by science will automatically and unquestioningly be accepted or assimilated as part of the nature of nature. People may simply deny that a finding of science is true and therefore not accept it as a part of what kinship 'is.' By the same token, some items in some people's inventories of the real, true, objective facts of nature may be those which scientific authority has long ago shown to be false" (1968: 24, note 3). Franklin writes, though, that Schneider "ignored the extent to which biology, even its traditional form, is about change. Biotechnology today is the matrix of unprecedented life-forms that have as little to do with the nature biology once depicted as they do with the biology portrayed by Schneider" (2001: 320).

64. Carsten (2000: 3). And see Strathern (1992), Thompson (2005), Pálsson (2007).

65. See Pennisi (1999), Woese (2000), Philippe and Douady (2003).

66. Ochman, Lawrence, and Groisman (2000: 304). See O'Malley and Dupré (2007) on implications of microbes for philosophy of evolution.

67. Ritvo (1998: xii). See Gilroy (2000) on race.

68. Hayles (1999).

69. Thacker (2004: 7).

70. Gogarten, Doolittle, and Lawrence (2002: 2227).

71. Ochman (2005).

72. Doolittle (1997: 12751, 12753).

73. See Foucault (2008). The question of whether genealogies can be revealed by making trees of traits standing in relations of origins and descent is at the heart of debates about cladistics, a field of classification that sorts creatures into ingroups and outgroups based on characteristics they share (or not) owing to common ancestry. Many biologists disagree that such sorting maps evolutionary history, seeing circular argumentation in designations of similarities and differences in the first place (see Bowker 2005: 132–34).

74. Wagner (1977).

75. Bapteste et al. (2004: 410).

76. On descent and alliance, see Fortes (1949) and Lévi-Strauss (1969). In descent theory, inheritance is often reckoned through gendered lineages, with women cast as safeguarding through marriage and chastity "legitimate" descendents for patrilineal or patriarchal kinship systems; in alliance theory, women are "exchanged" through marriage to secure intergroup affiliation (Rubin 1975). In both cases, women are symbolically significant for "reproduction." Reading vertical and horizontal kin through a feminist lens, "genes" are not only units of ontogenesis but, like women in traditional kinship theory, associated with reproduction of future relations. At the same time, as Franklin (1995) argues, the informatic gene can be seen as "unsexed" (or, alternatively, as a self-replicating paternal seed principle).

77. Bapteste et al. (2004: 409). The authors argue that "a framework for natural classification should be based on a true understanding of historical processes," though they also caution that "there might never be a perfectly natural classification" (409). See also O'Hara (1992).

78. Beer (2000: 159).

79. Darwin (1964: 490), emphasis added.

80. Discovery Institute, About Discovery, www.discovery.org/aboutFunctions.php, and Center for Science and Culture, About CSC, www.discovery.org/csc/aboutCSC.php.

81. Wells (n.d). To secular eyes, such a position looks like creationism. But advocates of intelligent design say they do not entertain any hypothesis about the designer's identity. Signs of design, they hold, can be discerned by measuring information content in DNA. This contention fails to recognize that information is a metaphor for (and an observer's imposition upon) biological phenomena. Design adherents participate in the same misplaced concreteness that biologists such as Woese risk when they assign an ontology of information processing to DNA. For an argument that intelligent design is creationism, see Fitelson, Stephens, and Sober (1999). On creation science, Toumey (1994).

82. From Luskin's "Design vs. Descent: A Contest of Predictions," www.ideacenter.org/contentmgr/showdetails.php/id/846. No date of composition is attached to this piece.

83. Dean (2005).

84. Dean (2005).

85. Beer (2000: 107)

86. Evolution and Christianity are hardly always at war, of course; the Catholic church accepts evolution so long as scientists do not pronounce upon the human soul, a point on which most are happy to demur. See Pope John Paul II (1998).

87. Falkowski (2002).

88. Caws (1993: 75).

89. Strathern (2005: 67).

90. Strathern (2005: 75).

91. Paul (1999).

92. Ochman, Lawrence, and Groisman (2000: 300–301). And see Nelson et al. (1999).

93. Bushman (2002: 2).

94. Jacob (1993: 324).

95. Pennisi (1999).

96. Pennisi (1999); see also Vogel (1999), Galtier, Tourasse, and Gouy (1999).

97. A collateral question was this: were genes with high guanine and cytosine (G+C) content particularly *likely* to be transferred in hydrothermal environments? Insofar as hyperthermophiles had high G+C to start with, some said yes; see Ochman, Lawrence, and Groisman (2000: 300).

98. This raises the question of whether the logical confusions that litter this seascape have more to do with extreme life forms or bioinformatics. Would taxonomic controversies look different for yeast? Yes. Even if the epistemological puzzles produced by bioinformatics would have something in common, the material possibilities opened by each case would differ.

99. Bateson (2000: 459).

100. Bapteste et al. (2004: 406).

101. Daubin, Moran, and Ochman (2003: 829).

102. Prak and Kazazian (2000). See also Doolittle (1998). Sapp (1994) calls this "evolution by association." See Salzberg et al. (2001) on "microbial genes in the human genome."

103. "Paradoxically, the alien's replicability extends a kind of invitation to resist the gravitational pull of the genetic family's imprimatur and bloodline. . . . the mobility of 'alien DNA' articulates the problem of a 'moveable' and 'removable script' of ourselves exteriorized to a site we cannot control" (Battaglia 2005a: 30, 22). Compare Franklin (2007: 31–32) on how Dolly, the cloned sheep, the product of somatic cell nuclear transfer, presses queries about "what happens to sex, breed, species, and reproduction when genealogy is retemporalized and respatialized."

104. See Bowker (2005: 30) on relational databases as nonnarrative memory.

105. Doyle (2003: 197).

106. Munn (2004: 89) writes, "This threshold corresponds to a 16S rRNA sequence similarity of 97%." See Bowker (2005) on numerical taxonomy.

Another strategy for labeling microbes is to discard the notion of classification in favor of case-by-case identification. See Barcode of Life Data Systems, www.barcodinglife.org.

107. Moore, Rocap, and Chisholm (1998).

108. Munn (2004: 85).

109. For analyses of how American political constructions of race may be resurrected in the genomic age, see Haraway (1997), Duster (2001b), Reardon (2005), Montoya (2007), Fullwiley (2007), Nelson (2008).

110. Omi and Winant (1994).

111. See DeLong (2004), Root (1967).

112. Doyle (2004) formulates this as a move from cryptographic to pragmatic approaches to genetic information.

113. Connery (1995: 292).

114. Melville (2001: 137). See Otter (1999) on Melville's cetology as a parody of nineteenth-century race classification, and Helmreich (2005) on a cetology for today. Compare to Melville's size-based categorization of whales the technically practical but naturalistically arbitrary classification of marine microbes by logarithmically diminishing size: microplankton (e.g., ciliates), nanoplankton (e.g., diatoms), picoplankton (e.g., bacteria), and femtoplankton (e.g., viruses).

115. Broad (1997: 280).

116. Agamben (1998: 3).

117. Agamben (1998: 1).

118. Rose (2001: 17). Thacker (2000) is the first mention of *molecular biopolitics.*

119. A report on deep seabed resources by United Nations University offers another version of transfer: "Bioinformatics is likely to change the way biotechnology research is conducted in the future, with trends suggesting that there is a decreasing dependence on physical transfers of biological materials in favor of electronic transfers" (Arico and Salpin 2005: 26).

120. Jones and Haraway (2006: 243).

121. Madigan (2001).

122. Center of Marine Biotechnology, Marine Pollution and Remediation, www.umbi.umd.edu/~comb/programs/biorem/bioreme.html; and COMB main page, www.umbi.umd.edu/~comb/.

123. Greg Bear, in *Vitals* (2002), imagines hydrothermal vents as fountains of youth, motivating his biotech thriller with speculation that vent microbes contain ancient secrets to prolonging life.

124. Robb (2000).

125. See BLAST: Basic Local Alignment Search Tool, www.ncbi.nlm.nih.gov/blast/.

126. This corroborates Haraway's sense that, in the genomic age, "Nature [has become] a genetic engineer that continually exchanges, modifies, and invents new genes across various barriers" (1997: 225).

127. Rabinow (1992: 241–42).

128. Schneider (1968: 23).

129. Rabinow (1999: 15).

130. Strathern (1992: 17).

131. Haraway (1997: 53).

132. Helmreich (2001).

133. A hypothesis about hyperthermophilic life I have not touched is Cornell geologist Thomas Gold's (1999) "deep hot biosphere." Gold suggests that microbial life extends into Earth's crust, living off primordial petroleum. His heterodox claim is that oil does not come from decayed prehistoric life but from this reservoir; that is, oil is not a fossil fuel.

134. Consult Weston (2002) on "unsexed," Bowker and Star (1999) on torque as a tension between bodies and their classifications.

3. BLUE-GREEN CAPITALISM

1. Franklin and Lock (2003: 8); see also Heller (2001).

2. Twenty-three participants came from Japan, eight from the Philippines, six each from Thailand and Korea, four from People's Republic of China, three each from India, Mexico, Italy, and Germany, and a handful from Hong Kong, Taiwan, Norway, Portugal, and Israel.

3. Engineering Research Centers Association, About the ERC Program, *BP Manual*, www.erc-assoc.org/manual/bp_ch1_2.htm.

4. Academic-industrial biotech hybrids became common in the United States after the Supreme Court in 1980 permitted the patenting of modified organisms in *Diamond v. Chakrabarty*.

5. COMB main page, www.umbi.umd.edu/~comb/. See National Academy of Sciences (2002).

6. "MarBEC research ranges from the discovery and screening of new organisms from diverse and extreme environments to the design of high-productivity cultivation systems for microalgae, *Bacteria, Archaea*, metazoa, extremophiles, and more complex life forms, and purification schemes for marine bioproducts derived from these organisms" (MarBEC 2003: v).

7. Harris skipped the military, Hawai'i's second-largest money source, 10 percent of state income.

8. Fortun (2008).

9. Haraway (1997).

10. Mestel (1999: 75).

11. Hayden (2003: 52). See Takacs (1996). See Thorne-Miller (1999) on marine biodiversity.

12. Colwell (1984: 3).

13. COMB main page, www.umbi,umd.edu/~comb/.

14. Yanagisako (2002: 21).

15. Merton (1973).

16. Mukerji (1989).

17. An account of sentiments animating China's marine biotechnology would follow connections between U.S. organizations such as the Society of

Chinese Bioscientists in America and researchers in China. See Meyer and Brown (1999) on scientific diasporas, Ong (1999) on transnational Chinese professionals.

18. Kuo (2005).

19. Takahashi (2003: 28–32).

20. Absent from the conference were papers about genetically modified fishes, at the center of debates about the Blue Revolution. Detractors call these "Frankenfish." For opponents of genetically modified foods, fishes symbolize the uncontrolled flow of engineered genes into the wild. See Stonich and Bailey (2000).

21. Kim and Mauborne (2004).

22. Blue-sky schemes are mentioned in a 1917 legal opinion of Justice McKenna of the U.S. Supreme Court, who censures "speculative schemes which have no more basis than so many feet of 'blue sky.'" Richard I. Alvarez, Esq., and Mark J. Astarita, Esq., Introduction to the Blue Sky Laws, *Securities Law*, www.seclaw.com/bluesky.htm.

23. The MarBEC final, fifth-year report (2004) records no patents awarded and none licensed.

24. See Rohrer (2006) on the history of the word *haole*.

25. MarBEC made big promises about deep creatures: "The deep-sea methanogen, *Methanococcus jannaschii*, has been found to produce novel squalenoids and C-35 isoprenoids. Using enzymatic cyclization reactions, compounds have been produced that show significant potential for the development of new anticholesterol agents" (MarBEC 2003: 2).

26. MarBEC (2003: 50).

27. MarBEC (2003: 3).

28. Franklin and Lock (2003: 7).

29. Oceanit, www.oceanit.com.

30. Aquasearch has reorganized as Mera Pharmaceuticals.

31. Center for Marine Microbial Ecology & Diversity, www.cmmed.hawaii.edu.

32. See Modis et al. (2004).

33. MarBEC (2003).

34. MarBEC (2003: 3).

35. Mintz (1985: 47).

36. The engineering of pharmaceutical biocapital, however, does direct us to clinical trials, to the mayor's mention of Hawai'i's "diverse human gene pool." To consider human populations test beds for medicines sees some people as drug-screening labor for others, workers in a pharmaceutical plantation system; see Petryna (2005).

37. Crane (2003) writes of the weave between military and counterculture in oceanography: "During the 1970s, 'big science' and the military tangled behind the laid-back surfer lifestyle" (33).

38. Mendola (2000). See Helmreich (2003) for an oral history.

39. Desmond (1999: 8).

40. See Broad (1997: 252).

41. Fortun (2008: 11).

42. MarBEC (2003: 11).

43. Mike Winslow, An Algae's Tale: Three Billion Years of Gas Bubbles, Oxygen and, More Recently, Dog Doo, *The Glory of Pond Scum*, www.uvm.edu/~empact/water/algae.php3.

44. Cox et al. (2005).

45. Waldby (2000: 33).

46. Marx (1976: 251).

47. Marx (1978: 228).

48. Franklin and Lock (2003: 8). And see Franklin (2007) on *breedwealth* in animal husbandry.

49. Marx, quoted in Franklin (2007: 106). See Thacker on "'molecular species being,' a species being in which labor power is cellular, enzymatic, and genetic" (2005: 40).

50. Landecker (2007).

51. Taussig (2004: 23).

52. Sunder Rajan (2006: 114, emphasis omitted). The contrast between science and capital may be overdrawn; science has trafficked in credibility as much as truth since the rise of the experimental way of life; see Shapin and Schaffer (1985).

53. In Franklin and Lock's terms, biocapital partakes of what Strathern (1992) identifies as a commitment in European kinship systems to the notion that mutation and recombination create "newness." Biomatter is not absent in Sunder Rajan's account—indeed, following Marx, he parses biocapital into industrial, commodity capital (therapeutic molecules) and speculative, commercial capital (stocks) (2006: 8–9); but rather than emphasize biological generativity he calls attention to the constructedness of biological facts upon which value is predicated.

54. The informal, nonmonetized exchange of biomatter among scientists is increasingly governed by MTAs too.

55. Such calibration can work the other way around, with lab practices in genomics informally producing new parsings of property; see Hilgartner (2004).

56. Cf. Silbey and Ewick (2003) on lab space regulation.

57. Casarino (2002: 91). Cf. Riles (2003) on "making white things white," legal moves that claim to recognize matters as they already are but in so doing formalize them.

58. Connery (1995: 289, 288). Steinberg writes that "first world capitalists have constructed the ocean in a manner that selectively reproduces and emphasizes its existence as a space apart from land-based capitalist society. . . . territoriality under capitalism became constructed in such a way as to support the concept of abstract,' 'emptiable' space" (2001: 25, 30). See also Walley (2004).

59. See Landecker (2007: 233).

60. Maurer (2000: 672).

61. It is this messiness that makes such stuff interesting to less product-minded scientists: "The mode of scientific existence peculiar to such entities [as the objects of molecular biology] derives precisely from their resistance, resilience, and recalcitrance rather than from their malleability in the framework of our constructive and purposive ends" (Rheinberger 2000: 272).

62. MarBEC (2003: 50).

63. See Karl, Bidigare, and Letelier (2001).

64. The word *aloha*, "the spirit of reciprocal giving and exchange between Hawaiians" (Halualani 2002: 6), is often taken up by Island visitors, who imagine that in saying it they become "Hawaiian at heart." Halualani argues that this usage bespeaks a desire of tourists to connect with native people outside economic inequality and colonial history; see Trask (1993).

65. Karl, Bidigare, and Letelier (2001).

66. Venture capitalists are interested: "Following the Kyoto conference on reduction of CO_2 emissions and global warming, the concept of a global market in the trading of 'carbon offset credits' has emerged. Entrepreneur Michael Markels has set up the commercial organization GreenSea Venture Inc., with the aim of profiting from ocean fertilization whilst mitigating the constant rise in atmospheric CO_2 levels" (Munn 2004: 158).

67. Pálsson (1998).

68. See Chisholm (2000).

69. Latour (1987).

70. See Marcus (1995).

71. See Hayden (2003: 10) on "slightly choppy" bioprospecting networks.

72. Diversa Corporation, Press Releases for 2002, Diversa Signs Biodiversity Access Agreement in Hawaii, www.diversa.com/presrele/2002/view_release.asp?id=20020604, accessed April 15, 2005.

73. UH's MarBEC Partners with Diversa, *Pacific Business News*, June 5, 2002, www.bizjournals.com/pacific/stories/2002/06/03/daily33.html.

74. Weird Science: Company Contracts with UH for Access to Strange DNA, *Pacific Business News*, June 7, 2002, www.bizjournals.com/pacific/stories/2002/06/10/story6.html.

75. See Brush (1999) and Moran, King, and Carlson et al. (2001).

76. See Shiva (1997).

77. Private property was introduced in 1848, when missionaries pressed King Kamehameha III to divide land into saleable plots.

78. Section 5(f) of An Act to Provide for the Admission of the State of Hawaii into the Union (Act of March 18, 1959, Pub L 86–3, 73 Stat 4), www.capitol.hawaii.gov/hrscurrent/Vol01_Ch0001–0042F/04-ADM/ADM_0005.htm.

79. Article XII, section 4, Constitution of the State of Hawaii, as Amended and in Force January 1, 2000, www.hawaii.gov/lrb/con/conorg.html.

80. Parker (1989: 165). This organization grew out of campaigns for reparations to Hawaiians from the United States, led by such organizations as ALOHA (Aboriginal Lands of Hawaiian Ancestry).

81. Ota (2002: 11).

82. Le'a Malia Kanehe, Testimony in Support of SB643. See also Harry and Kanehe (2005).

83. Le'a Malia Kanehe, Testimony in Support of SB643.

84. A second iteration of the bill in mid-February 2003 (SB643 SD1) added that "the Hawaiian people have customarily used Hawaii's biological resources in accordance with their traditional, cultural, and subsistence practices, including agriculture, fisheries, health, and horticulture. The legislature . . . finds that the Hawaiian people are traditional, indigenous knowledge holders with rights."

85. Mililani B. Trask, Indigenous Expert—Pacific Basin, United Nations Permanent Forum on Indigenous Issues, Testimony to: Senate: Water, Land & Agriculture, Senator Loriane Inouye/Senator Willie Espeno; House—Judiciary & Hawaiian Affairs, Senator Colleen Hanabusa/Senator Suzanne Chun Oakland; Re: Senate Bill 643—Re: Biodiversity, February 5, 2003. Hawai'i State Archives.

86. Conversation with Mililani Trask, July 9, 2003.

87. Le'a Malia Kanehe, Testimony in Strong Support of SCR 55. See Qanungo (2002) on marine bioprospecting.

88. Le'a Malia Kanehe, Testimony in Strong Support of SCR 55.

89. Jill Leilani Nunokawa, Strong Support for Senate Bill 643 Relating to Bioprospecting, February 6, 2003. Hawai'i State Archives.

90. Hayden (2003: 51). Cf. Zerner (1996).

91. Fredric J. Pashkow, M.D., representing BiophoriX, Pacific, Inc. Opposition to SB 643, which establishes a moratorium on bioprospecting and a temporary bioprospecting advisory commission, February 6, 2003. Hawai'i State Archives.

92. Dr. Will McClatchey, University of Hawai'i, Department of Botany, Some Comments on SB 643, February 5, 2003. Hawai'i State Archives.

93. See Hernández Castillo (2001) and Povinelli (2002) on burdens placed on aboriginal communities by multicultural nation-states demanding indigenous peoples be "authentic" to qualify for recognition.

94. Fredric J. Pashkow, M.D., representing BiophoriX, Pacific, Inc. Opposition to SB 643, which establishes a moratorium on bioprospecting and a temporary bioprospecting advisory commission, February 6, 2003. Hawai'i State Archives.

95. This position is partially articulated in a November 16, 2002, resolution from the Association of Hawaiian Civic Clubs reproduced in the Testimony of Charles Rose, President, Association of Hawaiian Civic Clubs on Senate Bill No. 643, Relating to Bioprospecting, February 6, 2003. Hawai'i State Archives. Note, too, that President Clinton's 1993 apology to Native Hawaiians for the illegal overthrow could open up a wider frame for contestation.

96. On their view, revivified Hawaiian citizenship would extend not just to Native Hawaiians but also to people descended from European and American citizens of the kingdom—a position other sovereignty advocates (Ka Lahui

Hawai'i) see as conservative, not so much out of any fantasy of ethnic purity but because foreign-born citizens dispossessed indigenous contemporaries; see Osorio (2003).

Repositioning discussion of the Senate bill on bioprospecting in the register of the historical kingdom might have offered other possibilities for drawing watery boundaries. One might have recalled King Kamehameha III's 1846 claim of the then standard marine league around each island (to which he added channels between islands), or his 1851 outlining of fishing boundaries: "The fishing ground extends from the beach to the reefs, or where there are no reefs one 'mile seaward of the beach at low watermark'" (Parker 1989: 124). One might even have called upon King Kamehameha IV's expansive 1856 claim of Johnson Atoll, 700 miles south of the archipelago, signaled by a ship sent to plant the Hawaiian flag. In a more presentist vein, an independent Hawaiian kingdom could through the Law of the Sea Convention claim an Exclusive Economic Zone of 200 miles (a space which, if one were worried about it, would reach to Station ALOHA). Other possibilities would include thinking back before Captain Cook, considering precontact classifications of ocean zones by the blueness of water, a parsing retrieved by nineteenth-century Native Hawaiian antiquarian David Malo; see Martin and Burke (1995: 190). Here, blueness does not stand for unbounded territory to be colonized with blue ocean dreams of plankton plantations but rather comes in shades representing shifting allocations of temporary social rights.

97. Kevin Kelly, University of Hawai'i, Director of Business Development, Marine Bioproducts Engineering Center (MarBEC), Comments on SB 643, February 5, 2003. Hawai'i State Archives.

98. The bill's definition of bioprospecting—"the collection, removal, purchase, sale, or use of biological and genetic resources of any organism, mineral, or other organic substance found within the ceded land trust for scientific research or commercial development"—encompasses almost every kind of field-based biological research.

99. OHA agreed, finding the definition of bioprospecting too wide, suggesting "that a proviso be added specifically exempting the cultivation of food products for consumption"; see Office of Hawaiian Affairs, Legislative Testimony, Measure Number: House Bill SB643, Committee: House Committee on Water, Land Use and Hawaiian Affairs, February 6, 2003. Hawai'i State Archives. Later versions substituted for "moratorium" "prohibition on the sale or transfer of biological resources and biological diversity on trust lands."

100. University of Hawai'i Testimony on SCR 55 SD1.

101. Jasanoff (2005: 237).

102. Parker (1989: 114).

103. The 2004 legislative season reopened the issue. The category of "public lands" substituted for "trust lands," eliminating OHA's claim for a 20 percent benefit from bioprospecting. More, language was added to one iteration of the bill specifying that nothing in the act could be construed as restricting Hawai'i's biotechnology industry. No wonder, then, that the university and the

Pineapple Growers Association of Hawai'i supported the bill while native groups jumped to a parallel version in the House. Neither passed. In 2005, two House committees pronounced that bioprospecting on organisms "neither indigenous nor endemic to the State shall not be prohibited, inhibited, or restricted," opening up everything else as a public domain; Hawai'i House Committees on Energy & Environmental Protection, Water, Land, & Ocean Resources Report No. 112, re: H.B. No. 247 H.D. 1., Regular Session of 2005. Another committee argued that what were at stake in benefit-sharing agreements were matters of "indigenous knowledge or intellectual property, rather than intellectual knowledge, of biological resources that are public natural resources held in trust by the State," implying that anything about which indigenous people did not know was fair game for bioprospecting; Hawai'i House Committee Higher Education, Report No. 1420, re: H.C.R. No. 146 H.D. 1., Regular Session of 2005. The first two House committees scolded the university, suggesting the school be prohibited "from entering into material transfer agreements that transfer title and ownership of the state's natural, biological, and genetic resources to any private entity unless expressly authorized by the Legislature through a Concurrent Resolution adopted by both houses"; Hawai'i House Committees on Energy & Environmental Protection, Water, Land, & Ocean Resources Report No. 112, re: H.B. No. 247 H.D. 1., Regular Session of 2005. The university argued that it never claimed *ownership* of state resources. 2005 ended in stalemate. By then, MarBEC was defunct.

104. See Haraway (1997: 137–43).

105. Hayden (2003: 4).

106. Merry and Brenneis (2003: 18).

107. Hayden reports a similar case, in which Diversa sought to buy bioprospecting rights for microbial biodiversity on public land in Mexico (2003: 97–98, 234–35).

108. Anthropologists have been complicit: "The largely landless and increasingly urbanized Hawaiians were understood anthropologically, until recently, only in their isolated villages and through a reconstruction of their past. This form of analysis ignored the ways colonialism shaped these villages, effacing its impact. This contributed to seeing Hawaiians as a residual, vanishing community" (Merry and Brenneis 2003: 27).

109. See Merry (2000).

110. Diversa Corporation, Technology Overview, Direct Evolution, www.diversa.com/techplat/innobiod/direevol.asp, accessed October 19, 2004.

111. Diversa Corporation, Press Releases for 2002, Diversa Signs Biodiversity Access Agreement in Hawai'i, www.diversa.com/presrele/2002/view_release.asp?id=20020604, accessed April 15, 2005.

112. Fortun (2008: 183).

113. Maurer (2005: 59).

114. Finney (1979, 2003). See Frake (1985) on chartless navigation.

115. Center for Microbial Oceanography, Education Partners, cmore.soest.hawaii.edu/education/partners.htm.

4. ALIEN SPECIES, NATIVE POLITICS

1. Staples and Cowie (2001).
2. Van Driesche and Van Driesche (2000).
3. Carlton (1996), Cox (1999), Mooney and Hobbs (2000), Pimentel (2002).
4. Fincham (1998). Clips available at Alien Ocean, http://sgnis.org/real/alien.htm.
5. Global Ballast Water Management Programme, The Problem, http://globallast.imo.org/index.asp?page=problem.htm&menu=true.
6. Global Ballast Water Management Programme, The New Convention, http://globallast.imo.org/index.asp?page=mepc.htm&menu=true. See Ruiz et al. (2000). Consult Colwell (2006) for discussion of how changing climate may affect cholera's movement across ecosystems.
7. Gro Harlem Brundtland, World Economic Forum Plenary Seminar, Addressing the Challenges of Unequal Distribution, Davos, January 29, 2001, www.who.int/director-general/speeches/2001/english/20010129_davosunequaldistr.en.html.
8. Falkner (2000: 3). Earlier introductions of organisms to San Francisco include import of oysters from Chesapeake Bay and transport of fauna on ships bringing people during the Gold Rush.
9. Godwin and Eldridge (2001). This report does not contain data about total volume of ballast water.
10. Although most taxonomists would say that blue-green algae and red algae are not related—cyanobacteria are single-celled prokaryotes and red algae multicellular seaweeds—others would point out that chloroplasts inside red algae descend from microbes.
11. Appadurai (1996).
12. Tsing (2004).
13. See Hearn (2002) for a history of currents and shipping paths.
14. Cf. Sarai Editorial Collective (2006).
15. As an example of the second tendency, consider the opinion of an editor at *Discover* who writes, "Alien species do not come from Mars; they are not 'other.' They are very much of us, by us; we are the main agent of their spread" (Burdick 2005: 40).
16. Collier and Ong (2005: 11).
17. The Nature Conservancy for the Division of Aquatic Resources, Department of Land and Natural Resources, State of Hawai'i, Aquatic Invasive Species (AIS) Management Plan, Division of Aquatic Resources Public Information, www.state.hi.us/dlnr/dar/pubs/ais_mgmt_plan_draft.pdf, accessed October 5, 2003.
18. Devine (1998: 262).
19. Strathern (1991).
20. See Staples and Cowie (2001: 109–10).
21. L. G. Eldredge and R. C. DeFelice, Checklist of the Marine Invertebrates of the Hawaiian Islands, www2.bishopmuseum.org/HBS/invert/list_home.htm.

22. McNeely (2000: 183).

23. Adapted from Staples and Cowie (2001: 3–4).

24. National Centre for Aquatic Biodiversity and Biosecurity (2002: 8).

25. Subramaniam (2001: 29–30, citations omitted).

26. Tsing (1995); see also Bright and Starke (1998).

27. Comaroff and Comaroff (2001: 650).

28. McNeely (2000: 172).

29. As biologist James Carlton puts it, "Not all invasions are created equal" (2000: 42). Cf. Simberloff (2003) on the varied politics that can travel alongside concerns about invasive species more generally.

30. Bishop Museum, About the Museum, www.bishopmuseum.org/about. html, accessed January 26, 2004.

31. Sahlins (1995).

32. Obeyesekere (1997).

33. Borofsky (1997).

34. Bulmer (1967).

35. Ritvo (1998).

36. A similar taxonomy animates Englund and Baumgartner (2000).

37. See Kauanui (2002).

38. Faxed to the Committee on Judiciary and Hawaiian Affairs and the Committee on Water, Land, and Agriculture, February 5, 2003.

39. Raffles (2002). And see Ingold (1990).

40. "The problem of human-induced invasive species is as old as our own species," writes conservationist James McNeely (2000: 187). See Kirch (2000) on pre-Cook agricultural intensification. Kirch argues that early terracing practices in Hawai'i were often harmful.

41. Rotman (2000: 63).

42. A pseudonym.

43. Devine (1998: 276).

44. See Chapter 107 of Title 13, Hawai'i Administrative Rules, adopted May 15, 1997, www.state.hi.us/dlnr/dofaw/rules/Chap107.pdf.

45. Hobsbawm and Ranger (1983).

46. Reichenbach (1938).

47. McNeely (2000: 175); and see Sprugel (1991).

48. Carlton (2000), Guggenheim (2006).

49. Devine (1998: 9–10); and see Speidel and Inn (1994).

50. Definitions of *alien* and *native* would be different if China's fifteenth-century forays into the Indian Ocean had inaugurated an age of Chinese maritime exploration; see Whitfield (1996: 34).

51. Cf. Warren (2007).

52. Devine (1998: 260–61).

53. See Finney (2003).

54. Meyer (2001: 126).

55. Unlike the names of other undergraduates in this book, Rebecca is a real name, used with approval.

56. Abbott (1999: 9).

57. Abbott (1999: 10).

58. Abbott (1999: 8).

59. Abbott (1999:10).

60. See Sagoff (2000). As Comaroff and Comaroff put it, "Alien vegetation . . . may, simultaneously, be one person's livelihood and another's apocalypse" (2001: 650).

61. K. Trask (1993), M. Trask (1998).

62. Abbott (1999: 8, references omitted).

63. Many Asian immigrants to Hawai'i—from China, Japan, the Philippines—came in the first half of the twentieth century, when the Islands were a U.S. territory. These people did not consider themselves Americans—not least because Asian exclusion acts prevented them from naturalizing as citizens, considering them "alien races." After Hawaiian statehood they and their descendants did not adopt the term Asian-American, popular among post-1965 immigrants to mainland states; see Okamura (1994).

64. Taste brings up questions of food traditions. In Hawai'i, seaweed is food for Native Hawaiians, Filipinos, Japanese, Chinese, Koreans, and haoles. Taxonomies by these groups—some *Gracilaria* are called *ogo* in Japanese—may not align with scientific or native terms; what is considered native or alien may also vary (see Abbott 1978). The privileging of Hawaiian names in the limu curriculum stakes an unmistakable political claim.

65. Abbott (1987).

66. Meyer (2001: 126).

67. Scheuer (2002), Turner (2002).

68. Quoted in Kapur (2005: 111, references omitted). See Puhipau and Lander (1996).

69. Kaipo Faris, Hihiwai Restocking Project: Makawai Stream Restoration Alliance, www.pixi.com/~isd/MakawaiHihiwai.html. The state of the stream also affects limu. Having less mixed fresh and saltwater—muliwai—in estuaries of Kāne'ohe Bay means less wāwae'iole, an important limu.

70. Fitzsimmons (2001).

71. See Boellstorff (2006) on the "archipelagic concept" organizing Indonesian nationalism.

72. To say nothing of rivers. "Rivers . . . are both guardians and betrayers of places. And, what's more, despite often being themselves the borders that make places, they are places too, as mobile as can be" (Raffles 2002: 182). See Orlove (2002) on lakes.

73. Celia Lowe writes of how the mobility of Sama seafarers in Indonesia has led the state to see them as "extraterrestrial others: both living beyond the land and alien" (2006: 81), decreasing their claims in the eyes of the territorially minded nation to political legitimacy.

74. Hau'ofa (1993). See Subramani (2001).

75. And see Maurer (2003: 780). We must adjust Russian linguist Vološinov's 1920s argument that "contexts do not stand side by side in a row,

as if unaware of one another, but are in a state of constant tension, or incessant interaction and conflict" (quoted in Raffles 2002: 30), to argue that context is a product of interaction.

5. ABDUCTING THE ATLANTIC

1. Venter (2004), Venter et al. (2004). See Venter et al. (2001) on the human genome.

2. Chisholm et al. (1988).

3. Shreeve (2004: 108).

4. Venter, on Voice of America's "Our World," March 17, 2007, transcript at www.voanews.com/english/archive/2007-03/2007-03-16-voa3.cfm?CFID=84736433&CFTOKEN=35525614.

5. Swyngedouw (1997); see also Smith (1992).

6. Peirce (1998: 299). See Doyle (2003: 25), which inspired me to use Peirce; see also Battaglia (2005b: 170).

7. Anderson (1983: 6).

8. Pollack (2004).

9. Carpine-Lancre (2001: 61).

10. See Haraway (1997); see Reardon (2005: 77).

11. Falkowski and de Vargas (2004: 59); see also DeLong (2005).

12. Binder (2002). I had to leave at Bermuda to fly to Seattle to join *Atlantis* (see chapter 6).

13. Bermuda Atlantic Time-Series Study, Overview of BATS, http://bats.bios.edu.

14. Columbus was delayed in the Sargasso, though he did not have an entirely bad time of it. He commented in 1492 that "the weather is like April in Andalusia" (quoted in Teal and Teal 1975: 13). See Butel (1999).

15. Mills (1989: 19).

16. Quoted in Mills (1989: 124).

17. Quoted in Mills (1989: 208).

18. Teal and Teal (1975: 200).

19. Fuhrman (1999: 546). See Hutchinson (1961). One of Hutchinson's best-known students is Donna Haraway.

20. See Giovannoni et al. (1990). In 2002, SAR-11 was cultured (Rappé et al. 2002) and given the name *Pelagibacter ubique*. The microbe features in the sci-fi novel *Fluke*, which postulates that whales are robots created by primordial deep-sea ooze to spy on ships. This ooze also "produced a minute bacterium that could spread throughout the oceans, be part of the great world ecosystem but could pass genetic information back to the source. We call the bacteria SAR-11" (Moore 2003: 253).

21. In the days before CTDs, marine biologists lowered bottles into the water, sending down a "messenger" to close the first bottle, triggering a chain reaction; see ZoBell (1941). See Goodwin (1995) on CTD sociology, Deacon (1971) on Hooke's 1660s attempts to measure salinity and temperature.

22. This is what we come up with:

Hey hey the CTD
A frame of steel and plastic
With its bottles twelve of PVC
It dives the deep fantastic

Hey hey the CTD
The college kids will heave it
With the Niskins firing 1, 2, 3
They'll cast it and retrieve it

In the nineteenth-century, Forbes offered "The Song of the Dredge" (quoted in Schlee 1973: 80):

Hurrah for the dredge, with its iron edge,
And its mystical triangle.
And its hided net with meshes set
Odd fishes to entangle!

See Hugill (1994) for a definitive collection of sea shanties.

23. Shapiro (1977: 976). This entire scientific article is in verse, with sheet music provided.

24. Karl, Bidigare, and Letelier (2001).

25. See Cambrosio and Keating (2000) on how flow cytometry defines cell populations in the act of bringing them into representation.

26. Chisholm (2004).

27. See Miyazake (2004) on the "method of hope" in financial and social analysis.

28. See Kwa (2002).

29. Whitehead, in Kwa (2002: 28).

30. See Redfield (1934).

31. Roger Revelle, Biographical Memoirs: Alfred C. Redfield, November 15, 1890–March 17, 1983, *National Academies Press*, www.nap.edu/readingroom/books/biomems/aredfield.html.

32. See Smil (2002).

33. DeLong (2003).

34. International Census of Marine Microbes (IcoMM), Project Descriptions: Census of Marine Life Portal, www.coml.org/descrip/icomm.htm.

35. Bowker (2005: 191).

36. Partensky, Hess, and Vaulot (1999: 119–20); see also Vaulot et al. (1995).

37. Partensky, Hess, and Vaulot (1999: 108).

38. Partensky, Hess, and Vaulot (1999: 120, 122).

39. Mills (1989: 125).

40. See Postgate (1994) on "microsenses."

41. Kull (1999); see also von Uexküll (1982).

42. See Agamben (2004: 40).

43. Boas (1938), quoted in Stocking (1974: 42); see also Sahlins (1976: 65).

44. Teal and Teal (1975: 201).

45. See Shapiro (2004).

46. Venter et al. (2004: 66).

47. See Sullivan, Waterbury, and Chisholm (2003).

48. Still, I shiver when I read *The Bermuda Virus* (O'Quinn 1995), a thriller about a disease that bleaches those it kills a deathly white.

49. See Ansell Pearson (1997).

50. Martin (1994).

51. Deleuze and Guattari (1987: 10).

52. Steward (1999).

53. See Sullivan, Waterbury, and Chisholm (2003). Fuhrman reports that "dissolved DNA is readily found in sea water, and it has been reported that viral lysis may be a major source mechanism" (1999: 547).

54. Fuhrman (1999: 546).

55. MacPhail (2004: 359, 364).

56. Fuhrman (1999: 546).

57. Fuhrman (1999: 546). See Rayl (2001) on "human viruses at sea."

58. Environmentalist Neil Evernden (1993) might put this down to his romantic argument that humanity is "the natural alien."

59. Fortun (2005).

60. Corfield (2003: 76).

61. Berlitz (1974: 103). And see North's 1955 sci-fi novel *Sargasso of Space*.

62. 2004 Boston University Honorary Degree Recipients, *B.U. Bridge: Boston University Community's Weekly Newspaper*, May 28, 2004, www.bu.edu/bridge/archive/2004/05–28/honorary.html.

63. Venter's genetically reductionist attitude is summed up in *Cracking the Ocean Code*, a Discovery Channel documentary in which he seeks to turn his cavalier attitude about context into a virtue. Speaking of a storm that hit his Sargasso cruise, he says (in Conover 2005), "These forces can really churn up the water and bring up even subsurface organisms to the top, which could give us, again, tremendous diversity with the sample we just took!"

64. Falkowski and de Vargas (2004: 60).

65. DeLong (2005). Venter's publication of the final results of his Global Ocean Sampling Expedition acknowledges the significance of microbial relations with environmental context, though he holds abductively to "the potential of using metagenomic data to tease out such relationships" (Rusch et al. 2007: 0416).

66. Maurer (2005: 98).

67. A phrase from Fuhrman (2003).

68. Kipling (1902). For a critique of leaps of logic in biological explanation, see Gould and Lewontin (1979).

69. Venter and collaborators have lately cautioned that environmental genomes might harbor imaginary creatures, warning against the production in databases of "computationally generated chimerism" (Rusch et al. 2007: 0411).

70. Ian Skoggard, Understanding Ethnography: An eHRAF Workbook for Introductory Anthropology Courses, *Teaching eHRAF,* www.yale.edu/hraf/workbook.htm.

71. What even the Census of Marine Microbes now calls "a global sampling of 'the marine microbial genome'" is only the latest cartographic practice to imagine "the" ocean as a unified thing.

72. Gupta and Ferguson (1992: 7).

73. Street (1993).

74. As Hastrup has it, "One gives in to an alien reality and allows oneself to change in the process" (1990: 50).

75. See Rainger (2004) on tensions at Scripps in the 1920s about seagoing fieldwork versus laboratory-based marine biology.

76. See Collier and Ong (2005).

77. That strategy fuses anthropology as cultural critique and as association; in 2006, I signed on as one of twenty-three project personnel of the Center for Microbial Oceanography, an NSF-sponsored science and technology center based at the University of Hawai'i, Oregon State University, MBARI, the University of California, Santa Cruz, Woods Hole, and MIT. In joining DeLong, Chisholm, Karl, and others in this endeavor, my aim will be less to undertake a second-order observation of scientists, legitimating their practice by adding a self-correcting reflexive dimension, than to generate discussion across science, culture, and politics. That, at least, is one abductive hope behind *Alien Ocean,* my contribution to mapping microbial oceanography.

78. Venter (2004). Transcription from Evolution of the Earth System conference DVD.

79. Verne (1962: 74).

80. Elizabeth Bravo, in Erosion, Technology, and Concentration Group (2004: 3).

81. See Haraway (1995).

82. Hughes (2005: 157, citation omitted).

83. Deleuze and Guattari (1987).

84. The tale would be less rosy if he'd modeled himself on Magellan, who perished during his circumnavigation (Zweig 1938).

85. J. Craig Venter Institute, IBEA Announces Sorcerer II Expedition, Global Expedition to Sample World's Oceans and Land to Characterize and Understand Microbial Populations Using Environmental DNA Sequencing, JCVI: Press/Press Releases/Full Text, www.jcvi.org/cms/press/press-releases/full-text/archive/2004//browse/1/article/ibea-researchers-publish-results-from-environmental-shotgun-sequencing-of-sargasso-sea-in-science-d/?tx_ttnews%5BbackPid%5D=67&cHash=f892088991.

86. On artificial life, see Helmreich (2000), which opens with an account of a 1994 lecture given in the same MIT hall where Venter spoke, a lecture in which biologist Tom Ray detailed a plan to disseminate a computer simulation of evolution into the Internet, creating a "biodiversity reserve for digital

organisms." Like Venter, Ray started small, on his personal computer, though he also imagined covering the globe.

Alain Pottage (2006: 137) argues that Venter's marine genome collection, which spans "evolutionary ecologies, database logics and programmable synthetic organisms," summons forth so many overlapping possible ownership claims—of states, scientists, companies—as to be formally undecidable. Microbial genes not only become alienable but emerge as entities not fitting into any one legal regime: alien.

87. Shreeve (2004: 007).

88. Shreeve (2004: 148).

89. Shreeve (2004: 149).

90. I learned later that Venter received permission to sample from the Bermuda Biological Station for Research (BBSR), a placing of Bermudan biodiversity into a public domain to which the Bermudan Ministry of the Environment would later object, perhaps because Diversa had "just patented a fluorescing protein extracted from coral collected under its agreement with the BBSR" (Pottage 2006: 152). "The director of the BBSR, writing to refute the allegations aired in *Nature,* observed that 'seawater moves quite fast off Bermuda,' and that as a result the collected samples were likely to contain bacteria 'from many Exclusive Economic Zones of many countries in the world,' making ownership 'a difficult and complicated issue'" (Pottage 2006: 152, references omitted). Note that the BBSR director's argument might clash with Venter's claim to be sequencing the Sargasso Sea.

91. See Tate (2003).

92. According to Carter (2003), the number was 5.4 percent as of June 24, 2003. Additionally, per Carter, a 2001 study by the *Journal of Blacks in Higher Education* ranked the University of Georgia last out of forty-nine flagship state universities in terms of the ratio of African American students to the African American population in the state. The university, in 2001, was 5.8 percent African American; the state of Georgia was 28.7 percent African American.

93. ASLO: Multiculturalism in the Aquatic Sciences, Minorities in the Aquatic Sciences, http://aslo.org/mas/history.html.

94. Du Bois (1989: xxxi).

95. Cuker (2001: 18).

96. It also overlooks practices such as African American subsistence fishing (see Corburn 2005) by focusing on swimming and recreation as paradigmatic modes of engaging the sea.

97. More recently, Brandon tells me, the university has decided to take the football team to Sapelo for field research; team demographics will mean more African American undergrad visitors.

98. Anthropologists have often nostalgically imagined the Sea Islands "to retain the most tangible African roots" of any rural black setting (Ebron 1998: 95; see Herskovits 1941, Maurer 2002). Bailey (2000) employs the notion of the African survival to make sense of her heritage, a project that has taken her to Sierra Leone. Such ideas are now being translated into the genomic; the

African American DNA Roots Project offers genetic genealogical services and promises "to reunite African-Americans with their ancestral roots in Africa." The project, whose logo is a slave ship trailing a double helix across the Atlantic, contrasts with Venter's imperial narrative even as it also writes a global tale in the hope-saturated language of genes.

99. Bailey (2000: 209–10).

100. Bailey (2000: 210).

101. See Sargent (2002).

102. Bailey (2000: 210–11).

103. See Fell and Newell (1998).

104. Indeed, researchers from the Marine Resources Research Institute in South Carolina have written of red tides, "The blooms appear to coincide with heavy spring rain events that produce increased run-off of terrestrial humic substances" (Lewitus et al. 2002: abstract).

105. Crook et al. (2003: 39).

106. Bailey (2000: 26).

107. See Gilroy (1993).

108. Thanks to Alondra Nelson for this point. As Gilroy has observed, seafaring could facilitate escape from slavery (Frederick Douglass's time as a fugitive involved passing as a sailor) and open paths toward new transnational connections; see Brown (2005). The Afrofuturist imagery of musical artists Sun Ra, Lee Perry, and George Clinton has slave ships reoutfitted as space-age starships of liberation, sci-fi versions of Marcus Garvey's early twentieth-century Black Star line, a shipping line premised on the dream of a return to Africa. In these stories, Afrodiasporic people, once alien abductees (see Dery 1994), become aliens, piloting UFOs that promise hopeful abductions into the utopian ocean of space; see Newitz (1993) on aliens as proxy whites, White (1994) on aliens as embodying tensions on race more broadly. The Detroit techno ensemble Drexciya offers a cybernetic version of this story, a musical mythology about a high-tech black Atlantis, founded by aquanaut descendents of people thrown overboard during the Middle Passage, recuperating the deadly sea into an ironic fable of an alien ocean populated by mutants who control the planet from a submarine command center; see Williams (2001).

109. Ebron (1998: 95).

110. Manning (1983: 78–79).

111. Just in Manning (1983: 175).

112. According to Hess (1995: 28), "Just criticized the nucleus-centered approach to genetics as 'a veritable decree of authoritarianism,'" joining a cadre of biologists critical of an emerging emphasis on cell nuclei as the key force behind development.

113. Just's descriptions of cell activation were musical: "The environment plays upon the ectoplasm and its delicate filaments as a player upon the strings of harp, giving them new forms and calling forth new melodies" (quoted in Manning 1983: 261).

114. Bauman (2000: 11).

115. Bauman (2000: 13).

116. Moore (2004).

117. Gilroy (1993: 4). Raffles adds a complementary insight: "Locality is both embodied and narrated and is, as a consequence, often highly mobile: places travel with the people through whom they are constituted" (1999: 324).

118. Stott (2003).

119. Venter's latest partnership is with people at the University of California, San Diego, to create the Community Cyberinfrastructure for Advanced Marine Microbial Ecology Research and Analysis (CAMERA), snapshotting the "planet's life system," as computer scientist Larry Smarr has it (Fikes 2006).

120. See Sapelo Island Microbial Observatory, http://simo.marsci.uga.edu/.

121. ASLO: Multiculturalism in the Aquatic Sciences, Profiles of Aquatic Scientists, http://aslo.org/mas/profiles.html. See Bolster (1997). Consider also Western Washington University's Minorities in Marine Science Undergraduate Program (www.hamptonu.edu/science/marine/hallbonner.htm). See Simpson (2006) on similar efforts at City, Queens and Lehman College in New York.

122. McKissack and McKissack (1999).

123. Microbiologist Bonita Johnson, employed by the Environmental Protection Agency in Georgia to study drinking water, remarks that, "oftentimes, when I attend meetings and conferences, I am the only African-American in attendance," a motivation for her mentoring work (ASLO: Minorities in the Aquatic Sciences, Bonita D. Johnson, http://aslo.org/mas/profiles/bjohnson.html). Although mentoring is to be lauded, the job marks a shift of responsibility for minority recruitment and retention from institutions—think affirmative action, increasingly downsized—onto the individual shoulders of often-overextended minority scientists.

124. See Bass (1999).

125. Duster (2001a).

126. Lorini (1998).

127. Thacker (2005: 48).

128. Law (2004: 18).

6. SUBMARINE CYBORGS

1. Williams (1990: 4).

2. Van Dover (1996: 16).

3. See Wiener (1985).

4. Van Dover (1996: 24–25).

5. Kelty (2003).

6. Mead (1968); see also Bateson (2000).

7. Clynes and Kline (1995: 30–31).

8. Woodward (2004), playing on Keller's (1983) "a feeling for the organism," develops a similar formulation to describe human relations to robots.

9. Crane (2003: 125).

10. Kaharl (1990: 194).

11. Crane reports that on voyages in the seventies women were warned about "showing too much skin" (2003: 61). Such sexism has attenuated and numbers of women have increased. NSF reports that 30.4 percent of Ph.D.s in oceanographic sciences went to women in 1994 and 44.4 percent in 2003 out of a steady yearly total of 120–130 (National Science Foundation, Science and Engineering Statistics, www.nsf.gov/statistics/).

12. Mindell (2002: 2).

13. Williams (1990: 7).

14. See Lewis (1967).

15. Comparing narratives about tropical ecosystems with stories of space, Haraway writes: "The extraterrestrial is coded to be fully general; it is about escape from the bounded globe into an anti-ecosystem called, simply, space. Space is not about 'man's' *origins* on earth but about 'his' *future*. . . . Space and the tropics are both utopian topical figures in western imaginations, and their opposed properties dialectically signify origins and ends for the creature whose mundane life is outside both: civilized man" (1989: 137). Submarine cyborg travelers in *Alvin*, as they move through vent fields, negotiate a site in which narratives of origin (hydrothermal beginnings of life) and future (outer space) coil around one another.

16. Kaharl (1990: 337).

17. Mody (2005: 176).

18. "like the waters of some alien ocean breaking about [one's] ears," to take a phrase from Samuel Delany's sci-fi novel *Trouble on Triton* (1996: 109).

19. The underwater realm is *not* a soundscape for people without prosthetic technologies. Humans cannot use underwater acoustic vibration to locate themselves in space: sound waves travel four times faster in water than in air, and human eardrums are too similar in density to water to interrupt most underwater vibrations. For humans, bones in the skull register underwater sound, making such vibrations seem omniphonic—coming from all directions at once—and immanent—coming from within one's own body.

20. A different circumstance obtains in saturation diving, which acclimatizes divers to pressures greater than one atmosphere, requiring extended decompression. To prevent oxygen poisoning, helium is sometimes added to air, causing divers' voices to rise, making them sound like the 1950s novelty act Alvin and the Chipmunks.

21. Subs depended upon sound for navigation before information theory. In 1901, the Submarine Signal Company of Boston imagined "a network of underwater bells whose sonorous gongs would carry through the water for great distances" (Schlee 1973: 246). The company built receivers to capture resonances for listeners onboard ships, though plans to use bells for Morse code were swamped by underwater turbulence.

22. Iselin and Ewing (1941).

23. By this time, temperature profiles of the world's waters were public; information about the area around Japan, of interest to the United States, had been published by the Japanese Hydrographic Department (Schlee 1973).

24. And it *was* a vision: "Cybernetic philosophy was premised on the opacity of the Other" (Galison 1994: 256), an opacity reinforced in sound ranging, premised on bouncing off exteriors, not penetrating interiors.

25. Kingsley (1863).

26. Sterne (2003: 15).

27. Hamilton-Paterson (1992: 21).

28. Stetten (1984).

29. Feld and Brenneis (2004: 462).

30. Corbin (1994: 164).

31. Hamilton-Paterson (1992: 110)

32. Hastrup (1990: 46).

33. Goldenweiser (1933: 349).

34. Swain and Lapkin (1983).

35. Mukerji (1989: 71).

36. Scripps Institution of Oceanography (1981).

37. Bull (2003).

38. Toop (1995: 271). Some composers attempt to fuse the immersively oceanic and musical. Redolfi's "Sonic Waters" (1989) was performed underwater, off the pier of Scripps.

39. Kaharl (1990: 273).

40. Théberge (2004).

41. The pilot, however, must be attentive to the rhythms of the sub, though more to the running of the engines than the spacing of sonar pings. Marine mammals outside, meanwhile, surely take our lullaby to be a racket. Stocker (2002/2003) reports on acoustic ecology as it pertains to marine mammals. The sea is no longer a "silent world" (Cousteau 1953).

42. Schwartz (2003).

43. Scuba divers are not sonic submarine cyborgs. They monitor equilibrium through *visual* checks of dive computer screens and surrounding space.

44. Crane (2003: 134).

45. See Van Dover et al. (1989).

46. Haraway (1991a: 149).

47. I use *transduced* instead of *translated* here in resonance with linguistic anthropologist Michael Silverstein, who frames the work of rendering meaning from one language into another as transduction: "We should think seriously of the underlying metaphor of the energy transducer that I invoke, such as a hydroelectric generator. Here, one form of organized energy [e.g., the gravitationally aided downstream and downward linear rush of water against turbine blades] is asymmetrically converted into another kind of energy at an energetic transduction site. . . . much of what goes into connecting an actual source-language expression to a target-language one is like such a transduction of energy" (2003: 83–84). Cf. Barad (2001).

48. Kaharl (1990: 308).

49. Van Dover (1996: 101).

50. Fabian (1983: 6).

51. The sea is louder in Kingsley's *The Water-Babies:* "Tom came to the white lap of the great Sea-mother, ten thousand fathoms deep. . . . as he walked along in the silence of the sea-twilight, . . . he was aware of a hissing and a roaring, and thumping, and a pumping, as of all the steam engines of the world at once. And, when he came near, the water grew boiling hot" (quoted in Kaharl 1990: xiii). "The Sea Priestess" by the British electronic musical outfit Coil is a 1999 echo of Kingsley:

> Her wizened mouthpiece whistles with silver fishes.
> Swirls of spider-crabs crackle like Wimshurst mechanicals.
> All around her, jellies are diaphanous.
>
> I had breakfast with the sea priestess,
> whose sibilant esses are escaping gas from the sea floor.
> . . .
> Gas fired from a gun, herbal hydrogen.
> If it goes any faster
> there'll be an astral disaster.

Louder is the alien ocean planet Kainui, in *Noise,* Hal Clement's sci-fi novel: "Noise from the ocean bottom was continuous, and deafening, and often deadly in overpressure" (2003: 26). See Crone et al. (2006) on research into vent sounds.

52. See Arico and Salpin (2005: 14).

53. Mukerji (1989: 153, 155).

54. Mukerji (1989: 146).

55. Mukerji (1989: 148).

56. Mukerji (1989: 153).

57. Van Dover (1996: 4).

58. Broad (1997: 104).

59. Kaharl (1990: 340).

60. Arico and Salpin (2005: 51).

61. Crane (2003: 106–7). Oreskes (2003) argues that military specifications for *Alvin's* depth rating also determined the presence of the sub in particular locales.

62. This fusion could be understood as following a tradition Steinberg identifies in *The Social Construction of the Ocean,* in which Americans "used the wildness of the sea to achieve liberation from the corruption and banality of land-based society, but . . . did this not through conquering the sea or submitting to the awesome powers that it represented. Rather, they achieved liberation by using science to become one with the sea" (2001: 121).

63. Myers (2007).

64. Haraway (1991a: 178).

65. For discussion of ethnographies that might be considered transductive, see Helmreich (2007).

66. Scientists and crew joke that being in *Alvin* is like being in a womb, listening to the heartbeat of the motherly sea outside. When described as a "she,"

Alvin partakes of a tradition of gendering ships feminine. Van Dover's image of working "inside her" follows this lead but suggests something else. Submerged, the submarine's femininity modulates into the maternal; this ship will take care of you. If one wanted to listen with a Freudian ear to scientists' and crew's jokes, the sounds of *Alvin* could be interpreted in line with maternal imagery. The sounds around *Alvin* become amniotic—the sea offering "a hydrologically filtered mother's voice promising the bliss of undifferentiation" (Kahn 1999: 257). A masculine gendering was playfully defended by one of the *Atlantis* crew, ordinary seaman Kevin Threadgold, who recited a verse he wrote about *Alvin* on our expedition's poetry night:

> *Alvin* is my favorite sub
> I'd like to take him to a pub
> We'd sit and drink our favorite beer
> and I'd say WHAT'S IT LIKE DOWN THERE?
> He'd smile and smoke a fat cigar
> and say I'M JUST A BIG WHITE CAR
> that drives around the ocean floor
> finds a rock and drives some more.

67. Weir (2001).

68. Shipboard e-mail has, students tell me, reduced the incidence of romantic liaisons at sea, since people can easily maintain e-contact with significant others ashore. It is no longer so true that "what happens at sea stays at sea."

69. Mindell (2005).

70. Van Dover (2000: 117).

71. Interestingly, "While Archaea and Bacteria are generally considered separate domains of life in the prokaryotic world, there are indications that many signal transduction systems in archaea have been acquired from bacteria through lateral gene transfer" (Galperin 2005).

72. Gould (1997).

73. Mark Wheeler, Signal Discovery? *Smithsonian.com*, March 1, 2004, www.smithsonianmag.si.edu/smithsonian/issues04/mar04/phenomena.html.

74. Roosth (2009).

75. This is not entirely clear. As a UN report states, "While scientific research is authorized and regulated within the [Endeavour] MPA [Marine Protected Area], the Management Plan and the Regulations remain silent regarding activities undertaken with a commercial purpose, such as bioprospecting, which seem to fall under the prohibition of Section 2 of the Regulations. . . . Implications of the Regulations and the Management Plan for expeditions involving both scientific research and bioprospecting remain unclear" (Arico and Salpin 2005: 52).

76. Broad (1997: 98).

77. Committee on the Implementation of a Seafloor Observatory Network for Oceanographic Research, Ocean Studies Board, Division on Earth and Life Sciences, National Research Council of the National Academies (2003).

78. Haraway (1991a: 162). Peter Watts's 1999 sci-fi novel *Starfish* imagines a dystopian vivification. A seismic monitoring network built on the Juan de Fuca Ridge achieves intelligence and schemes to trigger undersea quakes that will wipe out Pacific Rim humanity in a tsunami. The network, a compound of computer links and vat-grown brains—a cyborg—also seeks to disperse "apocalyptic microbes . . . from the deep sea" by infecting deep-diving humans.

79. The making of such a Neptunian instrumentarium will ironically require much more time at sea than a few cruises a year—and more people, drawn from engineers and ship's crew; it is not true, then, as my interlocutor on Sapelo saw it for genomics, that people "just don't have to *be* here anymore."

80. A sonic, cetacean premonition of this convergence appears in Ian Watson's 1975 sci-fi novel *The Jonah Kit* (2002) in which the ocean is a giant computer, operated by whales and dolphins that use echolocation to solve equations.

81. See Butterfield and Kelley (2004: 21).

82. NEPTUNE, www.neptune.washington.edu.

83. Poore (2003).

84. Haraway (1991b: 189).

85. Simondon (1992: 313). See Mackenzie (2002).

86. Maurer (2005: 5)

87. Dean (1998: 63).

88. Cf. Michael Rossi (2007) on the "virtual ocean" offered in the American Museum of Natural History's Hall of Ocean Life.

89. Arico and Salpin (2005: 14).

90. Glowka (1996: 158). According to Arico and Salpin (2005), of 125 known vents, sixty-one are within national jurisdiction, fifty-five outside, and the remainder contested or uncertain.

91. See Brush (1999).

92. Glowka (1996: 160).

93. Earle (1995: 314). See Hardin (1968) for "the tragedy of the commons."

94. Jacobson and Rieser (1998: 103).

95. Broad (1997: 276).

96. See Lawson and Downing (2002) for just this appropriative view.

97. Soares et al. (1998: 92–93).

98. Soares et al. (1998: 10), emphasis added.

99. Durrenberger and Pálsson (1987), McCay and Acheson (1987).

100. U.S. Cabinet (1999: 22–23).

101. Soares et al. (1998: 17).

102. See Seed (1995).

103. Arico and Salpin (2005).

104. Arico and Salpin (2005: 31).

105. Arico and Salpin (2005: 30).

106. Marcos Almeida, Commentary at Tapping the Oceans' Treasures: Bioprospecting in the Deep Seabed, June 9, 2005, United Nations Conference Room 6. Transcription mine.

107. Dando and Juniper (2001: 1).

108. Glowka (2001: 17).

109. Glowka (2001: 17).

110. Strathern (2004: 10).

111. Beck (1992).

112. Dando and Juniper (2001: 3–4).

113. The discussion of extraterritorial governance prompted by vent research is reminiscent of conversations that unfolded in the 1960s about the disposition of resources on the Moon. Debate about the United Nations' proposed Moon Treaty of 1967 was organized around similar divisions between "common heritage" and "open access" positions.

7. EXTRATERRESTRIAL SEAS

1. Sagan and Margulis (1993: 5). Some argue that burning fossil fuels is creating a more acidic sea (Kleypas et al. 2006).

2. Carson (1955). This area is an ecotone, a place where different ecosystems (microbial mats, salty ocean waters, dunes) converge. "Ecosystems," Margulis tells me, "are like nations: mail and goods travel faster within them than between them." See Traver (2006) on Sippewissett's history. Rachel Carson herself once frequented Sippewissett marsh, looking for eels.

3. Emma Taussig in "Extremophiles," an essay written for her first-grade class, suggests that these organisms "are amazing because they are so hard to find" (2005: 1). Thanks to Karen-Sue Taussig for sharing her daughter's essay.

4. From Early Biospheric Metabolisms to the Evolution of Complex Systems, *MBL Astrobiology Portal,* http://astrobiology.mbl.edu/overview.html.

5. The twenty-person workshop was convened by the Dibner Institute for the History of Science and Technology as its yearly Seminar in the History of Biology, traditionally held at the MBL. It was cosponsored by the NASA History Office.

6. Franklin and Lock (2003: 14). What Fischer (2003: 37) names "emergent forms of life," referring to biotechnologies and the social orders in which they reside, might also be called *divergent* forms of life.

7. Haraway (2008: 17–19) argues that *species* (from Latin *specere,* "to look") retains a reference to visual form and holds that mutual regard between species might manifest by seeing through species twice—to *respicere:* "to look (back) at," the root of *respect.* Apprehending subvisible biota, however, presents a complication to this ocular idiom (as microbes do to the species form). The auditory/tactile metaphor of transduction may offer another path toward human-microbial correspondence, one that can keep audible the technologies enabling translation across scales. Keeping in mind that *theory* and *species* derive from the Greek and Latin, respectively, for "to look," one might find oneself listening athwart species. Listening to the epistemological resonance of *species* and *specific,* Peter Galison's (2004) argument that we live in an age of "specific theory"—theory operating between the grand generalization and the

singular account—can be heard as not only as a claim about theory but also a query about how the *specific* should be demarcated.

8. Art historian Henri Focillon in 1934 argued that form always "suggests to us the existence of other forms" (1989: 34), which implies that, when form itself is at sea, so too might be the very intelligibility of otherness.

9. This future no longer looks so bright; NASA in 2006 cut funding in half for astrobiology (NAI, NSI History, http://nai.nasa.gov/about/timeline.cfm#2007).

10. McKay et al. (2002: 625).

11. Des Marais et al. (2002: 154).

12. Rothschild and Mancinelli (2001). On archaeal extremophiles, see DeLong (1998a).

13. MacElroy (1974).

14. From other organisms' points of view, humans and other oxygen-breathing animals (aerophiles) might be considered extremophiles since aerobic metabolism produces forms of oxygen that slowly damage tissue.

15. Mancinelli, White, and Rothschild (1998). Pavlov et al. (2006: 911) "propose that the radioresistance (tolerance to ionizing radiation) observed in several terrestrial bacteria has a martian origin."

16. Des Marais et al. (2002: 154).

17. "We want to explore observed features like dry riverbeds, ice in the polar caps and rock types that only form when water is present. We want to look for hot springs, hydrothermal vents or subsurface water reserves. We want to understand if ancient Mars once held a vast ocean in the northern hemisphere as some scientists believe" (NASA, The Mars Exploration Program, http://mars.jpl.nasa.gov/overview/).

18. See Committee on the Limits of Organic Life in Planetary Systems (2007).

19. Some of the earliest astronomical analogs to earthly oceans were lunar features believed to be seas. The Sea of Tranquility, where *Apollo 11* touched down, was thought by early astronomers to be a body of water (Mare Tranquillitatis). Such "seas," now called the Lunar maria (singular, *mare*), are plains formed by eruptions of basalt; less reflective than mountainous terrain that surrounds them, they have a dark cast. They exist in various sizes, designated by a descending scale of labels; thus Lacus Oblivionis (the Lake of Forgetfulness), Sinus Amoris (the Bay of Love), and Palus Putrednis (the Marsh of Rot).

20. Dick and Strick (2004: 83). Earth is a "complex entity involving the Earth's biosphere, atmosphere, oceans, and soil; the totality constituting a feedback or cybernetic system which seeks an optimal physical and chemical environment for life on this planet" (Lovelock quoted in Haraway 1995: xii).

21. Dick and Strick (2004: 83).

22. Des Marais et al. (2002: 156).

23. Haraway (1995: xiv).

24. Fuller (1969).

25. Lepselter (1997: 197).

26. Quoted in Dick and Strick (2004: 117).

27. Hoffman and Schrag (2000).

28. Bowker (2005: 19).

29. Quoted in Dick (1996: 69).

30. Wells (2001: 241).

31. Quoted in Dick (1996: 122)

32. Schulze-Makuch and Houtkooper (2007) have argued that Viking experiments, done before extremophile research, may have drowned or baked possible Martian microbes, which they hypothesize may thrive in a low-temperature water–hydrogen peroxide medium.

33. McKay et al. (1996: 928).

34. McKay et al. (1996: 929).

35. Wittgenstein (1958: 193). See Wittgenstein on Mars, www.george-coates.org/OnMars/.

36. Baker et al. (2006) claim to have found "nanoarchaea" in California mine drainage.

37. Krasnopolsky, Maillard, and Owen (2004: 537).

38. NASA, Goal 1: Determine in Life Ever Arose on Mars, http://mars.jpl.nasa.gov/science/life/index.html.

39. Mirmalek (2004). Vertesi (2008) documents how Pasadena personnel describe "seeing like a rover."

40. Grady (2001: 82).

41. Des Marais et al. argue that "spectral biosignatures can arise from organic constituents (e.g., vegetation) and/or inorganic products (e.g., atmospheric O_2)" (2002: 154). Martian meteorite researcher McKay offers with his colleagues a ranking of different kinds of biosignatures, based on their persuasiveness, arguing that "the reliability or usefulness of a biosignature is inversely proportional to how difficult it is to produce by non-biologic processes." Category I biosignatures—Grady's direct signatures—are "nearly indisputable evidence for life" and examples include "complex fossils such as trilobites, skeletons, and other forms with indisputable morphologies (extremely challenging with single-cell life)." Category II biosignatures—remote signatures—include the "presence of ozone and methane in a planetary atmosphere." Category III biosignatures embrace such items as "micrometer-size spherical or ovoid objects of appropriate composition"—which last describes the shapes found in ALH84001 and might be described as direct but iffy signatures (McKay et al. 2002: 625).

42. Markley (2004).

43. Derrida (1982: 328).

44. Thanks to Hillel Schwartz for this critique.

45. Des Marais et al. (2003: 233).

46. Markley (2005: 346).

47. Malin et al. (2006).

48. NASA, First-of-Its-Kind Antenna to Probe the Depths of Mars, http://mars.jpl.nasa.gov/express/spotlight/20050504.html.

49. There are problems with this theory: "No evidence of carbonates has yet been found anywhere on the planet. Carbonates are minerals that form readily when liquid water reacts with carbon dioxide in the atmosphere. If Mars had abundant liquid water in its past, carbonates should be detectable in the Martian rock record" (NASA, The Case of the Missing Mars Water, http://science.nasa.gov/headlines/y2001/ast05jan_1.htm).

50. Davies (2003: 675).

51. Davies (2003: 674–75).

52. Supporters of panspermia have been intrigued by (controversial) 2001 accounts of red rain in Kerala, India, which physicists Godfrey Louis and A. Santhosh Kumar (2006) say contained cell-like structures of cometary origin.

53. Davies (2003: 678). Harvard geneticist Gary Ruvkun suggests sending a PCR machine to Mars to look for genetic material familiar from Earth. See the abstract of his NASA-funded proposal, "An in situ PCR Detector For Life on Mars Ancestrally Related to Life on Earth," part of "Search for Extraterrestrial Genomes," Office of Space Science, NASA Announcements Opportunity, Astrobiology Science and Technology Instrument Development (ASTID): Abstracts of awarded proposals, http://research.hq.nasa.gov/code_s/nra/current/NRA-01-OSS-01-AST/winners.html.

54. PBS, Nova transcripts, Mars Dead or Alive, www.pbs.org/wgbh/nova/transcripts/3101_mars.html.

55. Untroubled by the trouble with trees, astrobiologist Peter Ward (2005: 255) suggests that "the universe is a forest . . . of trees of life of separate evolutionary creation," for which he proposes an arborcentric taxonomic designation above domain: Arborea. In the December 2007 *Scientific American*, Paul Davies asks whether more than one tree of life exists on Earth itself, whether life-as-we-know-it coexists with, for example, life made from left-handed coiling DNA or exotic amino acids. Putting a point on it, *Scientific American*'s cover asks "Are We Living with Alien Cells?"

56. Charles Carlson, The 'Other Agenda' in Exploring the Red Planet, *Florida Baptist Witness*, January 22, 2004, www.floridabaptistwitness.com/2036.article.

57. Stephen M. Yulish, Letter to the Editor, *Larry Taylor's Hyssop Chronicles, January–February 2004*, www.millennium-ark.net/Larry_Taylor/2004.Jan.Feb.html.

58. Cooper (2007: 32, references omitted).

59. Quoted in Wolfe (2002: 194).

60. Wolfe (2002: 195).

61. NASA, Solar System Exploration: Europa Orbiter, http://planetaryprotection.nasa.gov/pp/missions/planned/eurorb.htm.

62. Markley (2005: 8). The treatment of earthly ecologies as analogs for other worlds now sees astrobiologists transforming our planet into a patchwork of proxies for other places. Sites in the Nunavut territory of Canada now stand for Mars, and parts of Antarctica are surrogates for Europa. See Leane (2003) on "Antarctica as a Scientific Utopia."

63. Blumberg (2003: 467).

64. Quoted in Grinspoon (2003: 35).

65. Blumberg (2003: 470).

66. NASA, Goal 1: Determine in Life Ever Arose on Mars, http://mars.jpl.nasa.gov/science/life/index.html.

67. Des Marais et al. (2003: 234).

68. Committee on the Limits of Organic Life in Planetary Systems (2007: 9). Saturn's moon Titan, which may cover an ammonia-water ocean beneath an icy shell—itself topped by ethane or methane lakes and rivers—has attracted astrobiological speculation about novel life forms (Ward 2005), though see Stofan et al. (2007) for skepticism about a biogenic origin for methane.

Although the term *life-form* entered the twentieth century with a specific botanical meaning, a tool for classifying plants based on "the position of the buds relative to the soil surface during the unfavourable season" (OED), it can nowadays refer to morphological types, species, or even *alien life forms* (a phrase that enters science fiction around 1941; see *Science Fiction Citations*, full record for *alien life form n.*, www.jessesword.com/sf/view/12).

69. Given the jettisoning of *race* by most biologists, it is no wonder that scientifically minded science fiction has disposed of talk of alien races from other planets—a prominent framing in the 1950s and '60s. Instead, as illustrated by *Barlowe's Guide to Extraterrestrials,* a 1979 compendium of aliens from twentieth-century fiction, aliens are now marked by different "cultures."

70. Delaney (2001: 151). Saturn's moon Encedalus may also host liquid water beneath an icy shell.

71. Lipps (2003).

72. Cameron and Quale (2005).

73. The selection of a black woman as an emissary for humanity can be read a few ways. The most superficial reading would dismiss race and gender as irrelevant, hewing to a model of science as a zone of equal opportunity. Another might suggest that Figueroa functions as a role model for aspiring scientists of color (an interpretation Figueroa herself offered to *Ebony*; see Holloway 2005). Such an interpretation on its own deflects attention from institutional forces (underfunding of K–12 science education, the evisceration of affirmative action) that keep African Americans and other minorities at low numbers in marine biology. At a more symbolic level, putting the face of a person of color at the head of an apparatus of extraterrestrial exploration erases the colonial fantasies out of which such narratives issue. Meanwhile, within the frame of American political culture, a black woman ambassador to the stars might represent the culmination of the American promise of enfranchisement; here, Figueroa stands as the model citizen of earthly democracy, a once-Other reaching out to a new Other, promising harmony among the different-yet-equal. Thanks to Jake Dorman for conversations on this question.

74. Foucault (1986: 24).

75. Foucault in Casarino (2002: 12).

76. Compare other sci-fi treatments of alien ocean planets: In Werner Herzog's 2005 film *The Wild Blue Yonder,* astronauts fleeing an endangered

Earth arrive at a planet with a liquid helium ocean but, returning home to encourage humans to relocate, find humanity vanished. Arthur C. Clarke's novel *The Songs of Distant Earth* (1986) takes place on the water planet Thalassa and describes a less disturbing alien ocean. In this story, humans live on the surface while primitive lobster-like critters inaugurating their own cultural evolution scrabble around on the bottom.

77. See Ssorin-Chaikov (2006) on heterochrony.

78. Csicsery-Ronay (1985).

79. Such unruly complexity (Taylor 2005) might invite the conceit that encountering the unintelligible within serves a salve for self—that, as Julia Kristeva suggests, "We know that we are foreigners to ourselves, and it is with the help of that sole support that we can attempt to live with others" (1991: 170).

80. Margulis (2004: 123). Her contention extends not just to marsh spirochetes but also to that more famous spirochete, syphilis, which, she holds, still sleeps among us despite medicine's claim to have eliminated it. In "On Syphilis & Nietzsche's Madness: Spirochetes Awake!" Margulis revisits the story of Nietzsche's syphilitic dementia, offering that spirochetes in his brain came alive just like ones in microbial mats, causing him to go insane: "Nietzsche's brain on January 3, 1889 experienced a transformation like that of the microbial mat sample transferred into new fresh food" (2004: 125). And see Margulis, Navarrete, and Solé (1998).

81. Margulis and Sagan (2002).

82. Doyle (2003: 186).

83. See Myers (2007) on "liveliness" as a dynamic alternative to "life itself."

84. McMenamin and McMenamin (1993: 5).

85. Margulis and Sagan (1995).

86. Landecker (2007).

87. Doyle (2003: 21). Rotman (2000) builds a kindred case about the self, arguing that subjectivity in the digital age—distributed, multiple—has us becoming "beside ourselves." "Life itself," I claim, is becoming beside itself.

88. Lévi-Strauss (1963: 24).

89. He thought differently about kinship, which he *did* see as a universal form binding nature to culture, though later anthropologists would claim that kinship, too, lacked a central core that could be considered everywhere the same.

90. And, for scientists who simulate life *in silico,* so quotable.

91. Bamford (2007).

92. See Cleland and Chyba (2002).

93. Agamben (1998: 1).

94. Hayden (2003: 51).

95. Foucault (1978: 143).

96. Thacker (2004: 27).

97. Doyle (2003) suggests we might even be seeing the rise of the "postvital organism." Cf. *Nature* editorial (2007).

98. Foucault (1970: 387). His words forecast a circumstance desired by Lévi-Strauss, who wrote, "I believe the ultimate goal of the human sciences to be not to constitute, but to dissolve man" (1966: 247).

99. If marine microbiologists have insisted that the smallest drop of seawater is heavy with significance for earthly life, the Indian Ocean tsunami of December 26, 2004, was a reminder that the sea also exists and acts on another scale entirely. The permutation of the *alien ocean* that has the sea as other to humanity can materialize in massive manifestations, with sociopolitical effects different from the molecular and microbial—although, in the wake of the coastal deluges of Sri Lanka, Indonesia, India, Thailand, and other Indian Ocean nations, health workers did worry about the water-borne spread of microbially mediated diseases like cholera and malaria. Contamination of drinking water and overflow of sewage were collateral concerns. In southeastern India, a warning spread through text messaging advised that a "very dangerous Zulican virus is being spread through sea food," and that "fish are thriving on corpses of thousands of men, women and heads of cattle which were swept away that fateful Sunday and have not been fished out yet. Consuming these fish would lead to outbreak of diseases" (Tsunami Aftermath, www.textually.org/textually/archives/006575.htm, and Dangerous Virus in Seafood a Hurtful SMS Rumor, www.textually.org/textually/archives/006586.htm). India's national newspaper *The Hindu* reported that such rumors were unfounded; the director of the Centre for Advanced Study in Marine Biology in Portonovo, Tamil Nadu, posited that an organic boundary prevented transfers between humans and the sea: "Asked whether it would be safe to consume sea fish and lobsters, Mr. Balasubramanian said there would be no harm, because on-land viruses and bacteria would not survive in saline water. There existed a natural immune system in the ocean" (A. V. Ragunathan, Tsunami Has Not Affected Marine Life, *The Hindu*, www.hindu.com/2005/01/03/stories/2005010310720400.htm). The two views of the ocean in contention here—as predatory domain and as health-securing boundary between worlds—repeat some of the associations with the alien ocean mapped in this book. See Helmreich (2006) for more about the tsunami.

100. Public Radio International, "Living on Earth," www.loe.org/shows/shows.htm?programID=07-P13–00013#feature5. "If 'they're us' and 'we're them,'" writes Neil Badmington, "the opposition between the homely and unfamiliar can no longer hold, for the alien is no longer the wholly other" (2004: 30).

101. O'Malley and Dupré (2007).

102. Or *Homo microbis*. *Homo sapiens*, *Homo faber* ("man the maker"), and *Homo ludens* ("man the player") all press *nature* up against *culture*, whereas *Homo microbis* is a *nature-nature* hybrid emerging from the enmeshing of human and microbial cultures as well as the liquefaction of *culture* as a universal medium to describe either. The anthropology I offer in this book thus seeks less to synthesize biological and cultural anthropology—the mission of American four-field anthropology—than to read them athwart one another; cf. Segal and Yanagisako 2005.

103. Weiss (2005).

Bibliography

Abbott, Isabella. 1978. The Uses of Seaweed as Food in Hawaii. *Economic Botany* 32(4): 409–12.

———. 1987. There Are Aliens among the Algae, Too—or Limu malihini. *Hawaiian Botanical Society Newsletter* 26:60–63.

———. 1999. *Marine Red Algae of the Hawaiian Islands.* Honolulu: Bishop Museum Press.

Acheson, James M. 1981. Anthropology of Fishing. *Annual Review of Anthropology* 10: 275–316.

Agamben, Giorgio. 1998 [1995]. *Homo Sacer: Sovereign Power and Bare Life.* Trans. Daniel Heller-Roazen. Stanford, CA: Stanford University Press.

———. 2004 [2002]. *The Open: Man and Animal.* Trans. Kevin Attell. Stanford, CA: Stanford University Press.

Anderson, Benedict. 1983. *Imagined Communities: Reflections on the Origin and Spread of Nationalism.* London: Verso.

Ansell Pearson, Keith. 1997. *Viroid Life: Perspectives on Nietzsche and the Transhuman Condition.* London: Routledge.

Appadurai, Arjun. 1996. *Modernity at Large: Cultural Dimensions of Globalization.* Minneapolis: University of Minnesota Press.

Arico, Salvatore, and Charlotte Salpin. 2005. *Bioprospecting of the Genetic Resources in the Deep Seabed: Scientific, Legal and Policy Aspects.* Yokohama, Japan: United Nations University Institute of Advanced Studies Publication.

Azam, Farooq, and Alexandra Z. Worden. 2004. Microbes, Molecules, and Marine Ecosystems. *Science* 303:1622–24.

Badmington, Neil. 2004. *Alien Chic: Posthumanism and the Other Within.* London: Routledge.

Bailey, Cornelia Walker, with Christena Bledsoe. 2000. *God, Dr. Buzzard, and the Bolito Man: A Saltwater Geechee Talks about Life on Sapelo Island, Georgia.* New York: Anchor Books.

Baker, Graham, director. 1988. *Alien Nation.* Twentieth Century Fox.

Baker, J. Brett, Gene W. Tyson, Richard I. Webb, Judith Flanagan, Philip Hugenholtz, Eric E. Allen, and Jillian Banfield. 2006. Lineages of Acidophilic Archaea Revealed by Community Genomic Analysis. *Science* 314:1933–35.

Ball, Philip. 2001. *Life's Matrix: A Biography of Water*. Berkeley: University of California Press.

Bamford, Sandra. 2007. *Biology Unmoored: Melanesian Reflections on Life and Biotechnology*. Berkeley: University of California Press.

Bapteste, Eric, Yan Boucher, Jessica Leigh, and W. Ford Doolittle. 2004. Phylogenetic Reconstruction and Lateral Gene Transfer. *Trends in Microbiology* 12(9):406–11.

Barad, Karen. 2001. Performing Culture/Performing Nature: Using the Piezoelectric Crystal of Ultrasound Technologies as a Transducer between Science Studies and Queer Theories. In *Digital Anatomy*, ed. Christina Lammar, 98–114. Vienna: Turia and Kant.

Barlowe, Wayne Douglas, Ian Summers, and Beth Meacham. 1979. *Barlowe's Guide to Extraterrestrials*. New York: Workman.

Baross, John A. 1998. Do the Geological and Geochemical Records of the Early Earth Support the Prediction from Global Phylogenetic Models of a Thermophilic Cenancestor? In *Thermophiles: The Keys to Molecular Evolution and the Origin of Life?* ed. J. Wiegel and M. Adams, 3–18. London: Taylor and Francis.

Bass, Patrik Henry. 1999. Intimate Strangers. *BET Weekend Magazine*, November, 20.

Bateson, Gregory. 2000 [1972]. *Steps to an Ecology of Mind*. Chicago: University of Chicago Press.

Battaglia, Debbora. 2005a. Insider's Voices in Outerspaces. In *E.T. Culture: Anthropology in Outerspaces*, ed. Debbora Battaglia, 1–37. Durham: Duke University Press.

———. 2005b. "For Those Who Are Not Afraid of the Future": Raëlian Clonehood in the Public Sphere. In *E.T. Culture: Anthropology in Outerspaces*, ed. Debbora Battaglia, 149–79. Durham: Duke University Press.

Bauman, Zygmunt. 2000. *Liquid Modernity*. Cambridge: Polity Press.

Bear, Greg. 2002. *Vitals*. New York: Ballantine Books.

Beck, Ulrich. 1992. *Risk Society: Towards a New Modernity*. Trans. Mark Ritter. London: Sage.

Beer, Gillian. 2000. *Darwin's Plots: Evolutionary Narrative in Darwin, George Eliot and Nineteenth-Century Fiction*, 2d ed. Cambridge: Cambridge University Press.

Béjà, Oded, L. Aravind, Eugene V. Koonin, Marcelino T. Suzuki, Andrew Hadd et al. 2000. Bacterial Rhodopsin: Evidence for a New Type of Phototrophy in the Sea. *Science* 289:1902–6.

Benjamin, Walter. 1968 [1936]. Unpacking My Library. Trans. Harry Zohn. In *Illuminations*, ed. Hannah Arendt, 59–67. New York: Shocken Books.

Berlitz, Charles. 1974. *The Bermuda Triangle: The Incredible Saga of Unexplained Disappearances*. Garden City, NY: Doubleday.

Bernard, H. Russell 1976. Scientists and Mariners at Sea. *Marine Technology Society Journal* 10:21–30.

Bernard, H. Russell, and Peter D. Killworth. 1973. On the Social Structure of an Ocean-Going Research Vessel and Other Important Things. *Social Science Research* 2(2): 145–84.

Bestor, Theodore. 2004. *Tsukiji: The Fish Market at the Center of the World.* Berkeley: University of California Press.

Binder, Brian. 2002. *In Situ* Pico-Cyanobacterial Growth Rates in the Sargasso Sea Based on Cell-Specific rRNA Measurements. A Proposal Submitted 8/15/2002 to NSF Division of Ocean Sciences, Biological Oceanography Program.

Binns, Raymond A., and David L. Dekker. 1998. The Mineral Wealth of the Bismarck Sea. *Scientific American Presents* 9(3): 92–97.

Blumberg, Baruch S. 2003. The NASA Astrobiology Institute: Early History and Organization. *Astrobiology* 2(3): 463–70.

Boas, Franz. 1938. *The Mind of Primitive Man.* New York: Free Press.

Bock, Gregory, and Jamie Goode, eds. 1996. *Evolution of Hydrothermal Ecosystems on Earth (and Mars?).* New York: John Wiley and Sons.

Boellstorff, Tom. 2006. *The Gay Archipelago: Sexuality and Nation in Indonesia.* Princeton, NJ: Princeton University Press.

Bolster, W. Jeffrey. 1997. *Black Jacks: African American Seamen in the Age of Sail.* Cambridge, MA: MIT Press.

Bolter, Jay, and Richard Grusin. 1999. *Remediation: Understanding New Media.* Cambridge, MA: MIT Press.

Borges, Jorge Luis. 1962 [1956]. The Library of Babel. Trans. James E. Irby. In *Labyrinths: Selected Stories and Other Writings,* ed. Donald A. Yates and James E. Irby, 51–58. New York: New Directions.

Borges, Kim M., Shawna R. Brummet, Allison Bogert, Maria C. Davis, Kristine M. Hujer et al. 1996. A Survey of the Genome of the Hyperthermophilic Archaeon, *Pyrococcus furiosus. Genome Science and Technology* 1(2): 37–46.

Borofsky, Robert. 1997. Cook, Lono, Obeyesekere, and Sahlins. *Current Anthropology* 38(2): 255–82.

Bowker, Geoffrey C. 2005. *Memory Practices in the Sciences.* Cambridge, MA: MIT Press.

Bowker, Geoffrey C., and Susan Leigh Star. 1999. *Sorting Things Out: Classification and Its Consequences.* Cambridge, MA: MIT Press.

Bright, Chris, and Linda Starke. 1998. *Life Out of Bounds: Bioinvasion in a Borderless World.* New York: Norton.

Broad, William J. 1997. *The Universe Below: Discovering the Secrets of the Deep Sea.* New York: Simon and Schuster.

Brody, David. 2002. Ernst Haeckel and the Microbial Baroque. *Cabinet Magazine* 7:25–27.

Brown, Jacqueline Nassy. 2005. *Dropping Anchor, Setting Sail: Geographies of Race in Black Liverpool.* Princeton, NJ: Princeton University Press.

Brush, Stephen. 1999. Bioprospecting the Public Domain. *Cultural Anthropology* 14(4): 535–55.

Bull, Michael. 2003. Soundscapes of the Car: A Critical Study of Automobile Habitation. In *The Auditory Culture Reader,* ed. Michael Bull and Les Back, 357–74. Oxford: Berg.

Bulmer, Ralph. 1967. Why Is the Cassowary Not a Bird? A Problem of Zoological Taxonomy among the Karam of the New Guinea Highlands. *Man: The Journal of the Royal Anthropological Institute* 2(1): 5–25.

Burdick, Alan. 2005. The Truth about Invasive Species: How to Stop Worrying and Learn to Love Ecological Intruders. *Discover* 26(5): 34–41.

Bushman, Frederic. 2002. *Lateral DNA Transfer: Mechanisms and Consequences.* Cold Spring Harbor, NY: Cold Spring Harbor Laboratory Press.

Butel, Paul. 1999. *The Atlantic.* Trans. Iain Hamilton Grant. London: Routledge.

Butterfield, David, and Deborah Kelley. 2004. Status Report on the Endeavour ISS. *Ridge 2000 Events,* 4–5, 20–21.

Callon, Michel. 1986. Some Elements of a Sociology of Translation: Domestication of the Scallops and the Fishermen of St. Brieuc Bay. In *Power, Action and Belief: A New Sociology of Knowledge?* ed. John Law, 196–229. London: Routledge and Kegan Paul.

Cambrosio, Alberto, and Peter Keating. 2000. Of Lymphocytes and Pixels: The Techno-Visual Production of Cell Populations. *Studies in History and Philosophy of Biological and Biomedical Sciences* 31(2): 233–70.

Cameron, James, and Steven Quale, directors. 2005. *Aliens of the Deep.* Walt Disney Pictures and Walden Media.

Carlton, James T. 1996. Marine Bioinvasions: The Alteration of Marine Ecosystems by Non-Indigenous Species. *Oceanography* 9(1): 36–43.

———. 2000. Global Change and Biological Invasions in the Oceans. In *Invasive Species in a Changing World,* ed. Harold A. Mooney and Richard J. Hobbs, 31–53. Washington, DC: Island Press.

Carpine-Lancre, Jacqueline. 2001. Oceanographic Sovereigns: Prince Albert I of Monaco and King Carlos I of Portugal. In *Understanding the Oceans,* ed. Margaret Deacon, Tony Rice, and Colin Summerhayes, 56–68. London: University College London Press.

Carson, Rachel. 1951. *The Sea around Us.* New York: Oxford University Press.

———. 1955. *The Edge of the Sea.* Boston: Houghton Mifflin.

Carsten, Janet. 2000. Introduction: Cultures of Relatedness. In *Cultures of Relatedness: New Approaches to the Study of Kinship,* ed. Janet Carsten, 1–36. Cambridge: Cambridge University Press.

Carter, Kate. 2003. UGA, a School with a Low Black Student Ratio, Turns to Lawyers in Wake of Ruling. *SavannahNOW,* June 24. www.savannahnow.com/stories/062403/LOCaction.shtml.

Casarino, Cesare. 2002. *Modernity at Sea: Melville, Marx, Conrad in Crisis.* Minneapolis: University of Minnesota Press.

Cavanaugh, Colleen M., Stephen L. Gardiner, Meredith L. Jones, Holger W. Jannasch, and John B. Waterbury. 1981. Prokaryotic Cells in the

Hydrothermal Vent Tube Worm *Riftia pachyptila* Jones: Possible Chemoautotrophic Symbionts. *Science* 213:340–42.

Caws, Mary Ann. 1993. Baudelaire, Charles. In *The Johns Hopkins Guide to Literary Theory and Criticism,* ed. Michael Groden and Martin Kreiswirth, 74–76. Baltimore, MD: Johns Hopkins University Press.

Chisholm, Sallie W. 2000. Stirring Times in the Southern Ocean. *Nature* 407:685–87.

———. 2004. *Prochlorococcus:* How to Dominate the Oceans with 2000 Genes. Biological Engineering Seminar Series, Massachusetts Institute of Technology, Cambridge, April 22.

Chisholm, Sallie W., Robert J. Olson, Erik R. Zettler, Ralf Goericke, John B. Waterbury, and Nicholas A. Welschmeyer. 1988. A Novel Free-Living Prochlorophyte Occurs at High Cell Concentrations in the Oceanic Euphotic Zone. *Nature* 334:340–43.

Cipriano, Frank, and Stephen R. Palumbi. 1999. Rapid Genotyping Techniques for Identification of Species and Stock Identity in Fresh, Frozen, Cooked and Canned Whale Products. Report to the Scientific Committee, International Whaling Commission (SC/51/O9).

Clarke, Arthur C. 1986. *The Songs of Distant Earth.* New York: Ballantine Books.

Cleland, Carol E., and Christopher Chyba. 2002. Defining "Life." *Origins of Life and Evolution of the Biosphere* 32(4): 387–93.

Clement, Hal. 2003. *Noise.* New York: TOR.

Clynes, Manfred E., and Nathan S. Kline. 1995. Cyborgs and Space. In *The Cyborg Handbook,* ed. Chris Hables Gray, 29–34. New York: Routledge. [*Astronautics,* September, 1960, 26–27, 74–75]

Coleman, William. 1977. *Biology in the Nineteenth Century: Problems of Form, Function, and Transformation.* Cambridge: Cambridge University Press.

Colgan, Chuck, director. 2006. Sea of Genes. *Explorations* DVD 12(3). La Jolla, CA: Scripps Institution of Oceanography.

Collier, Stephen, and Aihwa Ong. 2005. Global Assemblages, Anthropological Problems. In *Global Assemblages: Technology, Politics, and Ethics as Anthropological Problems,* ed. Aihwa Ong and Stephen J. Collier, 3–21. Oxford: Blackwell.

Collins, Harry M. 1985. *Changing Order: Replication and Induction in Scientific Practice.* London: Sage.

Colwell, Rita R. 1984. The Industrial Potential of Marine Biotechnology. *Oceanus* 27(1): 3–12.

———. 2006. Global Climate and Health: Predicting Infectious Disease Outbreaks. *Innovations* 1(3): 19–23.

Comaroff, Jean, and John L. Comaroff. 2001. Naturing the Nation: Aliens, Apocalypse, and the Postcolonial State. *Journal of Southern African Studies* 27(3): 627–51.

Committee on the Implementation of a Seafloor Observatory Network for Oceanographic Research, Ocean Studies Board, Division on Earth and Life

Sciences, National Research Council of the National Academies. 2003. *Enabling Ocean Research in the 21st Century: Implementation of a Network of Ocean Observatories.* Washington, D.C.: National Academies Press.

Committee on the Limits of Organic Life in Planetary Systems, Committee on the Origins and Evolution of Life, National Research Council. 2007. *The Limits of Organic Life in Planetary Systems.* Washington, DC: National Academies Press.

Connery, Chris. 1995. The Oceanic Feeling and the Regional Imaginary. In *Global/Local: Cultural Production and the Transnational Imaginary,* ed. Rob Wilson and Wimal Dissanayake, 284–311. Durham: Duke University Press.

Conover, David, director. 2005. *Cracking the Ocean Code.* Silver Spring, MD: Discovery Communications.

Cooper, Melinda. 2007. Life, Autopoiesis, Debt: Inventing the Bioeconomy. *Distinktion* 14:25–43.

Corbin, Alain. 1994 [1988]. *The Lure of the Sea: The Discovery of the Seaside in the Western World 1750–1840.* Trans. Jocelyn Phelps. Berkeley: University of California Press.

Corburn, Jason. 2005. *Street Science: Community Knowledge and Environmental Health Justice.* Cambridge, MA: MIT Press.

Corfield, Richard. 2003. *The Silent Landscape: The Scientific Voyage of the HMS* Challenger. Washington, DC: Joseph Henry Press.

Corliss, John B., John A. Baross, and Sarah E. Hoffman. 1981. An Hypothesis Concerning the Relationship between Submarine Hot Springs and the Origin of Life on Earth. *Oceanologica Acta supplement,* 59–69.

Cousteau, Jacques, with Frédéric Dumas. 1953. *The Silent World.* New York: Harper and Brothers.

Cox, George W. 1999. *Alien Species in North America and Hawaii: Impacts on Natural Ecosystems.* Washington, DC: Island Press.

Cox, Paul A., Sandra A. Banack, Susan J. Murch, Ulla Rasmussen, Georgia Tien et al. 2005. Diverse Taxa of Cyanobacteria Produce β-N-methylamino-L-alanine, a Neurotoxic Amino Acid. *Proceedings of the National Academy of Sciences* 102(14): 5074–78.

Crane, Kathleen. 2003. *Sea Legs: Tales of a Woman Oceanographer.* Boulder: Westview.

Crawford, T. Hugh. 1997. Networking the (Non) Human: *Moby-Dick,* Matthew Fontaine Maury, and Bruno Latour. *Configurations* 5(1): 1–21.

Crone, Timothy J., William S. D. Wilcock, Andrew H. Barclay, and Jeffrey D. Parsons. 2006. The Sound Generated by Mid-Ocean Ridge Black Smoker Hydrothermal Vents. *PLoS ONE* 1(1): e133. doi: 10.1371/journal.pone. 0000133.

Crook, Ray, Cornelia Bailey, Norma Harris, and Karen Smith. 2003. *Sapelo Voices: Historical Anthropology and the Oral Traditions of Gullah-Geechee Communities on Sapelo Island, Georgia.* Carrollton: State University of West Georgia.

Csicsery-Ronay, Istvan, Jr. 1985. The Book Is the Alien: On Certain and Uncertain Readings of Lem's *Solaris*. *Science Fiction Studies* 12:6–21. http://fs6.depauw.edu:50080/~icronay/solaris.htm.

Cuker, Benjamin E. 2001. Steps to Increasing Minority Participation in the Aquatic Sciences: Catching up with Shifting Demographics. *ASLO Bulletin* 10(2): 17–21.

Dando, Paul, and S. Kim Juniper, conveners. 2001. *Management of Hydrothermal Vent Sites: Report from the InterRidge Workshop: Management and Conservation of Hydrothermal Vent Ecosystems.* Institute of Ocean Sciences, Sidney (Victoria), BC, Canada, September 28–30, 2000.

Darwin, Charles. 1964 [1859]. *On the Origin of Species.* Facsimile edition. Cambridge, MA: Harvard University Press.

Daubin, Vincent, Nancy A. Moran, and Howard Ochman. 2003. Phylogenetics and the Cohesion of Bacterial Genomes. *Science* 301:829–32.

Davidson, Osha Gray. 1998. *The Enchanted Braid: Coming to Terms with Nature on the Coral Reef.* New York: John Wiley and Sons.

Davies, Paul. 2001. The Origin of Life I: When and Where Did It Begin? *Science Progress* 8(1): 1–16.

———. 2003. Does Life's Rapid Appearance Imply a Martian Origin? *Astrobiology* 3(4): 673–79.

———. 2007. Are Aliens among Us? *Scientific American* 297(6): 62–69.

Davis, Susan. 1997. *Spectacular Nature: Corporate Culture and the Sea World Experience.* Berkeley: University of California Press.

Deacon, Margaret. 1971. *Scientists and the Sea 1650–1900: A Study of Marine Science.* New York: Academic Press.

Dean, Cornelia. 2005. A New Screen Test for IMAX: It's the Bible vs. the Volcano. *New York Times*, March 19, A1, A9. www.nytimes.com/2005/03/19/national/19imax.html.

Dean, Jodi. 1998. *Aliens in America: Conspiracy Cultures from Outerspace to Cyberspace.* Ithaca, NY: Cornell University Press.

De Kruif, Paul. 1996 [1926]. *Microbe Hunters*, with an Introduction by F. Gonzalez-Crussi. San Diego: Harcourt Brace and Company.

de la Cruz, Fernando, and Julian Davies. 2000. Horizontal Gene Transfer and the Origin of Species: Lessons from Bacteria. *Trends in Microbiology* 8(3): 128–33.

Delaney, John. 2001. Antarctica and the Europa Connection. In *National Geographic Atlas of the Ocean: The Deep Frontier*, ed. Sylvia Earle, 150–51. Washington DC: National Geographic Society.

Delany, Samuel. 1996 [1976]. *Trouble on Triton: An Ambiguous Heterotopia.* Middletown, CT: Wesleyan University Press.

Deleuze, Gilles, and Félix Guattari. 1987. *A Thousand Plateaus: Capitalism and Schizophrenia.* Trans. Brian Massumi. Minneapolis: University of Minnesota Press.

DeLong, Edward F. 1998a. Archaeal Means and Extremes. *Science* 280:542–43.

————. 1998b. Molecular Phylogenetics: New Perspective on the Ecology, Evolution and Biodiversity of Marine Organisms. In *Molecular Approaches to the Study of the Ocean*, ed. Keith E. Cooksey, 1–27. London: Chapman and Hall.

————. 2002. Microbial Population Genomics and Ecology. *Current Opinion in Microbiology* 5:520–24.

————. 2003. Integrating Perspectives on the Microbial World: From Nanosystems to Ecosystems. Civil and Environmental Engineering Distinguished Seminar Series, Massachusetts Institute of Technology, Cambridge, October 23.

————. 2004. Microbial Population Genomics and Ecology: A New Frontier. In *Microbial Genomics*, ed. C. M. Fraser, K. E. Nelson, and T. D. Read, 419–42. Totowa, NJ: Humana Press.

————. 2005. Microbial Community Genomics in the Ocean. *Nature Reviews Microbiology*, May 10, doi:10.1038/nrmicro1158.

DeLong, Edward F., and David M. Karl. 2005. Genomic Perspectives in Microbial Oceanography. *Nature* 437:336–42.

Derrida, Jacques. 1982. Signature Event Context. In *Margins of Philosophy*. Trans. Alan Bass, 309–30. Chicago: University of Chicago Press.

Dery, Mark. 1994. Black to the Future: Interviews with Samuel R. Delany, Greg Tate, and Tricia Rose. In *Flame Wars: The Discourse of Cyberculture*, ed. Mark Dery, 179–222. Durham: Duke University Press.

Des Marais, D. J., L. J. Allamandola, S. A. Benner, A. P. Boss, D. Deamer et al. 2003. The NASA Astrobiology Roadmap. *Astrobiology* 3(2): 219–35.

Des Marais, D. J., M. O. Harwit, K. W. Jucks, J. F. Kasting, D. N. C. Lin et al. 2002. Remote Sensing of Planetary Properties and Biosignatures on Extrasolar Terrestrial Planets. *Astrobiology* 2(2): 153–81.

Desmond, Jane. 1999. *Staging Tourism: Bodies on Display from Waikiki to Sea World*. Chicago: University of Chicago Press.

Devine, Robert S. 1998. *Alien Invasion: America's Battle with Non-Native Animals and Plants*. Washington, DC: National Geographic Society.

D'Hondt, Steven, Scott Rutherford, and Arthur J. Spivack. 2002. Metabolic Activity of Subsurface Life in Deep-Sea Sediments. *Science* 295:2067–70.

Dick, Steven J. 1996. *The Biological Universe: The Twentieth-Century Extraterrestrial Life Debate and the Limits of Science*. Cambridge: Cambridge University Press.

Dick, Steven J., and James E. Strick. 2004. *The Living Universe: NASA and the Development of Astrobiology*. New Brunswick, NJ: Rutgers University Press.

Doblin, Martina A., et al. 2003. Effects of Open-Ocean Exchange on Microbial Communities in Ships' Ballast Tanks. Paper presented at Third International Conference on Marine Bioinvasions, Convened at Scripps Institution of Oceanography, La Jolla, CA, March 16–19.

Doolittle, W. Ford. 1996. At the Core of the Archaea. *Proceedings of the National Academy of Sciences* 93:8797–99.

———. 1997. Fun with Genealogy. *Proceedings of the National Academy of Sciences* 94:12751–53.

———. 1998. You Are What You Eat: A Gene Transfer Ratchet Could Account for Bacterial Genes in Eukaryotic Nuclear Genomes. *Trends in Genetics* 14:307–11.

———. 1999. Phylogenetic Classification and the Universal Tree. *Science* 284:2124–28.

———. 2000. Uprooting the Tree of Life. *Scientific American* 282:90–95.

Doolittle, W. Ford, and James R. Brown. 1994. Tempo, Mode, the Progenote, and the Universal Root. *Proceedings of the National Academy of Sciences* 91:6721–28.

Doolittle, W. Ford, and John M. Logsdon Jr. 1998. Archaeal Genomics: Do Archaea Have a Mixed Heritage? *Current Biology* 8:R209–11.

Doyle, Richard. 1997. *On Beyond Living: Rhetorical Transformations of the Life Sciences.* Stanford, CA: Stanford University Press.

———. 2003. *Wetwares: Experiments in Postvital Living.* Minneapolis: University of Minnesota Press.

———. 2004. LSDNA: Consciousness Expansion and the Emergence of Biotechnology. In *Data Made Flesh: Embodying Information,* ed. Robert Mitchell and Phillip Thurtle, 103–20. New York: Routledge.

Drake, Lisa A., et al. 1999. Inventory of Microbes in Ballast Water of Ships Arriving in Chesapeake Bay. Paper presented at National Conference on Marine Bioinvasions, Massachusetts Institute of Technology, Cambridge, January 24–27.

Du Bois, W. E. B. 1989 [1903]. *The Souls of Black Folk,* with an Introduction by Henry Louis Gates Jr. New York: Bantam Books.

Durrenberger, E. Paul, and Gísli Pálsson. 1987. Ownership at Sea: Fishing Territories and Access to Sea Resources. *American Ethnologist* 14(3):508–22.

Duster, Troy. 2001a. Resolutions for the "Race in Science" Contradictions. Commentary presented at the meetings of the Society for the Social Study of Science, Cambridge, MA, November 1–4.

———. 2001b. The Sociology of Science and the Revolution in Molecular Biology. In *The Blackwell Companion to Sociology,* ed. Judith R. Blau, 213–26. Malden, MA: Blackwell.

Earle, Sylvia. 1995. *Sea Change: A Message of the Oceans.* New York: Fawcett Columbine.

Ebron, Paulla A. 1998. Enchanted Memories of Regional Difference in African American Culture. *American Anthropologist* 100(1): 94–105.

Ellis, Richard. 2001. *Aquagenesis: The Origin and Evolution of Life in the Sea.* New York: Viking.

———. 2003. *The Empty Ocean: Plundering the World's Marine Life.* Washington, DC: Island Press/Shearwater Books.

Englund, Ronald A., and Erin Baumgartner. 2000. A Fang-Toothed Blenny *Omobranchus ferox* from Pearl Harbor, O'ahu: A Probable Unintentional Introduction to the Hawaiian Islands. Records of the Hawaii Biological

Survey for 1999, Part I: Notes. *Bishop Museum Occasional Papers* 64:61–63.

Erosion, Technology, and Concentration Group. 2004. Playing God in the Galapagos: J. Craig Venter, Master and Commander of Genomics, on Global Expedition to Collect Microbial Diversity for Engineering Life. *ETC Communiqué*, no. 84.

Evernden, Neil. 1993. *The Natural Alien*, 2d ed. Toronto: University of Toronto Press.

Fabian. Johannes. 1983. *Time and the Other: How Anthropology Makes Its Object*. New York: Columbia University Press.

Falkner, Maurya. 2000. California's Ballast Water Management and Control Program Progress Report—September 2000. *Ballast Exchange* 3:2–5.

Falkowski, Paul G. 2002. The Ocean's Invisible Forest. *Scientific American* 287:54–61.

Falkowksi, Paul G., and Columban de Vargas. 2004. Shotgun Sequencing in the Sea: A Blast from the Past? *Science* 304:58–60.

Feld, Steven, and Donald Brenneis. 2004. Doing Anthropology in Sound. *American Ethnologist* 31(4): 461–74.

Fell, Jack W., and Steven Newell. 1998. Biochemical and Molecular Methods for the Study of Marine Fungi. In *Molecular Approaches to the Study of the Ocean*, ed. Keith E. Cooksey, 259–83. London: Chapman and Hall.

Feyerabend, Paul. 1975. *Against Method*. London: Verso.

Fikes, Bradley J. 2006. UCSD Venture Seeks to Map "Planet's Life System" of Marine Microbes. *North County Times,* January 17. www.nctimes.com/articles/2006/01/18/news/top_stories/22_23_391_17_06.txt.

Fincham, Michael W., director. 1998. *Alien Ocean* (30-minute videotape). Maryland Sea Grant College, NOAA.

Fink, Robert. 2005. *Repeating Ourselves: American Minimal Music as Cultural Practice*. Berkeley: University of California Press.

Finlayson, Alan Christopher. 1994. *Fishing for Truth: A Sociological Analysis of Northern Cod Stock Assessments from 1977 to 1990*. St. John's, Newfoundland: Institute of Social and Economic Research.

Finney, Ben. 1979. *Hokule'a: The Way to Tahiti*. New York: Dodd, Mead and Company.

———. 2003. *Sailing in the Wake of the Ancestors: Reviving Polynesian Voyaging*. Honolulu: Bishop Museum Press.

Fischer, Michael M. J. 2003. *Emergent Forms of Life and the Anthropological Voice*. Durham, NC: Duke University Press.

Fitelson, Branden, Christopher Stephens, and Elliott Sober. 1999. How Not to Detect Design. *Philosophy of Science* 66:472–88.

Fitzsimmons, John Michael. 2001. Hawaiian Stream Fishes and the Mauka-Makai Connection. Lecture at Valdosta State University, Valdosta, Georgia, April 5.

Flanagan, Richard. 2001. *Gould's Book of Fish*. New York: Grove Press.

Focillon, Henri. 1989 [1934]. *The Life of Forms in Art*. Trans. Charles Beecher Hogan and George Kubler. New York: Zone.

Fortes, Meyer. 1949. *The Web of Kinship among the Tallensi*. London: Oxford University Press.

Fortun, Mike. 2005. For an Ethics of Promising, or: A Few Kind Words about James Watson. *New Genetics and Society* 24(2): 157–73.

———. 2008. *Promising Genomics: Iceland and deCODE Genetics in a World of Speculation*. Berkeley: University of California Press.

Foucault, Michel. 1970 [1966]. *The Order of Things: An Archaeology of the Human Sciences* [Les mots et les choses]. New York: Random House.

———. 1986 [1967]. Of Other Spaces. Trans. Jay Miskowiec. *Diacritics* 16:22–27.

———. 1990 [1976]. *The History of Sexuality, Volume 1*. Trans. Robert Hurley. New York: Vintage.

———. 2008 [1973]. *This Is Not a Pipe*. Trans. James Harkness. Berkeley: University of California Press.

Frake, Charles. 1985. Cognitive Maps of Time and Tide among Medieval Seafarers. *Man: The Journal of the Royal Anthropological Institute* 20 (2): 254–70.

Franklin, Sarah. 1995. Romancing the Helix: Nature and Scientific Discovery. In *Romance Revisited*, ed. Lynne Pearce and Jackie Stacey, 63–77. London: Lawrence and Wishart.

———. 2001. Biologization Revisited: Kinship Theory in the Context of the New Biologies. In *Relative Values: Reconfiguring Kinship Studies*, ed. Sarah Franklin and Susan McKinnon, 302–25. Durham, NC: Duke University Press.

———. 2005. Stem Cells R Us: Emergent Life Forms and the Global Biological. In *Global Assemblages: Technology, Politics, and Ethics as Anthropological Problems*, ed. Aihwa Ong and Stephen J. Collier, 59–78. Oxford: Blackwell.

———. 2007. *Dolly Mixtures: The Remaking of Genealogy*. Durham, NC: Duke University Press.

Franklin, Sarah, and Margaret Lock. 2003. Animation and Cessation: The Remaking of Life and Death. In *Remaking Life and Death: Toward an Anthropology of the Biosciences*, ed. Sarah Franklin and Margaret Lock, 3–22. Santa Fe, NM: School of American Research Press.

Franklin, Sarah, and Susan McKinnon. 2000. New Directions in Kinship Study: A Core Concept Revisited. *Current Anthropology* 41(2): 275–78.

Fuhrman, Jed. 1999. Marine Viruses and their Biogeochemical and Ecological Effects. *Nature* 399:541–48.

———. 2003. Genome Sequences from the Sea. *Nature* 424:1001–2.

Fujimura, Joan. 1999. The Practices of Producing Meaning in Bioinformatics. In *The Practices of Human Genetics*, ed. Michael Fortun and Everett Mendelsohn, 49–87. Dordrecht: Kluwer Academic.

Fuller, Buckminster. 1969. *Operating Manual for Spaceship Earth*. Carbondale: Southern Illinois University Press.

Fuller, Matthew. 2005. *Media Ecologies: Materialist Energies in Art and Technoculture*. Cambridge, MA: MIT Press.

Fullwiley, Duana. 2007. The Molecularization of Race: Institutionalizing Human Difference in Pharmacogenetics Research. *Science as Culture* 16(1): 1–30.

Galison, Peter. 1994. The Ontology of the Enemy: Norbert Weiner and the Cybernetic Vision. *Critical Inquiry* 21(1): 228–66.

———. 2004. Specific Theory. *Critical Inquiry* 30(2): 379–83.

Galperin, Michael Y. 2005. A Census of Membrane-bound and Intracellular Signal Transduction Proteins in Bacteria: Bacterial IQ, Extroverts and Introverts. *BMC Microbiology* 5:35, June 14, doi:10.1186/1471-2180-5-35.

Galtier, Nicolas, Nicolas Tourasse, and Manolo Gouy. 1999. A Non-hyperthermophilic Common Ancestor to Extant Life Forms. *Science* 283: 220–21.

Geertz, Clifford. 1973. *The Interpretation of Cultures.* New York: Basic Books.

Giblett, Rod. 1996. *Postmodern Wetlands: Culture, History, Ecology.* Edinburgh: Edinburgh University Press.

Gilroy, Paul. 1993. *The Black Atlantic: Modernity and Double Consciousness.* Cambridge, MA: Harvard University Press.

———. 2000. *Against Race: Imagining Political Culture beyond the Color Line.* Cambridge, MA: Balknap/Harvard University Press.

Giovannoni, Stephen. J., Theresa B. Britschgi, Craig L. Moyer, and Katharine G. Field. 1990. Genetic Diversity in Sargasso Sea Bacterioplankton. *Nature* 345:60–63.

Glowka, Lyle. 1996. The Deepest of Ironies: Genetic Resources, Marine Scientific Research, and the Area. *Ocean Yearbook* 12:154–78.

———. 2001. Establishing the Legal Basis to Conserve and Sustainably Use Hydrothermal Vents and Their Biological Communities. In *Management of Hydrothermal Vent Sites: Report from the InterRidge Workshop: Management and Conservation of Hydrothermal Vent Ecosystems,* con-veners Paul Dando and S. Kim Juniper. Institute of Ocean Sciences, Sidney (Victoria), BC, Canada, September 28–30, 2000.

Godwin, L. S., and L. G. Eldridge. 2001. *South Oahu Marine Invasions Shipping Study.* Bishop Museum Technical Report no. 20, Honolulu.

Gogarten, J. Peter, W. Ford Doolittle, and Jeffrey G. Lawrence. 2002. Prokaryotic Evolution in Light of Gene Transfer. *Molecular and Biological Evolution* 19(12): 2226–38.

Gold, Thomas. 1999. *The Deep Hot Biosphere.* New York: Copernicus Books.

Goldenweiser, Alexander. 1933. Review of Paul Radin's *Social Anthropology. American Anthropologist* 35(2): 345–49.

Goodwin, Charles. 1995. Seeing in Depth. *Social Studies of Science* 25(2): 237–74.

Goonan, Kathleen Ann. 1996. *The Bones of Time.* New York: TOR.

Gordon, John Steele. 2002. *A Thread across the Ocean: The Heroic Story of the Transatlantic Cable.* New York: Walker.

Gould, Stephen Jay. 1997. Nonoverlapping Magisteria. *Natural History* 106:16–22.

Gould, Stephen J., and Richard C. Lewontin. 1979. The Spandrels of San Marcos and the Panglossian Paradigm: A Critique of the Adaptationist Programme. *Proceedings of the Royal Society of London, B* 205:581–98.

Grady, Monica. 2001. *Astrobiology.* Washington, DC: Smithsonian.

Grinspoon, David. 2003. *Lonely Planets: The Natural Philosophy of Alien Life.* New York: Ecco.

Guggenheim, Davis, director. 2006. *An Inconvenient Truth: The Planetary Emergency of Global Warming and What We Can Do about It.* Paramount Classics.

Gupta, Akhil, and James Ferguson. 1992. Beyond "Culture": Space, Identity, and the Politics of Difference. *Cultural Anthropology* 7(1): 6–23.

Gupta, Radhey, and Bohdan Soltys. 1999. Lateral Gene Transfer, Genome Surveys, and the Phylogeny of Prokaryotes. *Science* 286:1443a.

Hacking, Ian. 1983. *Representing and Intervening.* Cambridge: Cambridge University Press.

Haeckel, Ernst. 2005. *Art Forms from the Ocean: The Radiolarian Atlas of 1862.* Munich: Prestel Verlag.

Hall, Stuart. 1990. Cultural Identity and Diaspora. In *Identity: Community, Culture, Difference,* ed. Jonathan Rutherford, 222–37. London: Lawrence and Wishart.

Hallam, Steven, Nik Putnam, Christina M. Preston, John C. Detter, Daniel Rokhsar et al. 2004. Reverse Methanogenesis: Testing the Hypothesis with Environmental Genomics. *Science* 305:1457–62.

Halualani, Rona Tamiko. 2002. *In the Name of Hawaiians: Native Identities and Cultural Politics.* Minneapolis: University of Minnesota Press.

Hamblin, Jacob Darwin. 2005. *Oceanographers and the Cold War: Disciples of Marine Science.* Seattle: University of Washington Press.

Hamilton-Paterson, James. 1992. *The Great Deep: The Sea and Its Thresholds.* New York: Random House.

Handelsman, Jo. 2004. Metagenomics: Application of Genomics to Uncultured Microorganisms. *Microbiology and Molecular Biology Reviews* 68(4): 669–85.

Haraway, Donna. 1989. *Primate Visions: Gender, Race, and Nature in the World of Modern Science.* New York: Routledge.

———. 1991a. A Cyborg Manifesto: Science, Technology, and Socialist-Feminism in the Late Twentieth Century. In *Simians, Cyborgs, and Women: The Reinvention of Nature,* 149–82. New York: Routledge.

———. 1991b. Situated Knowledges: The Science Question in Feminism and the Privilege of Partial Perspective. In *Simians, Cyborgs, and Women: The Reinvention of Nature,* 183–201. New York: Routledge.

———. 1995. Cyborgs and Symbionts: Living Together in the New World Order. In *The Cyborg Handbook,* ed. Chris Hables Gray, xi–xx. New York: Routledge.

———. 1997. *Modest_Witness@Second_Millennium.FemaleMan©_Meets_OncoMouse™: Feminism and Technoscience.* New York: Routledge.

———. 2008. *When Species Meet*. Minneapolis: University of Minnesota Press.

Hardin, Garrett. 1968. The Tragedy of the Commons. *Science* 162:1243–47.

Harry, Debra, and Le'a Malia Kanehe. 2005. The BS in Access and Benefit Sharing (ABS): Critical Questions for Indigenous Peoples. In *The Catch: Perspectives on Benefit Sharing*, ed. Beth Burrows. Edmonds, WA: Edmonds Institute. www.ipcb.org/publications/other_art/bsinabs.html.

Hastrup, Kirsten. 1990. The Ethnographic Present: A Reinvention. *Cultural Anthropology* 5(1): 45–61.

Hau'ofa, Epeli. 1993. Our Sea of Islands. In *A New Oceania: Rediscovering Our Sea of Islands*, ed. Vijay Naidu, Eric Waddell, and Epeli Hau'ofa. Suva: School of Social and Economic Development, USP. [Reprinted in *Contemporary Pacific* 6.1 (1994): 147–61.

Hayden, Cori. 2003. *When Nature Goes Public: The Making and Unmaking of Bioprospecting in Mexico*. Princeton, NJ: Princeton University Press.

Hayles, N. Katherine. 1999. *How We Became Posthuman: Virtual Bodies in Cybernetics, Literature, and Informatics*. Chicago: University of Chicago Press.

Hayward, Eva. 2004. Jellyfish Optics: Immersion in Marine TechnoEcology. Paper presented at the meetings of the Society for Science and Literature, Durham, NC, October 14–17.

———. 2005. Enfolded Vision: Refracting *The Love Life of the Octopus*. *Octopus* 1:29–44.

Hearn, Chester G. 2002. *Tracks in the Sea: Matthew Fontaine Maury and the Mapping of the Oceans*. Camden, ME: McGraw Hill.

Heller, Chaia. 2001. McDonalds, MTV, and Monsanto: Resisting Biotechnology in the Age of Informational Capital. In *Redesigning Life? The Worldwide Challenge to Genetic Engineering*, ed. Brian Tokar, 405–19. London: Zed Books.

Helmreich, Stefan. 2000. *Silicon Second Nature: Culturing Artificial Life in a Digital World*, updated with a new preface (hardcover edition, 1998). Berkeley: University of California Press.

———. 2001. Kinship in Hypertext: Transubstantiating Fatherhood and Information Flow in Artificial Life. In *Relative Values: Reconfiguring Kinship Studies*, ed. Sarah Franklin and Susan McKinnon, 116–43. Durham, NC: Duke University Press.

———. 2003. A Tale of Three Seas: From Fishing through Aquaculture to Marine Biotechnology in the Life History Narrative of a Marine Biologist. *Maritime Studies* 2(2): 73–94.

———. 2005. Cetology Now. *Melville Society Extracts* 129:10–12.

———. 2006. Time and the Tsunami. *Reconstruction: Studies in Contemporary Culture* 6(3). http://reconstruction.eserver.org/063/contents.shtml.

———. 2007. An Anthropologist Underwater: Immersive Soundscapes, Submarine Cyborgs, and Transductive Ethnography. *American Ethnologist* 34(4): 621–41.

———. forthcoming. How Like a Reef. In *NatureCultures: Thinking with Donna Haraway*, Sharon Ghamari-Tabrizi, ed. Cambridge, MA: MIT Press.

Hernández Castillo, Rosalva Aída. 2001. *Histories and Stories from Chiapas: Border Identities in Southern Mexico*. Austin: University of Texas Press.

Herskovits, Melville. 1941. *The Myth of the Negro Past*. New York: Harper and Brothers.

Herzog, Werner, director. 2005. *The Wild Blue Yonder*. Werner Herzog Filmproduktion/Tetramedia/West Park Pictures/France 2.

Hess, David. 1995. *Science and Technology in a Multicultural World: The Cultural Politics of Facts and Artifacts*. New York: Columbia University Press.

Hilario, Elena, and Peter J. Gogarten. 1993. Horizontal Transfer of ATPase Genes: The Tree of Life Becomes a Net of Life. *Biosystems* 31(2–3): 111–19.

Hilgartner, Stephen. 2004. Mapping Systems and Moral Order: Constituting Property in Genome Laboratories. In *States of Knowledge: The Co-production of Science and Social Order*, ed. Sheila Jasanoff, 131–41. New York: Routledge.

Hobsbawm, Eric, and Terence Ranger, eds. 1983. *The Invention of Tradition*. Cambridge: Cambridge University Press.

Hoehler, Tori M., and Marc J. Alperin. 1996. Anaerobic Methane Oxidation by a Methanogen-sulfate Reducer Consortium: Geochemical Evidence and Biochemical Considerations. In *Microbial Growth on C1-Compounds*, ed. M. E. Lidstrom and F. R. Tabita, 326–33. Dordrecht: Kluwer Academic.

Hoffman, Paul F., and Daniel Schrag. 2000. Snowball Earth. *Scientific American* 282:68–75.

Höhler, Sabine. 2002. Depth Records and Ocean Volumes: Ocean Profiling by Sounding Technology, 1850–1930. *History and Technology* 18(2): 119–54.

Holloway, Lynette. 2005. Dijanna Figueroa: From the Lab to the Big Screen. *Ebony* 60(7): 118–20.

Hughes, David McDermott. 2005. Third Nature: Making Space and Time in the Great Limpopo Conservation Area. *Cultural Anthropology* 20(2): 157–84.

Hugill, Stan, collector. 1994. *Shanties from the Seven Seas: Shipboard Work-Songs and Songs Used as Work-Songs from the Great Days of Sail*. Mystic, CT: Mystic Seaport Museum.

Hutchins, Edwin. 1995. *Cognition in the Wild*. Cambridge, MA: MIT Press.

Hutchinson, G. Evelyn. 1961. The Paradox of the Plankton. *American Naturalist* 95:137–45.

Huxley, Thomas H. 1868. On Some Organisms Living at Great Depths in the North Atlantic Ocean. *Quarterly Journal of Microscopical Science* 8:202–12.

Ingold, Tim. 1990. An Anthropologist Looks at Biology. *Man* 25(2): 208–29.

Iselin, Columbus O'D, and Maurice Ewing. 1941. *Sound Transmission in Sea Water, a Preliminary Report*. Woods Hole, MA: Woods Hole Oceanographic Institution, for the National Defense Research Committee.

Jackson, Jeremy. 2006. Improvident Seas. Paper presented at Biodiversity in the Anthropocene: Perspectives on the Human Appropriation of the Natural World, Radcliffe Institute for Advanced Study, Harvard University, March 10.

Jacob, François. 1993 [1970]. *The Logic of Life: A History of Heredity*. Trans. Betty E. Spillman. Princeton, NJ: Princeton University Press.

Jacobson, Jon L., and Alison Rieser. 1998. The Evolution of Ocean Law. *Scientific American Presents* 9(3): 100–105.

James, C. L. R. 2001 [1953]. *Mariners, Renegades, and Castaways: The Story of Herman Melville and the World We Live In*. Hanover, NH: University Press of New England.

Jannasch, Holger. 1984. Chemosynthesis: The Nutritional Basis for Life at Deep-Sea Vents. *Oceanus* 27(3): 73–78.

Jasanoff, Sheila. 2005. *Designs on Nature: Science and Democracy in Europe and the United States*. Princeton, NJ: Princeton University Press.

Jenkins, Henry, and John Tulloch. 1995. *Science Fiction Audiences: Doctor Who, Star Trek, and Their Followers*. London: Routledge.

Johnson, J. C. 1996. Maritime Anthropology. In *The Encyclopedia of Cultural Anthropology*, vol. 3, ed. David Levinson and Melvin Ember, 726–28. New York: Henry Holt.

Jones, Caroline, and Donna Haraway. 2006. Zoon. In *Sensorium: Embodied Experience, Technology, and Contemporary Art*, ed. Caroline A. Jones, 241–45. Cambridge, MA: MIT Press.

Kaharl, Victoria A. 1990. *Water Baby: The Story of Alvin*. New York: Oxford University Press.

Kahn, Douglas. 1999. *Noise, Water, Meat: A History of Sound in the Arts*. Cambridge, MA: MIT Press.

Kaiser, David. 2006. Whose Mass Is It Anyway? Particle Cosmology and the Objects of Theory. *Social Studies of Science* 36(4): 533–64.

Kant, Immanuel. 1952 [1790]. *The Critique of Judgment*. Trans. J. C. Meredith. Oxford: Clarendon Press.

Kapur, Cari Costanzo. 2005. Race, Rights, and Resistance: Land and Indigenous (Trans)nationalism in Contemporary Hawai'i. Doctoral Dissertation, Department of Cultural and Social Anthropology, Stanford University.

Karl, David M. 1978. A Study of Microbial Biomass and Metabolic Activities in Marine Ecosystems: Development of Sensitive Techniques and Results of Selective Environmental Studies. Doctoral Dissertation, University of California, San Diego, Scripps Institution of Oceanography.

Karl, David. M., Robert R. Bidigare, and Ricardo M. Letelier. 2001. Long-term Changes in Plankton Community Structure and Productivity in the North Pacific Subtropical Gyre: The Domain Shift Hypothesis. *Deep Sea Research II* 48: 449–70.

Kauanui, J. Kehaulani. 2002. The Politics of Blood and Sovereignty in *Rice v. Cateyano*. *Political and Legal Anthropology Review* 25(1): 110–28.

Kay, Lily. 2000. *Who Wrote the Book of Life? A History of the Genetic Code*. Stanford, CA: Stanford University Press.

Keller, Catherine. 2003. *Face of the Deep: A Theology of Becoming*. London: Routledge.

Keller, Evelyn Fox. 1983. *A Feeling for the Organism: The Life and Work of Barbara McClintock*. New York: W. H. Freeman.

———. 1992. *Secrets of Life, Secrets of Death: Essays on Language, Gender and Science*. New York: Routledge.

———. 2000. *The Century of the Gene*. Cambridge, MA: Harvard University Press.

Kelty, Chris. 2003. Qualitative Research in the Age of the Algorithm: New Challenges in Cultural Anthropology. Lecture at the Research Libraries Group 2003 Annual Meeting, Rethinking the Humanities in a Global Age, Boston Public Library, Boston, May 5.

Kim, W. Chan, and Renée Mauborgne. 2004. Blue Ocean Strategy. *Harvard Business Review* 82(10): 76–85.

Kingsley, Charles. 1863. *The Water-Babies, A Fairy Tale for a Land Baby*. London: Macmillan.

Kipling, Rudyard. 1902. *Just So Stories for Little Children*. New York: Doubleday.

Kirch, Patrick V. 2000. *Historical Ecology in the Pacific Islands*. New Haven, CT: Yale University Press.

Klapisch-Zuber, Christiane. 1991. The Genesis of the Family Tree. *I Tatti Studies: Essays in the Renaissance* 4(1): 105–29.

Kleypas, J. A., R. A. Feely, V. J. Fabry, C. Langdon, C. L. Sabine, and L. L. Robbins. 2006. Impacts of Ocean Acidification on Coral Reefs and Other Marine Calcifiers: A Guide for Future Research. Report from a workshop sponsored by the National Science Foundation, the National Oceanic and Atmospheric Administration, and the U.S. Geological Survey.

Knapp, Steven, and Walter Benn Michaels. 1982. Against Theory. *Critical Inquiry* 8(4): 723–42.

Knorr-Cetina, Karin. 1999. *Epistemic Cultures: How the Sciences Make Knowledge*. Cambridge, MA: Harvard University Press.

Kohler, Robert E. 2002. *Landscapes and Labscapes: Exploring the Lab-Field Border in Biology*. Chicago: University of Chicago Press.

Krasnopolsky, V. A., J. P. Maillard, and T. C. Owen. 2004. Detection of Methane in the Martian Atmosphere: Evidence for Life. *Icarus* 172(2): 537–47.

Kristeva, Julia. 1991. *Strangers to Ourselves*. Trans. Leon Roudiez. New York: Columbia University Press.

Kroeber, Alfred L. 1952. *The Nature of Culture*. Chicago: University of Chicago Press.

Kull, Kalevi. 1999. On the History of Joining *Bio* to *Semio*: F. S. Rothschild and the Biosemiotic Rules. *Sign System Studies* 27:128–38.

Kuo, Wen-Hua. 2005. Japan and Taiwan in the Wake of Bio-globalization: Drugs, Race, and Standards. Doctoral Dissertation, Program in History, Anthropology, and Science, Technology, and Society, Massachusetts Institute of Technology.

Kuper, Adam. 1983. *Anthropology and Anthropologists: The Modern British School,* rev. ed. London: Routledge.

Kwa, Chunglin. 2002. Romantic and Baroque Conceptions of Complex Wholes in the Sciences. In *Complexities: Social Studies of Knowledge Practices*, ed. John Law and Annemarie Mol, 23–52. Durham, NC: Duke University Press.

Landecker, Hannah. 2007. *Culturing Life: How Cells Became Technologies*. Cambridge, MA: Harvard University Press.

Langewiesche, William. 2004. *The Outlaw Sea: A World of Freedom, Chaos, and Crime*. New York: North Point Press.

Laporte, Dominique. 2000 [1978]. *History of Shit*. Trans. Nadia Benabid and Rodolphe el-Khoury. Cambridge, MA: MIT Press.

Latour, Bruno. 1987. *Science in Action: How to Follow Scientists and Engineers through Society*. Cambridge, MA: Harvard University Press.

———. 1988. *The Pasteurization of France*. Trans. Alan Sheridan and John Law. Cambridge, MA: Harvard University Press.

———. 1999. *Pandora's Hope: Essays on the Reality of Science Studies*. Cambridge, MA: Harvard University Press.

Latour, Bruno, and Steve Woolgar. 1986. *Laboratory Life: The Construction of Scientific Facts*, 2d ed. Princeton, NJ: Princeton University Press.

Law, John. 2004. And If the Global Were Small and Non-Coherent? Method, Complexity and the Baroque. *Environment and Planning D: Society and Space* 22: 13–26.

Lawson, Charles, and Susan Downing. 2002. It's Patently Absurd—Benefit Sharing Genetic Resources from the Sea under UNCLOS, the CBD and TRIPs. *International Journal of Wildlife Law and Policy* 5:211–33.

Leach, Edmund. 1964. Anthropological Aspects of Language: Animal Categories and Verbal Abuse. In *New Directions in the Study of Language*, ed. Eric H. Lenneberg, 23–63. Cambridge, MA: MIT Press.

Leane, Elizabeth. 2003. Antarctica as a Scientific Utopia. *Foundation: The International Review of Science Fiction* 89:27–35.

Lem, Stanislaw. 1970 [1961]. *Solaris*. Trans. Joanna Kilmartin and Steve Cox. East Rutherford, NJ: Berkley Publishing.

Lenček, Lena, and Gideon Bosker. 1998. *The Beach: The History of Paradise on Earth*. New York: Penguin USA.

Lenoir, Timothy. 1997. *Instituting Science: The Cultural Production of Scientific Disciplines*. Stanford, CA: Stanford University Press.

———. 2002. Science and the Academy of the 21st Century: Does Their Past Have a Future in an Age of Computer-Mediated Networks? In *Ideale Akademie: Vergangene Zukunft oder konkrete Utopie?* ed. Wilhelm Vosskamp, 113–29. Berlin: Berlin Akademie der Wissenschaften.

Lepselter, Susan. 1997. From the Earth Native's Point of View: The Earth, the Extraterrestrial, and the Natural Ground of Home. *Public Culture* 9:197–208.

Lévi-Strauss, Claude. 1963 [1962]. *Totemism*. Trans. Rodney Needham. Boston: Beacon Press.

———. 1966 [1962]. *The Savage Mind*. Chicago: University of Chicago Press.

————. 1969 [1949]. *The Elementary Structures of Kinship*. Trans. J. Bell and J. von Sturmer. Boston: Beacon Press.

Lewis, Flora. 1967. *One of Our H-Bombs Is Missing*. New York: McGraw-Hill.

Lewitus, Alan, Jennifer Wolny, Jason Kempton, and Amy Ringwood. 2002. Identity, Physiological Ecology, and Toxicity of the Red Tide Dinoflagellate, *Kryptoperidinium* sp. Current Research Projects, Baruch Marine Field Laboratory, North Inlet-Winyah Bay, National Estuarine Research Reserve, University of South Carolina.

Lilly, John. 1961. *Man and Dolphin: Adventures on a New Scientific Frontier*. New York: Doubleday.

Lipps, Jere. 2003. Life on Ice-Covered Worlds: Antarctica, Snowball Earth, and Europa. Lecture presented at Moss Landing Marine Labs, Moss Landing, CA, February 28.

Lorini, Alessandra. 1998. The Cultural Wilderness of Canadian Water in the Ethnography of Franz Boas. *Cromohs* 3:1–7.

Louis, Godfrey, and A. Santhosh Kumar, 2006. The Red Rain Phenomenon of Kerala and Its Possible Extraterrestrial Origin. *Astrophysics and Space Science* 302:175–87.

Lovecraft, H. P. 2001 [1936]. At the Mountains of Madness. In *The Thing on the Doorstep and Other Weird Stories*. London: Penguin.

Low, Stephen, director. 2003. *Volcanoes of the Deep Sea*. Stephen Low Company and Rutgers University.

Lowe, Celia. 2006. *Wild Profusion: Biodiversity Conservation in an Indonesian Archipelago*. Princeton, NJ: Princeton University Press.

Lynch, Michael. 1992. Extending Wittgenstein: The Pivotal Move from Epistemology to the Sociology of Science. In *Science as Practice and Culture*, ed. Andrew Pickering, 215–65. Chicago: University of Chicago Press.

Lynch, William. 2005. The Ghost of Wittgenstein: Forms of Life, Scientific Method and Cultural Critique. *Philosophy of the Social Sciences* 35(2): 139–74.

MacElroy, R. D. 1974. Some Comments on the Evolution of Extremophiles. *Biosystems* 6:4–75.

Mackenzie, Adrian. 2002. *Transductions: Bodies and Machines at Speed*. London: Continuum.

————. 2003. Bringing Sequences to Life: How Bioinformatics Corporealizes Sequence Data. *New Genetics and Society* 22(3): 315–32.

MacPhail, Theresa. 2004. The Viral Gene: An Undead Metaphor Recoding Life. *Science as Culture* 13(3): 325–45.

Madigan, Michael. 2001. Extremophile Diversity and Ecology. Paper presented at Extremophile Research: Theory and Techniques. Center of Marine Biotechnology, Baltimore, MD, July 23.

Malin, Michael C., Kenneth S. Edgett, Liliya V. Posiolova, Shawn M. McColley, and Eldar Z. Noe Dobrea. 2006. Present-Day Impact Cratering Rate and Contemporary Gully Activity on Mars. *Science* 314:1573–77.

Malinowski, Bronislaw. 1984 [1922]. *Argonauts of the Western Pacific: An Account of Native Enterprise and Adventure in the Archipelagoes of Melanesian New Guinea.* Prospect Heights, IL: Waveland Press.

Mancinelli, R. L., M. R. White, and L. J. Rothschild. 1998. Biopan-Survival I: Exposure of the Osmophiles *Synechococcus sp.* (Nägeli) and *Haloarcula sp.* to the Space Environment. *Advances in Space Research* 22(3): 327–34.

Mann, David. 1972. Pioneers of Moss Landing. . . , and History of Moss Landing. *Game and Gossip* 17(8): 2–4, 22–23, 28–29; 10–11, 30–31.

Manning, Kenneth R. 1983. *Black Apollo of Science: The Life of Ernest Everett Just.* Oxford: Oxford University Press.

Manovich, Lev. 2001. *The Language of New Media.* Cambridge, MA: MIT Press.

MarBEC. 2003. Marine Bioproducts Engineering Center Year Four Annual Report. Honolulu: MarBEC.

———. 2004. Marine Bioproducts Engineering Center Final Report. Honolulu: MarBEC.

Marciniak, Katarzyna. 2006. *Alienhood: Citizenship, Exile, and the Logic of Difference.* Minneapolis: University of Minnesota Press.

Marcus, George E. 1995. Ethnography in/of the World System: The Emergence of Multi-Sited Ethnography. *Annual Review of Anthropology* 24:95–117.

Marcus, George E., and Michael M. J. Fischer. 1986. *Anthropology as Cultural Critique: An Experimental Moment in the Human Sciences.* Chicago: University of Chicago Press.

Margulis, Lynn. 2004. On Syphilis & Nietzsche's Madness: Spirochetes Awake! *Daedalus,* Fall: 118–25.

Margulis, Lynn, Antoni Navarrete, and Mónica Solé. 1998. Cosmopolitan Distribution of the Large Composite Microbial Mat Spirochete *Spirosymplokos deltaeiberi. International Microbiology* 1:27–34.

Margulis, Lynn, and Dorion Sagan. 1995. *What Is Life?* Berkeley: University of California Press.

———. 2002. *Acquiring Genomes: A Theory of the Origins of Species.* New York: Basic Books.

Markley, Robert. 2004. Methane on Mars: Message and Materiality. Paper presented at the meetings of the Society for Literature and Science, Durham, NC, October 14–17.

———. 2005. *Dying Planet: Mars in Science and the Imagination.* Durham, NC: Duke University Press.

Martin, Elizabeth Pa, and John Kekoa Burke. 1995. Ocean Governance Strategies: Governance in Partnership with Na Keiki O Ke Kai, the Children of the Sea. In *Ocean Governance for Hawai'i.* The Law of the Sea Institute Special Publication No. 3, chairman Thomas A. Mensah, 173–90. Honolulu: Law of the Sea Institute.

Martin, Emily. 1994. *Flexible Bodies: Tracking Immunity in American Culture from the Days of Polio to the Age of AIDS.* Boston: Beacon.

Marx, Karl. 1976 [1867]. *Capital,* vol. 1. Trans. Ben Fowkes. London: Penguin.

————. 1978 [1857–58]. The *Grundrisse*. Excerpted in *The Marx-Engels Reader*, 2d ed., ed. Robert C. Tucker, 221–93. New York: W. W. Norton.

Masco, Joe. 2004. Mutant Ecologies: Radioactive Life in Post-Cold War New Mexico. *Cultural Anthropology* 19(4): 517–50.

Maurer, Bill. 2000. A Fish Story: Rethinking Globalization on Virgin Gorda, British Virgin Islands. *American Ethnologist* 27(3): 670–701.

————. 2002. Fact and Fetish in Creolization Studies: Herskovits and the Problem of Induction, or, Guinea Coast, 1593. *New West Indian Guide/ Nieuwe West-Indisches Gids* 76 (1–2): 5–22.

————. 2003. Got Language? Law, Property, and the Anthropological Imagination. *American Anthropologist* 105(4): 775–81.

————. 2005. *Mutual Life, Limited: Islamic Banking, Alternative Currencies, Lateral Reason*. Princeton, NJ: Princeton University Press.

McCay, Bonnie J., and James M. Acheson, eds. 1987. *The Question of the Commons: The Culture and Ecology of Communal Resources*. Tucson: University of Arizona Press.

McGraw, Donald J. 2002. Claude ZoBell, Hadal Bacteria, and the "Azoic Zone." In *Oceanographic History: The Pacific and Beyond*, ed. Keith R. Benson and Philip F. Rehbock, 259–70. Seattle: University of Washington Press.

McKay, D. S., S. Clemett, K. Thomas-Keprta, and E. K. Gibson. 2002. Recognizing and Interpreting Biosignatures, Abstract # 12873 (Oral Presentation)—The Classification of Biosignatures. *Astrobiology* 2(4): 625–26.

McKay, D. S., Everett K. Gibson Jr., Kathie L. Thomas-Keprta, Hojatollah Vali, Christopher S. Romanek et al. 1996. Search for Past Life on Mars: Possible Relic Biogenic Activity in Martian Meteorite ALH84001. *Science* 273: 924–30.

McKissack, Patricia C., and Fredrick L. McKissack. 1999. *Black Hands, White Sails: The Story of African-American Whalers*. New York: Scholastic Press.

McLuhan, Marshall. 1964. *Understanding Media: The Extensions of Man*. New York: McGraw-Hill.

McLuhan, Marshall, and Quentin Fiore. 1967. *The Medium Is the Massage: An Inventory of Effects*. Coord. Jerome Age. New York: Random House.

McMenamin, Mark A. S., and Dianna L. S. McMenamin. 1993. *Hypersea: Life on Land*. New York: Columbia University Press.

McNeely, Jeffrey A. 2000. The Future of Alien Invasive Species: Changing Social Views. In *Invasive Species in a Changing World*, ed. Harold A. Mooney and Richard J. Hobbs, 171–89. Washington, DC: Island Press.

Mead, Margaret. 1968. Cybernetics of Cybernetics. In *Purposive Systems: Proceedings of the Fifth Annual Symposium of the American Society for Cybernetics*, ed. Heinz von Foerster, John D. White, Larry J. Peterson, and John K. Russell, 1–11. New York: Spartan Books.

Melville, Herman. 2001 [1851]. *Moby-Dick*, ed. Harrison Hayford, Hershel Parker, and G. Thomas Tanselle. Evanston, IL: Northwestern University Press.

Mendola, Dominick. 2000. Aquacultural Production of Bryostatin 1 and Ecteinascidin 743. In *Drugs from the Sea,* ed. Hobuhiro Fusetani, 120–33. Basel: Karger.

Merry, Sally Engle. 2000. *Colonizing Hawai'i: The Cultural Power of Law.* Princeton, NJ: Princeton University Press.

Merry, Sally Engle, and Donald Brenneis. 2003. Introduction. In *Law and Empire in the Pacific: Fiji and Hawai'i,* ed. Sally Engle Merry and Donald Brenneis, 3–34. Santa Fe, NM: School of American Research Press.

Merton, Robert. 1973 [1942]. *The Sociology of Science: Theoretical and Empirical Investigations.* Chicago: University of Chicago Press.

Mestel, Rosie. 1999. Drugs from the Sea. *Discover* 20(3): 70–75.

Meyer, Jean-Baptiste, and Mercy Brown. 1999. Scientific Diasporas: A New Approach to the Brain Drain. Management of Social Transformations, Discussion Paper no. 41, Prepared for the World Conference on Science, UNESCO-ICSU, Budapest, Hungary, June 26–July 1. www.unesco.org/most/meyer.htm.

Meyer, Manulani Aluli. 2001. Our Own Liberation: Reflections on Hawaiian Epistemology. *Contemporary Pacific* 13(1): 124–48.

Mills, Eric. 1989. *Biological Oceanography: An Early History, 1870–1960.* Ithaca, NY: Cornell University Press.

Mindell, David A. 2002. *Between Human and Machine: Feedback, Control, and Computing before Cybernetics.* Baltimore, MD: Johns Hopkins University Press.

———. 2005. Between Human and Machine. *Technology Review,* February. www.technologyreview.com/articles/05/02/issue/news_myview.asp?p=1.

Mintz, Sidney. 1985. *Sweetness and Power: The Place of Sugar in Modern History.* New York: Penguin Books.

Mirmalek, Zara. 2004. A Martian Ethnography. Paper presented at the meetings of the Society for the Social Study of Science, Paris, France, August 25–28.

Miyazaki, Hirokazu. 2004. *The Method of Hope: Anthropology, Philosophy, and Fijian Knowledge.* Stanford, CA: Stanford University Press.

Modis, Yorgo, Steven Ogata, David Clements, and Stephen C. Harrison. 2004. Structure of the Dengue Virus Envelope Protein after Membrane Fusion. *Nature* 427:313–19.

Mody, Cyrus C.M. 2005. The Sounds of Science: Listening to Laboratory Practice. *Science, Technology, and Human Values* 30(2): 175–98.

Montoya, Michael. 2007. Bioethnic Conscription: Genes, Race, and Mexicana/o Ethnicity in Diabetes Research. *Cultural Anthropology* 22(1): 94–128.

Mooney, Harold A., and Richard J. Hobbs, eds. 2000. *Invasive Species in a Changing World.* Washington, DC: Island Press.

Moore, Christopher. 2003. *Fluke, or, I Know Why the Winged Whale Sings.* New York: William Morrow.

Moore, Henrietta L. 2004. Global Anxieties: Concept-metaphors and Pre-theoretical Commitments in Anthropology. *Anthropological Theory* 4(1): 71–88.

Moore, Lisa R., Gabrielle Rocap, and Sallie W. Chisholm. 1998. Physiology and Molecular Phylogeny of Coexisting *Prochlorococcus* Ecotypes. *Nature* 393:464–67.

Moran, Katy, Steven R. King, and Thomas J. Carlson. 2001. Biodiversity Prospecting: Lessons and Prospects. *Annual Review of Anthropology* 30:505–26.

Morgan, C. Lloyd. 1878. Physiography. *American Naturalist* 12(10): 665–82.

Morgan, Elaine. 1984. The Aquatic Hypothesis. *New Scientist* 102(1405): 11–13.

Morgan, Lewis Henry. 1871 [1866]. *Systems of Consanguinity and Affinity of the Human Family*. Smithsonian Contributions to Knowledge, vol. 17. Washington, DC: Smithsonian Institute.

Morrison, Toni. 1992. *Playing in the Dark: Whiteness and the Literary Imagination*. Cambridge, MA: Harvard University Press.

Mukerji, Chandra. 1989. *A Fragile Power: Scientists and the State*. Princeton, NJ: Princeton University Press.

Munn, Colin B. 2004. *Marine Microbiology: Ecology and Applications*. London: BIOS Scientific Publishers/Taylor and Francis Group.

Myers, Natasha. 2007. Modeling Proteins, Making Scientists: An Ethnography of Pedagogy and Visual Cultures in Contemporary Structural Biology. Doctoral Dissertation, Program in History, Anthropology, and Science, Technology, and Society, Massachusetts Institute of Technology.

National Academy of Sciences. 2002. *Marine Biotechnology in the Twenty-first Century: Problems, Promise and Products*. Washington, DC: National Academies Press.

National Centre for Aquatic Biodiversity and Biosecurity, New Zealand. 2002. *Aquatic Biodiversity and Biosecurity Newsletter* 1.

National Oceanic and Atmospheric Administration. 1998. *National Ocean Conference Discussion Papers*. www.yoto98.noaa.gov/papers.htm.

National Research Council, Committee on the Ocean's Role in Human Health. 1999. *From Monsoons to Microbes: Understanding the Ocean's Role in Human Health*. Washington, DC: National Academies Press.

Nature. 2007. Meanings of "Life." *Nature* 447:1031–32.

Needham, Cynthia, Mahlon Hoagland, Kenneth McPherson, and Bert Dodson. 2000. *Intimate Strangers: Unseen Life on Earth*. Washington, DC: American Society of Microbiology Press.

Nelson, Alondra. forthcoming, 2008. Bio Science: Genetic Genealogy Testing and the Pursuit of African Ancestry. *Social Studies of Science* 38.

Nelson, Diane. 2003. A Social Science Fiction of Fevers, Delirium, and Discovery: *The Calcutta Chromosome*, the Colonial Laboratory, and the Postcolonial New Human. *Science Fiction Studies* 30:246–66.

Nelson, Karen E., Rebecca A. Clayton, Steven R. Gill, Michelle L. Gwinn, Robert J. Dodson et al. 1999. Evidence for Lateral Gene Transfer between Archaea and Bacteria from Genome Sequence of *Thermotoga maritima*. *Nature* 399:323–29.

Newitz, Annalee. 1993. Alien Abductions and the End of White People. *Bad Subjects* 5. http://bad.eserver.org/issues/1993/06/newitz.html.

North, Andrew. 1955. *Sargasso of Space*. New York: Ace.

Nye, David. 2003. *America as Second Creation: Technology and Narratives of New Beginnings*. Cambridge, MA: MIT Press.

Obeyesekere, Gananath. 1997 [1992]. *The Apotheosis of Captain Cook: European Mythmaking in the Pacific*. Princeton, NJ: Princeton University Press.

Ochman, Howard. 2005. Genomic Insights into Bacterial Species. Lecture at Microbial Sciences Symposium 2005, Harvard University, Cambridge, MA, April 16.

Ochman, Howard, Jeffrey G. Lawrence, and Eduardo A. Groisman. 2000. Lateral Gene Transfer and the Nature of Bacterial Innovation. *Nature* 405:299–304.

O'Hara, Robert J. 1992. Telling the Tree: Narrative Representation and the Study of Evolutionary History. *Biology and Philosophy* 7:135–60.

Okamura, Jonathan Y. 1994. Why There Are No Asian Americans in Hawai'i: The Continuing Significance of Local Identity. *Social Process in Hawai'i* 35:161–78.

O'Malley, Maureen A., and John Dupré. 2007. Size Doesn't Matter: Towards a More Inclusive Philosophy of Biology. *Biology and Philosophy* 22:155–91.

Omi, Michael, and Howard Winant. 1994. *Racial Formation in the United States: From the 1960s to the 1990s*, 2d ed. New York: Routledge.

Ong, Aihwa. 1999. *Flexible Citizenship: The Cultural Logics of Transnationality*. Durham, NC: Duke University Press.

O'Quinn, Bob. 1995. *The Bermuda Virus*. Hamilton, Bermuda: Bermudian Publishing.

Oreskes, Naomi. 1999. *The Rejection of Continental Drift: Theory and Method in American Earth Science*. Oxford: Oxford University Press.

———. 2003. A Context of Motivation: US Navy Oceanographic Research and the Discovery of Sea-Floor Hydrothermal Vents. *Social Studies of Science* 33(5): 697–742.

Orlove, Benjamin. 2002. *Lines in the Water: Nature and Culture in Lake Titicaca*. Berkeley: University of California Press.

Orphan Victoria J., Christopher H. House, Kai-Uwe Hinrichs, Kevin D. McKeegan, and Edward F. DeLong. 2002. Multiple Archaeal Groups Mediate Methane Oxidation in Anoxic Cold Seep Sediments. *Proceedings of the National Academy of Sciences* 99(11): 7663–68.

Osorio, Jonathan Kamakawiwo'ole. 2003. Kū'ē and Kū'oko'a: History, Law and Other Faiths. In *Law and Empire in the Pacific: Fiji and Hawai'i*, ed. Sally Engle Merry and Donald Brenneis, 213–37. Santa Fe, NM: School of American Research Press.

Ota, Charles. 2002. "Water Everywhere, Not a Drop to Drink." *Wai Ola o OHA* 19(9): 11. www.hawaii.edu/hivandaids/Ka_Wai_Ola_o_OHA_newspaper__September_2002.pd.

Otter, Samuel. 1999. *Melville's Anatomies.* Berkeley: University of California Press.

Oyama, Susan. 2000. *The Ontogeny of Information. Developmental Systems and Evolution,* 2d ed. Durham, NC: Duke University Press.

Pace, Norman, David A. Stahl, David J. Lane, and Gary J. Olsen. 1986. The Analysis of Natural Microbial Populations by Ribosomal RNA Sequences. *Advances in Microbial Ecology* 9:1–55.

Pálsson, Gísli. 1991. *Coastal Economies, Cultural Accounts: Human Ecology and Icelandic Discourse.* Manchester: Manchester University Press.

———. 1998. The Birth of the Aquarium: The Political Ecology of Icelandic Fishing. In *The Politics of Fishing,* ed. Tim Gray, 209–27. London: Macmillan.

———. 2007. *Anthropology and the New Genetics.* Cambridge: Cambridge University Press.

Parker, Linda S. 1989. *Native American Estate: The Struggle over Indian and Hawaiian Lands.* Honolulu: University of Hawai'i Press.

Partensky, Fred, Wolfgang R. Hess, and Daniel Vaulot. 1999. *Prochlorococcus,* a Marine Photosynthetic Prokaryote of Global Significance. *Microbiology and Molecular Biology Reviews* 63(1): 106–27.

Paul, John. 1999. Microbial Gene Transfer: An Ecological Perspective. *Journal of Molecular Microbiology and Biotechnology* 1(1): 45–50. www.jmmb. net/v1n1/07/07.html.

Pauly, Philip. J. 2000. *Biologists and the Promise of American Life: From Meriwether Lewis to Alfred Kinsey.* Princeton, NJ: Princeton University Press.

Pavlov, Anatoly, K. Vitaly L. Kalinin, Alexei N. Konstantiniov, Vladimir N. Shelegedin, and Alexander A. Pavlov. 2006. Was Earth Ever Infected by Martian Biota? Clues from Radioresistant Bacteria. *Astrobiology* 6(6): 911–18.

Paxson, Heather. 2008. Post-Pasteurian Cultures: The Microbiopolitics of Raw-Milk Cheese in the United States. *Cultural Anthropology* 23(1): 15–47.

Peirce, Charles S. 1998 [1893–1913]. *The Essential Peirce. Selected Philosophical Writings,* vol. 2, ed. Peirce Edition Project. Bloomington and Indianapolis: Indiana University Press.

Pennisi, Elizabeth. 1999. Is It Time to Uproot the Tree of Life? *Science* 284:1305–07. www.sciencemag.org.

Petryna, Adriana. 2005. Ethical Variability: Drug Development and Globalizing Clinical Trials. *American Ethnologist* 32(2): 183–97.

Philbrick, Nathaniel. 2003. *Sea of Glory: America's Voyage of Discovery, The U.S. Exploring Expedition, 1838–1842.* New York: Viking.

Philippe, Hervé, and Christophe J. Douady. 2003. Horizontal Gene Transfer and Phylogenetics. *Current Opinion in Microbiology* 6:498–505.

Pimentel, David, ed. 2002. *Biological Invasions: Economic and Environmental Costs of Alien Plant, Animal, and Microbe Species.* Boca Raton, FL: CRC Press.

Pollack, Andrew. 2003. A New Kind of Genomics, with an Eye on Ecosystems. *New York Times,* October 21, D1, D6.

——. 2004. Groundbreaking Gene Scientist Is Taking His Craft to the Oceans. *New York Times,* March 5.

Poore, Barbara. 2003. Blue Lines: Water, Information, and Salmon in the Pacific Northwest. Doctoral Dissertation, Geography, University of Washington.

Pope John Paul II. 1998. Message to the Pontifical Academy of Sciences. In *Evolutionary and Molecular Biology: Scientific Perspectives on Divine Action,* ed. Robert John Russell, William R. Stoeger, S. J., and Francisco J. Ayala, 2–9. Vatican City State: Vatican Observatory Publications.

Postgate, John. 1994. *The Outer Reaches of Life.* Cambridge: Cambridge University Press.

Pottage, Alain. 2006. Too Much Ownership: Bio-prospecting in the Age of Synthetic Biology. *BioSocieties* 1(2): 137–58.

Povinelli, Elizabeth. 2002. *The Cunning of Recognition: Indigenous Alterities and the Making of Australian Multiculturalism.* Durham, NC: Duke University Press.

Powdermaker, Hortense. 1966. *Stranger and Friend: The Way of an Anthropologist.* New York: Norton.

Prak, Eline T. Luning, and Haig H. Kazazian Jr. 2000. Mobile Elements and the Human Genome. *Nature Reviews, Genetics* 1:134–44.

Proctor, Lita M., and David M. Karl. 2007. Introduction to "A Sea of Microbes" Special Issue. *Oceanography* 20(2): 14–15.

Puhipau, and Joan Lander, directors. 1996. *Stolen Waters.* Nāʻālehu Hawaiʻi: Nā Maka o ka ʻAina Video Productions.

Qanungo, Kushal. 2002. Time for a New Deal on Marine Bioprospecting. *SciDev.Net,* January 17. www.scidev.net/Opinions/index.cfm?fuseaction= readOpinions&itemid=37&language=1.

Raban, Jonathan, ed. 1993. *The Oxford Book of the Sea.* Oxford: Oxford University.

Rabinow, Paul. 1992. Artificiality and Enlightenment: From Sociobiology to Biosociality. In *Incorporations,* ed. Jonathan Crary and Sanford Kwinter, 234–52. New York: Zone.

——. 1999. *French DNA: Trouble in Purgatory.* Chicago: University of Chicago Press.

——. 2003. *Anthropos Today: Reflections on Modern Equipment.* Princeton, NJ: Princeton University Press.

Raffles, Hugh. 1999. "Local Theory": Nature and the Making of an Amazonian Place. *Cultural Anthropology* 14(3): 323–60.

——. 2002. *In Amazonia: A Natural History.* Princeton, NJ: Princeton University Press.

Rainger, Ronald. 2004. Oceanography and Fieldwork: Geopolitics and Research at the Scripps Institution. *Proceedings of the California Academy of Sciences* 55 (supplement 1) (8): 185–208.

Rapp, Rayna. 1999. *Testing Women, Testing the Fetus: A Social History of Amniocentesis in America.* New York: Routledge.

Rappé, Michael S., Stephanie A. Connon, Kevin L. Vergin, and Stephen J. Giovannoni. 2002. Cultivation of the Ubiquitous SAR11 Marine Bacterioplankton Clade. *Nature* 418:630–33.

Rayl, A. J. S. 2001. Human Viruses at Sea: Researchers Confirm Presence of Adenovirus in Pacific Waters off Southern California. *Scientist* 16(6): 19.

Reardon, Jenny. 2005. *Race to the Finish: Identity and Governance in an Age of Genomics.* Princeton, NJ: Princeton University Press.

Redfield, Alfred Clarence. 1934. On the Proportions of Organic Derivatives in Sea Water and their Relation to the Composition of Plankton. In *James Johnstone Memorial Volume,* ed. R. J. Daniel, 176–92. Liverpool: University Press of Liverpool.

Redfield, Peter. 2000. *Space in the Tropics: From Convicts to Rockets in French Guiana.* Berkeley: University of California Press.

Rediker, Marcus. 1987. *Between the Devil and the Deep Blue Sea: Merchant Seamen, Pirates, and the Anglo-American Maritime World, 1700–1750.* Cambridge: Cambridge University Press.

Redolfi, Michel. 1989. *Sonic Waters #2 (Underwater Music) 1983–1989.* Therwil, Switzerland: Hat Hut Records.

Reeburgh, William S. 2003. Global Methane Biogeochemistry. In *Treatise on Geochemistry,* vol. 4: *The Atmosphere,* ed. Ralph. F. Keeling, 65–89. Oxford: Elsevier Pergamon.

Rehbock, Philip F. 1975. Huxley, Haeckel, and the Oceanographers: The Case of *Bathybius haeckelii. Isis* 66:504–33.

———, ed. 1992. *At Sea with the Scientifics: The Challenger Letters of Joseph Matkin.* Honolulu: University of Hawai'i Press.

Reichenbach, Hans. 1938. *Experience and Prediction: An Analysis of the Foundations and the Structure of Knowledge.* Chicago: University of Chicago Press.

Rheinberger, Hans-Jörg. 2000. Cytoplasmic Particles: The Trajectory of a Scientific Object. In *Biographies of Scientific Objects,* ed. Lorraine Daston, 270–94. Chicago: University of Chicago Press.

Rhys, Jean. 1982 [1966]. *Wide Sargasso Sea.* New York: Norton.

Ricketts, Edward F., and Jack Calvin. 1952. *Between Pacific Tides: An Account of the Habits and Habitats of Some Five Hundred of the Common, Conspicuous Seashore Invertebrates of the Pacific Coast between Sitka, Alaska, and Northern Mexico,* 3d ed. Stanford, CA: Stanford University Press.

Riles, Annelise. 2000. *The Network Inside Out.* Ann Arbor: University of Michigan Press.

———. 2003. Making White Things White: An Ethnography of Legal Knowledge. Paper presented at Constance International Conference on Social Studies of Finance, Konstanz, Germany, May 16–18. www.uni-konstanz.de/ssf-conference/Riles.pdf.

Ritvo, Harriet. 1998. *The Platypus and the Mermaid, and Other Figments of the Classifying Imagination.* Cambridge, MA: Harvard University Press.

Robb, Frank. 2000. Marine Extremophiles: An Opportunity for Gene and Enzyme Discovery. Keynote lecture given at International Conference on Marine Biotechnology 2000, Townsville, Queensland, Australia, September 29–October 4.

Rohrer, Judy 2006. "Got Race?" The Production of Haole and the Distortion of Indigeneity in the *Rice* Decision. *Contemporary Pacific* 18(1): 1–31.

Roosth, Sophia. 2009. Screaming Yeast: Sonocytology, Cytoplasmic Milieu, and Cellular Subjectivities. *Critical Inquiry* 35.

Root, Richard B. 1967. The Niche Exploitation Pattern of the Blue-Grey Gnatcatcher. *Ecological Monographs* 37:317–50.

Rose, Nikolas. 2001. The Politics of Life Itself. *Theory, Culture and Society* 18(6): 1–30.

———. 2007. *The Politics of Life Itself: Biomedicine, Power, and Subjectivity in the Twenty-First Century.* Princeton, NJ: Princeton University Press.

Rossi, Michael. 2007. Kitsch and the Meaning(s) of Life. Paper presented at *Mutamorphosis: Challenging Arts and Sciences*, November 8–10, Prague, Czech Republic. www.mutamorphosis.org/index.php?lang=en&node=120 &catid=108&id=48.

Rothschild, Lynn J., and Rocco Mancinelli. 2001. Life in Extreme Environments. *Nature* 409:1092–1101.

Rotman, Brian. 2000. Going Parallel. *SubStance* 91:56–79.

Rozwadowski, Helen M. 1996. Small World: Forging a Scientific Maritime Culture for Oceanography. *Isis* 87(3): 409–29.

———. 2005. *Fathoming the Ocean: The Discovery and Exploration of the Deep Sea.* Cambridge, MA: Harvard University Press.

Rubin, Gayle. 1975. The Traffic in Women: Notes On a Political Economy of Sex. In *Toward an Anthropology of Women*, ed. Rayna Reiter, 157–210. New York: Monthly Review Press.

Ruggieri, George, and Norman David Rosenberg. 1969. *The Healing Sea: A Voyage into the Alien World Offshore.* New York: Dodd, Mead.

Ruiz, Gregory M., Tonya K. Rawlings, Fred C. Dobbs, Lisa A. Drake, Timothy Mullady et al. 2000. Global Spread of Microorganisms by Ships: Ballast Water Discharged from Vessels Harbours a Cocktail of Potential Pathogens. *Nature* 408:49–50.

Rusch, Douglas B., et al. 2007. The *Sorcerer II* Global Ocean Sampling Expedition: Northwest Atlantic through Eastern Tropical Pacific. *PLoS Biology* 5(3): e77. doi:10.1371/journal.pbio.0050077.

Safina, Carl. 1997. *Song for the Blue Ocean: Encounters along the World's Coasts and beneath the Seas.* New York: Holt.

Sagan, Dorian, and Lynn Margulis. 1993. *Garden of Microbial Delights: A Practical Guide to the Subvisible World.* Dubuque: Kendall/Hunt.

Sagoff, Mark. 2000. Why Exotic Species Are Not as Bad as We Fear. *Chronicle of Higher Education* 46(42): B7.

Sahlins, Marshall. 1976. *Culture and Practical Reason.* Chicago: University of Chicago Press.

———. 1995. *How "Natives" Think: About Captain Cook, for Example.* Chicago: University of Chicago Press.

Salinas Californian. 1998. Delegates Agree: Sea Is Life. *Salinas Californian,* June 12, 1.

Salzberg, Steven L., Owen White, Jeremy Peterson, and Jonathan A. Eisen. 2001. Microbial Genes in the Human Genome: Lateral Transfer or Gene Loss. *Science* 292:1903–6.

Sapp, Jan. 1994. *Evolution by Association: A History of Symbiosis.* Oxford: Oxford University Press.

Sarai Editorial Collective. 2006. In Turbulence. In *Sarai Reader 06: Turbulence,* ed. Monica Narula, Shuddhabrata Sengupta, Ravi Sundaram, Awedhendra Sharan, Jeebesh Bagchi, and Geert Lovink, vii–ix. Delhi: Centre for the Study of Developing Societies.

Sargent, William. 2002. *Crab Wars: A Tale of Horseshoe Crabs, Bioterrorism, and Human Health.* Hanover, NH: University Press of New England.

Scheuer, Jonathan L. 2002. Water and Power in Hawai'i: The Waiahole Water Case and the Future of the Islands. Doctoral Dissertation, Environmental Studies, University of California, Santa Cruz.

Schlee, Susan. 1973. *The Edge of an Unfamiliar World: A History of Oceanography.* New York: Dutton.

Schneider, David. 1968. *American Kinship: A Cultural Account.* Chicago: University of Chicago Press.

Schrader, Astrid. 2006. Phantomatic Species Ontologies: Untimely Re/productions of Toxic Dinoflagellates. Paper presented at the meetings of the Society for the Social Study of Science, Vancouver, BC, Canada, November 2–4.

Schrödinger, Erwin. 1944. *What Is Life? The Physical Aspect of the Living Cell.* Cambridge: Cambridge University Press.

Schultz-Makuch, Dirk, and J.M. Houtkooper. 2007. Life on Mars? Reinterpretation of the Viking Life Detection Experiments: A Possible Biogenic Origin of Hydrogen Peroxide. Paper presented at the meetings of the American Astronomical Society, Seattle, WA, January 5–10.

Schwartz, Hillel. 2003. The Indefensible Ear. In *The Auditory Culture Reader,* ed. Michael Bull and Les Back, 487–501. Oxford: Berg.

Sconce, Jeffrey. 2000. *Haunted Media: Electronic Presence from Telegraphy to Television.* Durham, NC: Duke University Press.

Scripps Institution of Oceanography. 1981. The Deep Sea Hydrothermal Vents: The Hottest Thing in Oceanography. Video recording of UCSD Symposium, October 21.

Seed, Patricia. 1995. *Ceremonies of Possession in Europe's Conquest of the New World, 1492–1640.* Cambridge: Cambridge University Press.

Segal, Dan, and Sylvia Yanagisako. 2005. Introduction. In *Unwrapping the Sacred Bundle: Reflections on the Disciplining of Anthropology*, ed. Dan Segal and Sylvia Yanagisako, 1–23. Durham, NC: Duke University Press.

Shapin, Steven, and Simon Schaffer. 1985. *Leviathan and the Air-Pump: Hobbes, Boyle, and the Experimental Life*. Princeton, NJ: Princeton University Press.

Shapiro, Howard. 1977. Fluorescent Dyes for Differential Counts by Flow Cytometry: Does Histochemistry Tell Us Much More than Cell Geometry? *Journal of Histochemistry and Cytochemistry* 25(8): 976–89.

———. 2000. Microbial Analysis at the Single-Cell Level: Tasks and Techniques. *Journal of Microbiological Methods* 42:3–16.

———. 2004. "Cellular Astronomy"—A Foreseeable Future in Cytometry. www.shapirolab.com/CellularAstronomyWebPreprint.htm.

Sheridan, Cormac. 2005. It Came from beneath the Sea. *Nature Biotechnology* 12(10): 1199–1201.

Shiva, Vandana. 1997. *Biopiracy: The Plunder of Nature and Knowledge*. Boston: South End Press.

Shor, Elizabeth Noble. 1978. *Scripps Institution of Oceanography: Probing the Oceans, 1936 to 1976*. San Diego: Tofua Press.

Shreeve, James. 2004. Craig Venter's Epic Voyage of Discovery. *Wired*, August, 104–13, 146–51.

Silbey, Susan S., and Patricia Ewick. 2003. The Architecture of Authority: The Place of Law in the Space of Science. In *The Common Place of Law*, ed. Austin Sarat, Lawrence Douglas, and Martha Umphrey, 75–108. Ann Arbor: University of Michigan Press.

Silverstein, Michael. 2003. Translation, Transduction, Transformation: Skating "Glossando" on Thin Semiotic Ice. In *Translating Cultures: Perspectives on Translation and Anthropology*, ed. Paula G. Rubel and Abraham Rosman, 75–105. Oxford: Berg.

Simberloff, Daniel. 2003. Confronting Introduced Species: A Form of Xenophobia? *Biological Invasions* 5: 179–92.

Simondon, Gilbert. 1992. The Genesis of the Individual. Trans. Mark Cohen and Sanford Kwinter. In *Incorporations*, ed. Jonathan Crary and Sanford Kwinter, 296–319. New York: Zone.

Simpson, April. 2006. The Lure of the Sea, and Science, for Minority Students. *New York Times*, June 30, B6.

Smil, Vaclav. 2002. *The Earth's Biosphere: Evolution, Dynamics, and Change*. Cambridge, MA: MIT Press.

Smith, Neil. 1992. Geography, Difference, and the Politics of Scale. In *Postmodernism and the Social Sciences*, ed. Joe Doherty, Elspeth Graham, and Mo Malek, 57–79. London: Macmillan.

Soares, Mário, and members of the Independent World Commission on the Oceans. 1998. *The Ocean, Our Future: The Report of the Independent World Commission on the Oceans*. Cambridge: Cambridge University Press.

Sommerlund, Julie. 2006. Classifying Microorganisms: The Multiplicity of Classifications and Research Practices in Molecular Microbial Ecology. *Social Studies of Science* 36(6): 909–28.

Speidel, Gisela E., and Kristina Inn. 1994. The Ocean Is My Classroom. *Kamehameha Journal of Education* 5: 11–23.

Sprugel, Douglas G. 1991. Disturbance, Equilibrium, and Environmental Variability: What Is "Natural" Vegetation in a Changing Environment? *Biological Conservation* 58:1–18.

Ssorin-Chaikov, Nikolai. 2006. On Heterochrony: Birthday Gifts to Stalin, 1949. *Journal of the Royal Anthropological Institute* 12(2): 355–75.

Staples, George W., and Robert H. Cowie. 2001. *Hawai'i's Invasive Species.* Honolulu: Mutual Publishing and Bishop Museum Press.

Stein, Roger B. 1975. *Seascape and the American Imagination.* New York: Clarkson N. Potter.

Steinbeck, John. 1995 [1941]. *The Log from the Sea of Cortez.* Introduction by Richard Astro. New York: Penguin Books.

Steinberg, Philip E. 2001. *The Social Construction of the Ocean.* Cambridge: Cambridge University Press.

Sterne, Jonathan. 2003. *The Audible Past: Cultural Origins of Sound Reproduction.* Durham, NC: Duke University Press.

Stetten, George D. 1984. *Alvin's* Memory. *Oceanus* 27(3): 44–46.

Steward, Grieg. 1999. Viruses in the Sea: Distribution, Dynamics, and Consequences. Monterey Bay Aquarium Research Institute Seminar Series, Friday, March 26. Abstract online, www.mbari.org/seminars/1999/mar26_steward.html.

Stewart, Kathleen, and Susan Harding. 1999. Bad Endings: American Apocalypsis. *Annual Review of Anthropology* 28:285–310.

Stocker, Michael. 2002/2003. Ocean Bio-Acoustics and Noise Pollution: Fish, Mollusks and Other Sea Animals' Use of Sound, and the Impact of Anthropogenic Noise on the Marine Acoustic Environment. *Soundscape: The Journal of Acoustic Ecology* 3(2)/4(1): 16–29.

Stocking, George W., Jr. 1982 [1968]. From Physics to Ethnology. In *Race, Culture, and Evolution: Essays in the History of Anthropology,* 133–60. Chicago: University of Chicago Press.

———. 1974. *The Shaping of American Anthropology, 1883–1911: A Franz Boas Reader.* New York: Basic Books.

Stofan, Ellen R., C. Elachi, J. I. Lunine, R. D. Lorenz, B. Stiles et al. 2007. The Lakes of Titan. *Nature* 445:61–64.

Stonich, Susan C., and Conner Bailey. 2000. Resisting the Blue Revolution. *Human Organization* 59(1): 23–36.

Stott, Rebecca. 2003. *Darwin and the Barnacle: The Story of One Tiny Creature and History's Most Spectacular Breakthrough.* New York: Norton.

Strathern, Marilyn. 1991. *Partial Connections.* ASAO special publications 3. Lanham, MD: Rowman and Littlefield.

———. 1992. *Reproducing the Future: Anthropology, Kinship, and the New Reproductive Technologies.* New York: Routledge.

———. 2004. *Commons and Borderlands: Working Papers on Inter-disciplinarity, Accountability and the Flow of Knowledge.* Wantage, Oxon: Sean Kingston Publishing.

———. 2005. *Kinship, Law and the Unexpected: Relatives Are Always a Surprise*. Cambridge: Cambridge University Press.

Strayer, Eric P. 2002. *Community at the Crossroads: A California Fishing Village Struggles for Control in the Information Era*. Master of Arts thesis, Sociology, University of New Mexico.

Street, Brian. 1993. Culture Is a Verb: Anthropological Aspects of Language and Cultural Process. In *Language and Culture*, ed. D. Graddol, L. Thompson, and M. Byram, 23–43. Clevedon, Avon: British Association for Applied Linguistics in association with Multilingual Matters.

Subramani. 2001. The Oceanic Imaginary. *Contemporary Pacific* 13(1): 149–62.

Subramaniam, Banu. 2001. The Aliens Have Landed! Reflections on the Rhetoric of Biological Invasions. *Meridians: Feminism, Race, Transnationalism* 2(1): 26–40.

Sullivan, Matthew B., John B. Waterbury, and Sallie W. Chisholm. 2003. Cyanophages Infecting the Oceanic Cyanobacterium *Prochlorococcus*. *Nature* 424:1047–51.

Sunder Rajan, Kaushik. 2006. *Biocapital: The Constitution of Postgenomic Life*. Durham, NC: Duke University Press.

Swain, Merrill, and Lapkin, Sharon. 1983. *Evaluating Bilingual Education: A Canadian Case Study*, Acton, ON: Scholarly Book Services.

Swyngedouw, Erik. 1997. Neither Global nor Local: "Glocalization" and the Politics of Scale. In *Spaces of Globalization: Reasserting the Power of the Local*, ed. Kevin Cox, 137–66. New York: Guilford Press.

Takacs, David. 1996. *The Idea of Biodiversity: Philosophies of Paradise*. Baltimore, MD: Johns Hopkins University Press.

Takahashi, Patrick. 2003. Energy from the Sea: The Potential and Realities of Ocean Thermal Energy Conversion (OTEC). Intergovernmental Oceanographic Commission Anton Bruun Memorial Lecture, June 30, *IOC Technical Series 66*, UNESCO.

Tate, Greg, ed. 2003. *Everything but the Burden: What White People Are Taking from Black Culture*. New York: Broadway Books.

Taussig, Emma. 2005. Extremophiles. Essay for first grade class, Minneapolis Public Schools.

Taussig, Michael. 2004. *My Cocaine Museum*. Chicago: University of Chicago Press.

Taylor, Michael. 1999. *Dark Life: Martian Nanobacteria, Rock-Eating Cave Bugs, and Other Extreme Organisms of Inner Earth and Outer Space*. New York: Scribner.

Taylor, Patricia L. 1995. *Science without Borders: Scripps Institution of Oceanography and Baja California*. San Diego, UCSD-TV videorecording.

Taylor, Peter. 2005. *Unruly Complexity: Ecology, Interpretation, Engagement*. Chicago: University of Chicago Press.

Teal, John, and Mildred Teal. 1975. *The Sargasso Sea*. Boston: Little, Brown.

Thacker, Eugene. 2000. The Post-Genomic Era Has Already Happened. *Biopolicy Journal* 3, https://tspace.library.utoronto.ca/bitstream/1807/91/1/py00001.pdf.

———. 2004. *Biomedia.* Minneapolis: University of Minnesota Press.

———. 2005. *The Global Genome: Biotechnology, Politics, and Culture.* Cambridge, MA: MIT Press.

Théberge, Paul. 2004. The Network Studio: Historical and Technological Paths to a New Ideal in Music Making. *Social Studies of Science* 34(5): 759–81.

Thompson, Charis. 2005. *Making Parents: The Ontological Choreography of Reproductive Technologies.* Cambridge, MA: MIT Press.

Thomson, Frank. K, III, et al. 2003. Patterns of Antibiotic Resistance in Cholera Bacteria Isolated from Ships' Ballast Water. Paper presented at Third International Conference on Marine Bioinvasions, Convened at Scripps Institution of Oceanography, La Jolla, CA, March 16–19.

Thorne-Miller, Boyce. 1999. *The Living Ocean: Understanding and Protecting Marine Biodiversity,* 2d ed. Washington, DC: Island Press.

Thurtle, Phillip, and Robert Mitchell. 2004. Data Made Flesh: The Material Poiesis of Informatics. In *Data Made Flesh: Embodying Information,* ed. Robert Mitchell and Phillip Thurtle, 1–23. New York: Routledge.

Toop, David. 1995. *Ocean of Sound: Aether Talk, Ambient Sound, and Imaginary Worlds.* London: Serpent's Tail.

Toumey, Christopher. 1994. *God's Own Scientists: Creationists in a Secular World.* New Brunswick, NJ: Rutgers University Press.

Trask, Haunani-Kay. 1993. *From a Native Daughter: Colonialism and Sovereignty in Hawai'i.* Monroe, ME: Common Courage Press.

Trask, Mililani. 1998. Advocacy and Resistance in Hawai'i. *Resist Newsletter* 7(7). resistinc.org/newsletter/issues/1998/09/arh.html, accessed March 6, 2004.

Traver, Tim. 2006. *Sippewissett, or, Life on a Salt Marsh.* White River Junction, VT: Chelsea Green.

Tsing, Anna Lowenhaupt. 1995. Empowering Nature, or: Some Gleanings in Bee Culture. In *Naturalizing Power: Essays in Feminist Cultural Analysis,* ed. Sylvia Yanagisako and Carol Delaney, 113–43. New York: Routledge.

———. 2004. *Friction: An Ethnography of Global Connection.* Princeton, NJ: Princeton University Press.

Turner, Tom. 2002. *Justice on Earth: Earthjustice and the People It Has Served.* White River Junction, VT: Chelsea Green.

Turner, Victor. 1967. *The Forest of Symbols: Aspects of Ndembu Ritual.* Ithaca, NY: Cornell University Press.

U.S. Cabinet. 1999. *Turning to the Sea: America's Ocean Future.* Washington, DC: National Oceanic and Atmospheric Administration.

Van Dover, Cindy Lee. 1996. *Deep-Ocean Journeys: Discovering New Life at the Bottom of the Sea.* Redwood City, CA: Addison-Wesley.

———. 2000. *The Ecology of Deep-Sea Hydrothermal Vents.* Princeton, NJ: Princeton University Press.

Van Dover, C. L., E. Z. Szuts, S. C. Chamberlain, and J. R. Cann. 1989. A Novel Eye in "Eyeless" Shrimp from Hydrothermal Vents of the Mid-Atlantic Ridge. *Nature* 337:458–60.

Van Driesche, Jason, and Roy Van Driesche. 2000. *Nature out of Place: Biological Invasions in the Global Age.* Washington, DC: Island Press.

Vaulot, Daniel, Dominique Marie, Robert J. Olson, and Sallie W. Chisholm. 1995. Growth of *Prochlorococcus,* a Photosynthetic Prokaryote, in the Equatorial Pacific Ocean. *Science* 268:1480–82.

Venter, J. Craig. 2004. Whole Environment Shotgun Sequencing: The Sargasso Sea. Paper presented at WAS . . . IS . . . MIGHT BE . . .: Perspectives on the Evolution of the Earth System, Massachusetts Institute of Technology, Cambridge, March 8–9.

Venter, J. Craig, et al. 2001. The Sequence of the Human Genome. *Science* 291:1304–51.

———. 2004. Environmental Genome Shotgun Sequencing of the Sargasso Sea. *Science* 304:66–74.

Verne, Jules. 1962 [1870]. *20,000 Leagues under the Sea.* Trans. Anthony Bonner, Introduction by Ray Bradbury. New York: Bantam Books.

Vertesi, Janet. 2008. Seeing Like a Rover: Embodied Experience on the Mars Exploration Rover Mission. Paper presented at 2008 Conference on Human Factors in Computing Systems, Florence, Italy, April 5–10. www.chi2008.org/altchisystem/submissions/submission_jvertesi_1.pdf.

Vogel, Gretchen. 1999. RNA Study Suggests Cool Cradle of Life. *Science* 283:155a.

von Uexküll, Jakob. 1982. The Theory of Meaning. *Semiotica* 42(1): 25–82.

Wagner, Roy. 1977. Analogic Kinship: A Daribi Example. *American Ethnologist* 4(4): 623–42.

Wakeford, Tom. 2001. *Liaisons of Life: From Hornworts to Hippos, How the Unassuming Microbe Has Driven Evolution.* New York: Wiley.

Waldby, Catherine. 2000. *The Visible Human Project: Informatic Bodies and Posthuman Medicine.* London: Routledge.

Walley, Christine. 2004. *Rough Waters: Nature and Development in an East African Marine Park.* Princeton, NJ: Princeton University Press.

Ward, Peter. 2005. *Life as We Do Not Know It: The NASA Search for (and Synthesis of) Alien Life.* New York: Viking.

Warren, Charles R. 2007. Perspectives on the "Alien" versus "Native" Species Debate: A Critique of Concepts, Language and Practice. *Progress in Human Geography* 31(4): 427–46.

Watson, Ian. 2002 [1975]. *The Jonah Kit.* London: Gollancz.

Watts, Peter. 1999. *Starfish.* New York: TOR.

Weir, Gary. 2001. *An Ocean in Common: American Naval Officers, Scientists, and the Ocean Environment.* College Station: Texas A&M University Press.

Weiss, Martin. 2005. The Body of Phenomenology: Unforeseen Phenomenological Outcomes of Biotechnologies. In *On the Future of Husserlian Phenomenology,* an ongoing Internet project organized by the Husserl Archives at the New School for Social Research in Memory of Alfred Schutz. www.newschool.edu/gf/phil/husserl/Future/Part%20Two/Part TwoFrames/PartTwo.html.

Wells, H. G. 2001 [1898]. *The War of the Worlds*. Ed. Leon Stover. Jefferson, NC: McFarland.

Wells, Jonathan. n.d. After Darwin, What? http:1164.82.73.94/Publications/Dialogue/AfterDarwin.rtf.

Wertheim, Margaret. 2007. Figuring Life: Redrawing Darwin's Evolutionary Tree. *Cabinet Magazine* 27: 54–59.

Weston, Kath. 2002. *Gender in Real Time: Power and Transience in a Visual Age*. New York: Routledge.

White, Luise. 1994. Alien Nation. *Transition* 63:24–33.

Whitfield, Peter. 1996. *The Charting of the Oceans: Ten Centuries of Maritime Maps*. Rohnert Park, CA: Pomegranate Artbooks.

Wiener, Norbert. 1985 [1961]. *Cybernetics: or, Control and Communication in the Animal and the Machine*, 2d ed. Cambridge, MA: MIT Press.

Williams, Ben. 2001. Black Secret Technology: Detroit Techno and the Information Age. In *Technicolor: Race, Technology, and Everyday Life*, ed. Alondra Nelson and Thuy Linh N. Tu with Alicia Headlam Hines, 154–76. New York: NYU Press.

Williams, Rosalind. 1990. *Notes on the Underground: An Essay on Technology, Society, and the Imagination*. Cambridge, MA: MIT Press.

Wittgenstein, Ludwig. 1958 [1953]. *Philosophical Investigations*, 3d ed. Trans. G. E. M Ascombe. New York: Macmillan.

Woese, Carl R. 1967. *The Genetic Code: The Molecular Basis for Genetic Expression*. New York: Harper and Row.

———. 2000. Interpreting the Universal Phylogenetic Tree. *Proceedings of the National Academy of Sciences* 97:8392–96.

Woese, Carl R., and George E. Fox. 1977. Phylogenetic Structure of the Prokaryotic Domain: The Primary Kingdoms. *Proceedings of the National Academy of Sciences* 74:5088–90.

Woese, Carl R., Otto Kandler, and Mark L. Wheelis. 1990. Towards a Natural System of Organisms: Proposal for the Domains Archaea, Bacteria, and Eucarya. *Proceedings of the National Academy of Sciences* 87:4576–79.

Wolfe, Audra J. 2002. Germs in Space: Joshua Lederberg, Exobiology, and the Public Imagination, 1958–1964. *Isis* 93:183–205.

Woodard, Colin. 2000. *Ocean's End: Travels through Endangered Seas*. New York: Basic Books.

Woodward, Kathleen. 2004. A Feeling for the Cyborg. In *Data Made Flesh: Embodying Information*, ed. Robert Mitchell and Phillip Thurtle, 181–97. New York: Routledge.

Woodward, Lillian. 1983. *Lillian Woodward's Moss Landing*. Carmel, CA: Woodward Publishing.

Yanagisako, Sylvia. 2002. *Producing Culture and Capital: Family Firms in Italy*. Princeton, NJ: Princeton University Press.

Yanagisako, Sylvia, and Carol Delaney. 1995. Naturalizing Power. In *Naturalizing Power: Essays in Feminist Cultural Analysis*, ed. Sylvia Yanagisako and Carol Delaney, 1–22. New York: Routledge.

Zerner, Charles. 1996. Telling Stories about Biological Diversity. In *Valuing Local Knowledge: Indigenous People and Intellectual Property Rights*, ed. Stephen B. Brush and Doreen Stabinsky, 68–101. Washington, DC: Island Press.

ZoBell, Claude. 1941. Apparatus for Collecting Water Samples from Different Depths for Bacteriological Analysis. *Journal of Marine Research* 4:173–88.

———. 1946. *Marine Microbiology.* New York: Reinhold.

———. 1949. The Influence of Hydrostatic Pressure on the Growth and Viability of Terrestrial and Marine Bacteria. *Journal of Bacteriology* 57:179–89.

———. 1959. My Mother, Stella Davis ZoBell, 1881–1958: A Story of Her Life and Loved Ones. Self-printed and distributed.

Zuckerkandl, Emile, and Linus Pauling. 1965. Molecules as Documents of Evolutionary History. *Journal of Theoretical Biology* 8(2): 357–66.

Zweig, Stefan. 1938. *Conquerer of the Seas: The Story of Magellan.* Trans. Eden and Cedar Paul. New York: Literary Guild of America.

Index

Note: page numbers in *italic type* indicate photographs or illustrations

Text: 10/13 Aldus
Display: Aldus
Compositor: International Typesetting and Composition
Indexer: Ellen Sherron